GENETICS - RESEARCH AND ISSUES

METABOLOMICS: METABOLITES, METABONOMICS, AND ANALYTICAL TECHNOLOGIES

GENETICS - RESEARCH AND ISSUES

Additional books in this series can be found on Nova's website
under the Series tab.

Additional E-books in this series can be found on Nova's website
under the E-book tab.

GENETICS - RESEARCH AND ISSUES

METABOLOMICS: METABOLITES, METABONOMICS, AND ANALYTICAL TECHNOLOGIES

JUSTIN S. KNAPP

AND

WILLIAM L. CABRERA

EDITORS

Nova Science Publishers, Inc.

New York

For permission to use material from this book please contact us:
Telephone 631-231-7269; Fax 631-231-8175
Web Site: http://www.novapublishers.com

LIBRARY OF CONGRESS CATALOGING-IN-PUBLICATION DATA

Metabolomics : metabolites, metabonomics, and analytical technologies / editors, Justin S. Knapp and William L. Cabrera.
p. ; cm.
Includes bibliographical references and index.
ISBN 978-1-61668-006-0 (hardcover)
1. Metabolism--Regulation. 2. Physiological genomics. I. Knapp, Justin S. II. Cabrera, William L.
[DNLM: 1. Metabolomics. 2. Metabolism. 3. Models, Statistical. 4. Nutrigenomics. QU 120 M5873 2009]
QP171.M3823 2009
612.3'9--dc22

2009050743

Published by Nova Science Publishers, Inc. † New York

CONTENTS

PREFACE

Metabolomics is the logical progression of the study of genes, transcripts and proteins. Nutrients, gut microbial metabolites and other bioactive food constituents interact with the body at system, organ, cellular and molecular levels, and effect the expression of genome at several levels, and subsequently, the production of metabolites. This book presents an overview of nutrigenomics and metabolomics tools, and their perspective in livestock health and production. In addition, this book describes how lists of masses (molecular ions) and mass unit bins of interest are searched within online databases for compound identification, the extra biochemical data required for metabolite confirmation, how data are visualized and what the putative and protein sequences are associated with observed metabolic changes. Moreover, environmental metabolomics is the application of metabolomics to the investigation of both free-living organisms directly obtained from the natural environment or laboratory conditions. This book outlines some of the advances made in areas of plant environmental metabolomics. The applications of microbial metagenomics, the use of genomics techniques to the study of communities of directly in their diverse natural environments, are explored as well. Other chapters examine the abnormalities in metabolism of cancer cells, which could play a strategic role in tumour initiation and behavior.

As explained in Chapter 1, metabolism represents a complex system characterized by a high variability in metabolites' structures, concentrations and regulation ratios. Metabolic information can be stored in and analysed from metabolomic matrix consisting of concentrations of different metabolites analysed in different individuals (subjects). From such a matrix, different relationships can be highlighted between metabolites through a correlation analysis between their levels. When the set of all the metabolites are considered, their levels can be converted into ratios representing their metabolic regulations by reference to their metabolic profile. The complexity of network resulting from all the metabolic profiles can be structured by classifying the different profiles into different homogeneous groups representing different metabolic trends. Beyond the correlations between metabolites and their associations to different metabolic trends, a third variability can be observed consisting of atypical or original profiles in the population due to atypical values for some metabolites. Such cases provide information on extreme states in the studied population or on new emergent populations. Extreme cases are detected by combining analysis of variables with that of profiles leading to the outlier diagnostics. These three statistical aspects of variability analysis of metabolomic datasets are detailed in this chapter by different numerical examples and illustrations. Additionally to these correlation and distance matrices-based approaches, the chapter gives a background on different other metabolomic approaches based on other criteria/constraints/information stored in other types of matrices. According to the context, such matrices can contain (a) binary codes formulating the adjacencies between metabolites, (b) stoichiometric coefficients of metabolic reactions, (c) transition probabilities between different metabolic states, (d) partial derivatives of the system according to small perturbations, (e) contributions of different metabolic pathways, etc. Such matrices are used to describe/handle the complex structures, processes and evolutions of metabolic systems. General applications and interests of these different matrix-based approaches are illustrated in a first general section of the chapter, followed by a second detailed section on the correlation and distance-based analyses.

As discussed in Chapter 2, during the last decades compelling evidence has accumulated indicating that abnormalities in metabolism of cancer cells could play a strategic role in tumour initiation and behaviour. Abnormalities in metabolism are likely a consequence of several alterations in the complex network of signal transduction pathways, which may be caused by both genetic and epigenetic factors. An aberrant energy metabolism was recognized as one of the prominent features of the malignant phenotype, since the pioneering work of Warburg. It is now well established that the majority of tumours is characterized by a high glucose consumption, even under aerobic conditions, in absence of the Pasteur Effect, i.e. the lack of inhibition of glycolysis when cancer cells are exposed to normal oxygen consumption. Several investigators provided experimental data in support of a specific structure of the metabolic network in cancer cells. The 'tumour metabolome' has been defined as the metabolic tumour profile characterized by high glycolytic and glutaminolytic capacity and a high channelling of glucose carbons toward synthetic processes.

Despite no archetypal cancer cell genotype exists, facing the wide genotypic heterogeneity of each tumour cell population, some malignant features (i.e. invasion, uncontrolled growth, apoptosis inhibition, metastasis spreading) are virtually shared by all cancers. This paradox of a common clinical behaviour despite marked both genotypic and epigenetic diversity needs to be investigated by a Systems Biology approach and suggests that cancer phenotype should be considered as a sort of "attractor" in a specific space phase defined by thermodynamic and kinetic constraints. This is not the only phase space cancer cells are embedded into: in principle cancer cells, like any living entity travel along an integrated set of genetic, epigenetic or metabolomic parameters. A fractal dimension formalism can be used in a prospective reconstruction of cancer attractors. Studies conducted on MCF-7 and MDA-MB-231 breast cancer cells, exposed to different morphogenetic fields, show that metabolomic profile correlates to cell shape: modification of cell shape and/or architectural characteristics of the cancer- tissue relationships, induced through manipulation of environmental cues, are followed by significant modification of the cancer metabolome as well as of the fractal dimensions at both single cell and cell population level. These results suggest how metabolomic shifts in cancer cells need to be considered as an adaptive modification adopted by a complex system under environmental constraints defined by the non-linear thermodynamic of the specific attractor occupied by the system. Indeed, characterization of cancer cells behaviour by means of both metabolomic and fractal parameters could be used to build an operational and meaningful space phase, that could help in evidencing the transitions boundaries as well as the singularities of cancer behaviour. Hence, by revealing tumour-specific metabolic shifts in tumour cells, metabolic profiling enables drug developers to identify the metabolic steps that control cell proliferation, thus aiding the identification of new anti-cancer targets and screening of lead compounds for anti-proliferative metabolic effects.

As discussed in Chapter 3, molecular biology has recently concentrated on the determination of multiple gene-expression changes at the RNA level (transcriptomics), and into determination of multiple protein expression changes (proteomics). Similar developments have been taking place at metabolite small-molecule level, leading to the increasing expansion in studies now termed metabolomics. This approach can be used to provide comprehensive and simultaneous systematic profiling of metabolite levels in biofluids and tissues, and their systematic and temporal changes. Analysis of metabolites is not a new field; long prior to the development of the various ''omics'' approaches, the simultaneous analysis of the plethora of metabolites seen in biological fluids had been carried out largely, but historically it has been limited to relatively small numbers of target analytes. However, the realization that metabolic pathways do not act in isolation but rather as part of an extensive network has led to the need for a more holistic approach to metabolite analysis.

The main analytical techniques employed for metabolomics studies are based on NMR spectroscopy and mass spectrometry (MS), that, in turn, can be considered complementary each other. Neverthless, MS measurement following chromatographic separation offers the best combination of sensitivity and selectivity, so it is central to most metabolomics approaches. Either gas chromatography after chemical derivatization, or liquid chromatography (LC), with the newer method of ultrahigh-performance LC being used increasingly, can be adopted. Capillary electrophoresis coupled to MS has also shown some promises. Analyte detection by MS in complex mixtures is not as universal as for NMR and quantitation can be impaired by variable ionization and ion-suppression effects. A LC chromatogram is generated with MS detection, usually using electrospray ionization (ESI), and both positive- and negative-ion chromatograms can be recorded. The utilization of nano-ESI can reduce ionization suppression effects due to the increased ionization efficiency. Mass analyzer able to produce high mass resolution, mass accuracy, and tandem MS, such as quadrupole-time-of-flight (Q-TOF) or high-resolution ion trap instruments, are employed.

Direct infusion (DI)-MS/MS using Fourier transform ion cyclotron resonance mass spectrometers provides a sensitive, high-throughput method for metabolic fingerprinting. Unfortunately, DI-MS analysis is particularly susceptible to ionization suppression arising from competitive ionization. In metabolomics, matrix assisted laser desorption-ionization (MALDI) has largely been confined to the targeted analysis of high-molecular weight metabolites due to the substantial signals generated by the matrix in the low-molecular-weight region (<1,000 *m/z*). Recent advancements in laser desorption techniques include desorption-ionization MS from porous silicon chips and matrices that have minimal background signals in the low-molecular-weight region. These offer new opportunities for the utilization of MALDI ionization in metabolite screening and fingerprinting employing MALDI-TOF/TOF. However, the technique is still subject to ion suppression and yields poor quantitative detection. Desorption ESI (DESI), a new ambient, soft-ionization technique that combines features from both ESI and desorption-ionization methods, allows the direct analysis of animal and plant tissues. However, DESI experimental conditions typically require optimization for each sample type, so time must be invested initially in optimizing the experimental parameters.

It was quoted in 1953 at the 'Changing flora of Britain' conference that 'we should mobilize a team which could tackle the problems, genetical, cytological, physiological, ecological and chemical, and see whether out of the available mass of material we can not only reach a settled nomenclature... but make a serious contribution to the problems of evolution' (Raven 1953). Nearly 60 years later, we are now starting to assemble such genomic and post-genomic teams with the appropriate infrastructure, technology and bioinformatic power to answer questions in plant ecology and evolution. Of course, the chemical component of the team can now be termed environmental metabolomics and is progression of the study of genes (genomics), mRNA (transcriptomics) and proteins (proteomics).

The main intention of plant metabolomics research is to provide an unbiased assessment of metabolism across multiple pathways. Ideally, all plant metabolites should be identified and quantified at a relevant temporal and spatial scale by untargeted metabolomic fingerprinting using mass spectrometry or NMR or by targeted, quantitative metabolite profiling; to provide a comprehensive view of metabolism. Such global screening of the metabolites has been termed biochemical, or metabolic phenotyping. This approach builds on the much valid work carried out by plant biologists such as Richard Dixon and Jeffrey Harborne to name but a very few. However, the ease of application and software to analyse results, alongside the increase in interdisciplinary science, has opened up such technology to more research fields to answer a wider range of questions.

Chapter 4 will outline some of the advances made in such areas of plant environmental metabolomics.

As explained in Chapter 5, despite enormous advancements in microbial culturing methods, more than 95% of the global microbial diversity still remains cryptic. Microbial metagenomics- the applications of modern genomics techniques to the study of communities of microbes directly in their diverse natural environments, bypassing the need for isolation, is changing our comprehension of the biosphere. Advances in technologies designed to access this wealth of genetic information through environmental nucleic acids extraction and analysis have provided the means of overcoming the limitations of conventional culture-dependent microbial exploitation. Further developments and applications of these methods promise to provide opportunities to link distribution and identity of gut microbes in their natural habitats, and explore their use for promoting livestock health and industrial biotechnological applications.

Nutrition exhibits the most important life-long environmental impact on health. Nutrients, gut microbial metabolites and other bioactive food constituents interact with body at system, organ, cellular and molecular levels, and affect the expression of genome at several levels, and subsequently, the overall production of metabolites. Direct measurement of cellular metabolites is essential for the study of biological processes, and may allow causes of disease, toxicological progression, and novel disease-biomarkers to be identified. Advances in analytical techniques and the algorithms for management of the data has allowed a precise and global analysis of biological substances such as DNA (genomics), RNA (transcriptomics), proteins (proteomics) and smaller molecules (metabolomics). Holistic "**omics**" approaches are indispensable to cover the complex nutrient-cell and gut microbial-host interactions. Chapter 6 presents an overview of nutrigenomics and metabolomics tools with reference to their perspective in livestock health and production.

Metabolomics is a rapidly growing field with the goal of measuring and interpreting the complex time and condition dependent concentration, activity or flux of metabolites in cells, tissues and other biosamples. On the other side, the integrated approach to studying biological systems in Systems Biology has led to significant improvement of our understanding of such systems. Since biological circuits are hard to model and simulate, many efforts are being made to develop computational models that can handle their intrinsic complexity. However, a large part of the biological networks remains unknown and hard to understand and Metabolomics technology that allows simultaneous acquisition of many metabolite measurements can lead to further analysis for discovering novel pathway components and unknown network relationships. Metabolic networks are structurally complex and behave in a stochastic fashion. In Chapter 7 the authors describe how symbolic-statistical machine learning techniques can be used to reconstruct metabolic networks from metabolic profiling data. The authors show that symbolic machine learning methods have the power to model structural and relational complexity while statistical machine learning ones provide principled approaches to uncertainty modeling. They apply a symbolic-statistical learning framework to analyze sequences of reactions for biologically active paths in metabolic networks. The authors show through experiments that their approach provides a robust methodology for machine reconstruction of metabolic networks from metabolomic data.

As discussed in Chapter 8, generally, a large proportion of the genes in any genome encode enzymes of primary and specialized (secondary) metabolism [1]. Not all primary metabolites, those that are found in all or most species, have been identified and only a small portion of the estimated hundreds of thousand specialized metabolites, those found only in restricted lineages, have been studied in any species [1]. Fridman and Pichersky [1] noted that the correlative analysis of extensive metabolic profiling and gene expression profiling had proven a powerful approach for the identification of candidate genes and enzymes, particularly those in secondary metabolism [2]. It is rapidly becoming possible to measure hundreds or thousands of metabolites in small samples of biological fluids or tissues. Arita [3] said that metabolomics, a comprehensive extension of traditional targeted metabolite analysis, had recently attracted much attention as the biological missing pieces that can complement transcriptome and proteome analysis. Metabolic profiling applied to functional genomics (metabolomics) is in an early stage of development [4]. Fridman and Pichersky [1] said that the final characterization of substrates, enzymatic activities, and products requires biochemical analysis, which had been most successful when candidate proteins have homology to other enzymes of known function. To facilitate the analysis of experiments using post-genomic technologies, new concepts for linking the vast amount of raw data to a biological context have to be developed [5]. Visual representations of pathways help biologists to understand the complex relationships between components of metabolic network [5].

Organ function can only be completely understood through knowledge of molecular and cellular processes within the constraints of structure-function relations at the tissue level [6]. Knowledge on integrative computational physiology is required. Cellular components interact with each other to form networks that process information and evoke biological responses [7]. Today different database systems for molecular structures (genes and proteins) and metabolic pathways are available. All these systems are characterized by the static data representation [8]. For progress in biotechnology the dynamic representation of this data is important. The metabolism can be characterized as a complex biochemical network [8]. A deep understanding of the behavior of these networks requires the development and analysis of mathematical models [7]. Computer modeling of metabolic networks can help better understand complex metabolism [9 - 10]. As previously mentioned, mathematical modeling is one of the key methodologies of metabolic engineering [11]. Based on a given metabolic model different computational tools for the simulation, data evaluation, systems analysis, prediction, design and optimization of metabolic systems have been developed [11]. More details on mathematical modeling can be seen in another specific chapter in this book. In additional to mathematical model, graph-based analysis of metabolic networks is another widely used technique in metabolomics [12].

Various types of evidence have implicated estrogens in the etiology of human breast cancer [1-8]. They are generally thought to cause proliferation of breast epithelial cells through estrogen receptor-mediated processes [4]. Rapidly proliferating cells are susceptible to genetic errors during DNA replication, which, if uncorrected, can ultimately lead to malignancy. While receptor-mediated processes may play an important role in the development and growth of tumors, accumulating evidence suggests that specific oxidative metabolites of estrogens, if formed, can be endogenous ultimate carcinogens that react with DNA to cause the mutations leading to initiation of cancer [6-9]. Thus, estrogen metabolites, specifically catechol estrogen-3,4-quinones, are hypothesized to be endogenous initiators of breast, prostate and other human cancers.

Several lines of evidence, including metabolism and carcinogenicity studies by Liehr and coworkers, led to the recognition that the 4-hydroxylated estrogens play a major role in the genotoxic properties of estrogens [1-3]. In Chapter 9, the authors have hypothesized that the estrogens estrone (E_1) and estradiol (E_2) initiate breast and other human cancers by reaction of their electrophilic metabolites, catechol estrogen-3,4-quinones [E_1(E_2)-3,4-Q], with DNA to form depurinating adducts [5-8]. These adducts generate apurinic sites leading to mutations that may initiate breast, prostate and other human cancers [6-9].

In: Metabolomics: Metabolites, Metabonomics... ISBN: 978-1-61668-006-0
Editors: J.S. Knapp and W.L. Cabrera, pp. 1-85

Chapter 1

CORRELATIONS- AND DISTANCES-BASED APPROACHES TO STATIC ANALYSIS OF THE VARIABILITY IN METABOLOMIC DATASETS. APPLICATIONS AND COMPARISONS WITH OTHER STATIC AND KINETIC APPROACHES

Nabil Semmar[*]

ISSBAT, Institut Supérieur des Sciences Biologiques Appliquées de Tunis, Tunisia.
Laboratoire de Pharmacocinétique et Toxicocinétique,
Pharmacy School of Marseilles, France

Abstract

Metabolism represents a complex system characterized by a high variability in metabolites' structures, concentrations and regulation ratios. Metabolic information can be stored in and analysed from metabolomic matrix consisting of concentrations of different metabolites analysed in different individuals (subjects). From such a matrix, different relationships can be highlighted between metabolites through a correlation analysis between their levels. When the set of all the metabolites are considered, their levels can be converted into ratios representing their metabolic regulations by reference to their metabolic profile. The complexity of network resulting from all the metabolic profiles can be structured by classifying the different profiles into different homogeneous groups representing different metabolic trends. Beyond the correlations between metabolites and their associations to different metabolic trends, a third variability can be observed consisting of atypical or original profiles in the population due to atypical values for some metabolites. Such cases provide information on extreme states in the studied population or on new emergent populations. Extreme cases are detected by combining analysis of variables with that of profiles leading to the outlier diagnostics. These three statistical aspects of variability analysis of metabolomic datasets are detailed in this chapter by different numerical examples and illustrations. Additionally to these correlation and distance matrices-based approaches, the chapter gives a background on different other metabolomic approaches based on other criteria/constraints/information stored in other types of matrices.

[*] E-mail address: nabilsemmar@yahoo.fr. (Corresponding author)

According to the context, such matrices can contain (a) binary codes formulating the adjacencies between metabolites, (b) stoichiometric coefficients of metabolic reactions, (c) transition probabilities between different metabolic states, (d) partial derivatives of the system according to small perturbations, (e) contributions of different metabolic pathways, etc. Such matrices are used to describe/handle the complex structures, processes and evolutions of metabolic systems. General applications and interests of these different matrix-based approaches are illustrated in a first general section of the chapter, followed by a second detailed section on the correlation and distance-based analyses.

I. Introduction

Metabolomics aims at unbiased and comprehensive analysis of the biosynthesis, regulation, distribution and control processes of the metabolites in cells, tissues or organisms (Figure 1) (Goodacre et al., 2004; Sumner et al., 2003; Kell, 2004; Sweetlove and Fernie, 2005; Fernie et al., 2004). It is a multidisciplinary field including many approaches which analyse the metabolites' content of a biological system in relation to several biological factors (genome, proteome, physiology, environment) leading to a better understanding of the organization, behaviour and control of metabolic networks (Olivier et al., 1998; Roessner et el., 2001; Nicholson et al., 1999; Kell, 2002; Ott et al., 2003; Weckwerth, 2003).

Metabolism represents a complex system characterized by a great variability of chemical structures, biosynthesis levels, regulation ratios and flux distributions of metabolites (Kacser and Burns, 1973; Savageau, 1976; Atkinson, 1977; Hayashi and Sakamoto, 1986; Fell, 1996; Heinrich and Schuster, 1996). Such complex variability can be observed from continuums of metabolic profiles in which the metabolites vary qualitatively and quantitatively the ones in favour or at the expense of others. Subsequently, statistical methods are needed to detect, quantify, classify and associate different kinds of variations at metabolite and at metabolic pathway levels.

Statistically, the metabolic variability is analysed from a dataset or matrix consisting of *n* rows (or *n* profiles) and *p* columns (*p* metabolites). Therefore, three kinds of variability can be analysed, viz. along the rows, along the columns and by associating rows and columns (Nicholson et al., 1999; Semmar et al., 2001, 2005a, 2007, 2008; Lindon et al., 2007; Denkert et al., 2008):

Column analysis is closely linked to a correlation screening between variables. The set of different correlations between metabolites (variables) helps to detect different trends that can be interpreted as different metabolic pathways in the metabolic network. Row analysis aims to quantify similarities between individual profiles on the basis of distances or similarity indices calculus. The resulting calculated distance or similarity matrix can be used to classify profiles into different groups that can be interpreted in terms of different polymorphim poles. Association analysis between rows and columns provides complementary information concerning original or atypical profiles due to relatively high (or low) values for some metabolites. Such analysis is closely linked to outlier diagnostics which use different distance kinds to detect atypical profiles according to different statistical criteria. The application of different outlier diagnostic criteria allows to check if atypical profiles are confirmed by different criteria or particularly highlighted by only one criterion

Apart from these three basic statistical analyses (column-, row-, and association-analyses), helping to describe the variability of metabolic datasets under correlation, classification and outlier diagnostic aspects, the metabolomics includes other approaches requiring different matricial formulations. Such matrix-based approaches offer static and kinetic analyses of the variability in metabolic network. Static approaches include connectivity, stoichiometric and combined patterns analyses which are based on adjacency, stoichiometric and Scheffe mixture matrices, respectively (Ivanciuc et al., 1993; Ponce, 2004; Yanai et al., 2008; González-Díaz et al., 2007; Todeschini and Consonni, 2000; Llaneras and Picó, 2008; Steuer, 2007; Papin et al., 2003; Papin et al., 2004; Calik and Ozdamar, 2002; Semmar et al., 2007; Eide, 1996; Pattarino et al., 1993; Nyieredy et al., 1985; Glajch et al., 1982). Kinetic or temporal approaches include stability analysis and stochastic analysis based on Jacobian and Markov transition probability matrices, respectively (Yang et al., 2004; Steuer, 2007; Crampin et al., 2004; Fall et al., 2005; Cruz-Monteagudo et al., 2008a, b; Gonzalez-Diaz et al., 2005, 2008). These different matrix-based approaches will be briefly presented in the first part of this chapter to give a general background on metabolomic approaches. The second part of this chapter presents details and illustrations on the principles and applications of the three basic statistical methods consisting of row-, column- and association analyses, on the basis of different correlation and distance matrices.

II. Diversity and Intrinsic Variability of Metabolomic Datasets

II.1. Presentation of Metabolomic Datasets

A metabolomic dataset consists of several individuals (patients/animals/plants) in whom/which the concentrations of several metabolites were measured. The set of concentrations of p metabolites analysed in n individuals is stored into a matrix (n rows $\times$ p columns); the rows represent individual profiles, each one containing p metabolites (p variables) which are stored in columns (Figure 3). Each row of the concentration dataset represents initially a chemical profile; such a profile can be converted into a metabolic profile by dividing the concentration C_j of each metabolite j by the sum of concentrations all the metabolites (Figure 4).

A metabolomic dataset can be static or kinetic whether its n rows are measured at a one time or at different times (Figure 3). In the second case, the n profiles of p metabolites can be grouped *a priori* into q subsets (for each metabolite separately) representing successive q time-dependent profiles of the metabolite in the q studied subjects (e.g. q patients).

II.2. Repeated Experiments for Highlighting of Metabolic States

Metabolic systems (biological systems) are complex because of the high number of their components, the multiple interactions between them, and the numerous internal and external variability sources which result in several different states of the system. Because of such complexity and variability, single measurements are not sufficient to extract reliable information on system backbone. Therefore, repeated measurements (or replicates) are needed to gain information on the variability and the most probable (or the average) state of the system (Figure 2).

Even under approximately constant experimental conditions, metabolism is a highly dynamic system, responding to small factor (stimuli) variations. For example, slight differences in enzyme concentrations or metabolic oscillations (among other factors) contribute to variability in metabolite levels. The results are metabolic fluctuations which propagate through metabolic reaction chains and ultimately induce an emergent and experimentally observable pattern of metabolites (Steuer et al., 2003a, b; Weckwerth, 2003; Weckwerth et al., 2004 a, b; Morgenthal et al., 2005, 2006).

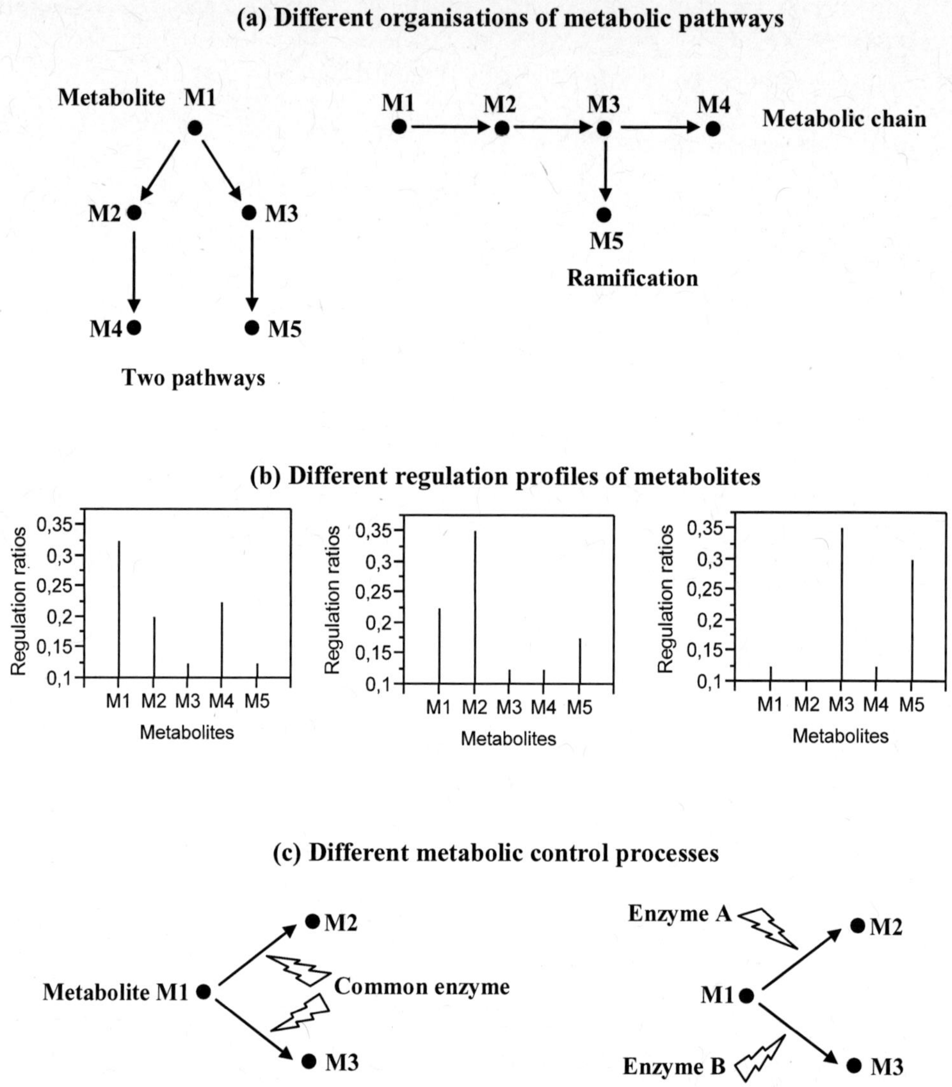

Figure 1. Schematic representations of different objectives in metabolomics. Analysis of pathways' organization (a), phenotypic expressions (b) and control processes (c) of metabolic networks.

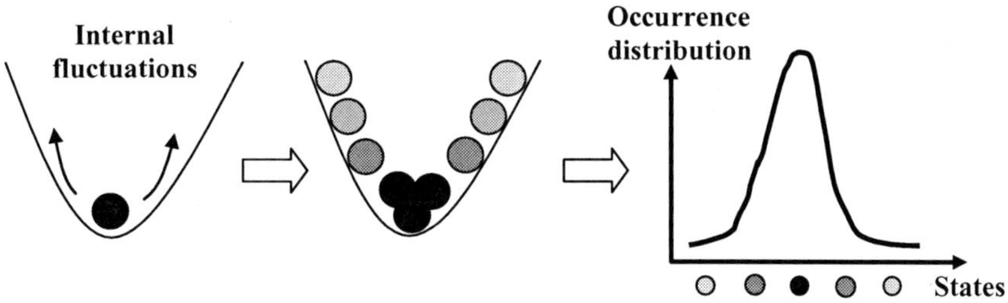

Figure 2. Internal fluctuations of a system resulting in a characteristic distribution of its different possible states, and making replications to be required for its reliable analysis.

			Metabolites					
Subject	**Time (h)**	**Profiles**	M1	M2	…	Mj	…	Mp
1	0.5	1	C_{11}	C_{12}	…	C_{1j}	…	C_{1p}
1	1	2	C_{21}	C_{22}	…	C_{2j}	…	C_{2p}
1	2	3	C_{31}	C_{32}	…	C_{3j}	…	C_{3p}
1	3	4	C_{41}	C_{42}	…	C_{4j}	…	C_{4p}
1	4	5	C_{51}	C_{52}	…	C_{5j}	…	C_{5p}
2	0.5	6	C_{61}	C_{62}	…	C_{6j}	…	C_{6p}
2	1	7	C_{71}	C_{72}	…	C_{7j}	…	C_{7p}
2	2	8	C_{81}	C_{82}	…	C_{8j}	…	C_{8p}
2	3	9	C_{91}	C_{92}	…	C_{9j}	…	C_{9p}
2	4	10	C_{101}	C_{102}	…	C_{10j}	…	C_{10p}
:	:	:	:	:	:	:	:	:
:	:	:	:	:	:	:	:	:
:	:	i	C_{i1}	C_{i2}	…	C_{ij}	…	C_{ip}
:	:	:	:	:	:	:	:	:
:	:	:	:	:	:	:	:	:
q	:	n	C_{n1}	C_{n2}	…	C_{nj}	…	C_{np}

Kinetic profile of metabolite *p* in subject 1

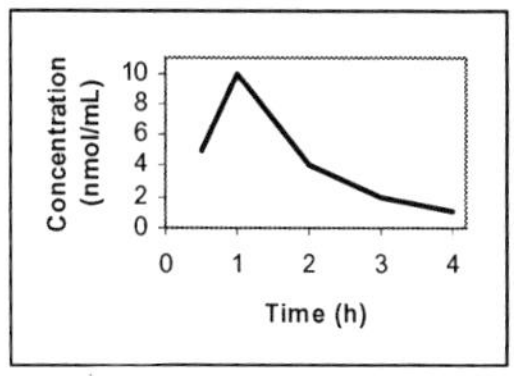

Concentration value of metabolite *j* in the profile *i*

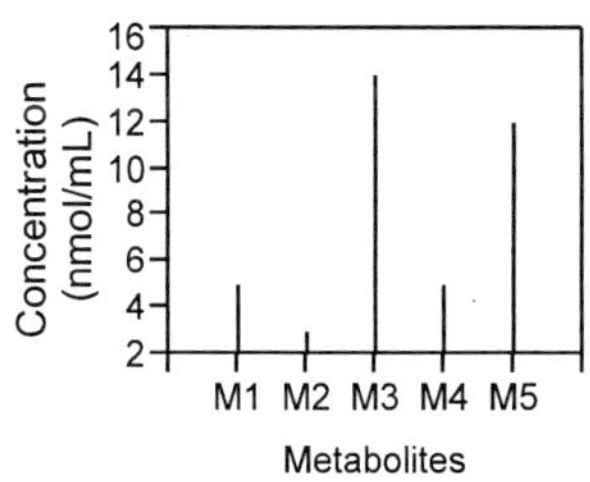

Concentration profile *i*

Figure 3. Representation of a metabolomic dataset (*n* profiles × *p* metabolites) with its different parameters. Concentration and kinetic profiles are read along rows and column, respectively.

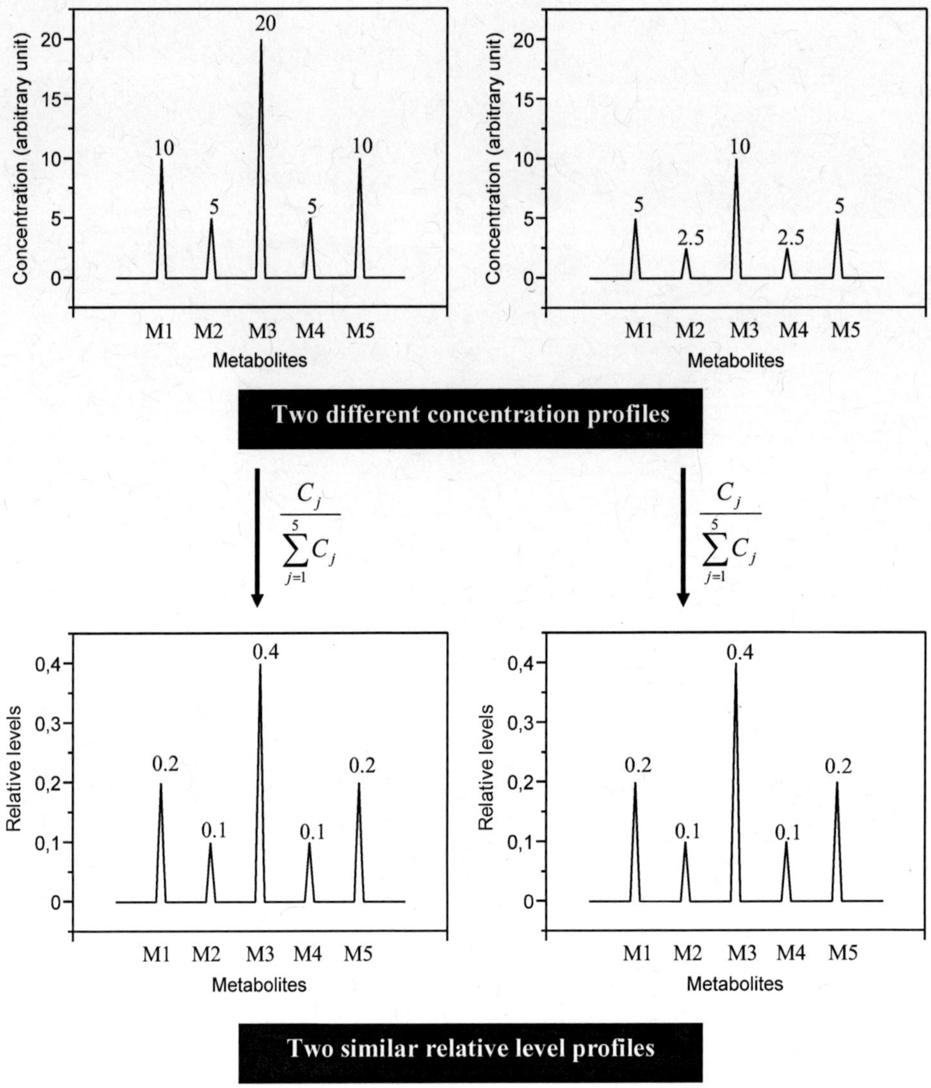

Figure 4. Standardization of concentration profiles giving relative level (or regulation) profiles.

III. General presentation of Different Metabolomic Approaches and Parameters

III.1. Classification of Metabolomic Approaches Based on Different Criteria

Metabolomic approaches can be classified according to different criteria depending on the goal, dataset, matrix formulation, etc. . Under the goal criterion, one can distinguish descriptive and predictive approaches. The first ones tend to describe complex structures of metabolic systems through different variability trends; the second ones aim to predict the behaviour of the system subjected to different controllability factors (Figure 5). In other words, the descriptive approaches aim identification of different variability trends/states of metabolic systems; for that, metabolomic datasets are analyzed in order to highlight how units or individuals separate the ones from the others leading to multidirectional behaviours within the system. This helps to identify substructures or system components from which the biological (metabolic) complexity can be described. However in predictive approaches, the steps and the aim are inverted: different variability factors are combined in order to estimate precisely what internal state could be acquired by the system. This helps to identify the most significant factors which control the system.

Metabolomic approaches can be also classified according to the type of datasets. Several classifications are considered, one of the most classical consists in separating static from kinetic datasets. These two kinds of datasets differ by the fact that the variable time is not considered or considered, respectively. In the first case (static), a dataset is treated as a whole block to obtain a global picture on the components or states of the system. In the second case (kinetic), a dataset is undertaken as succession of different subsets varying in time leading to analyze a serial of small and successive pictures representing a sequence of the system behaviour (Figure 6).

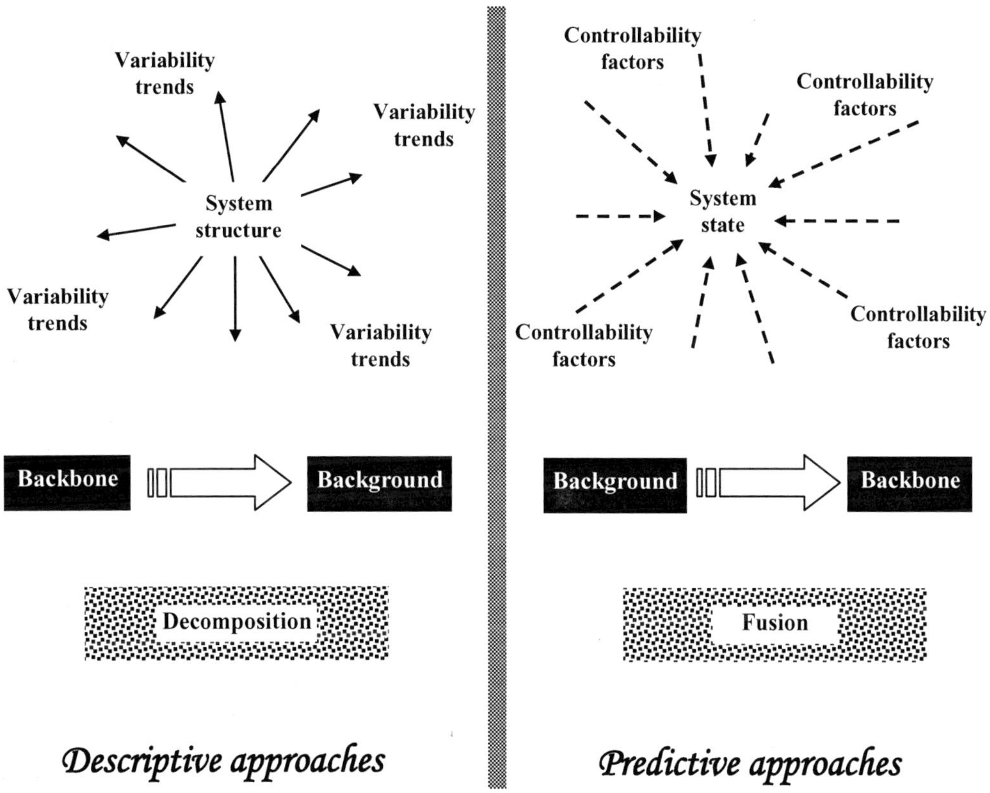

Figure 5. Schematic representation of the general goals of descriptive and predictive approaches.

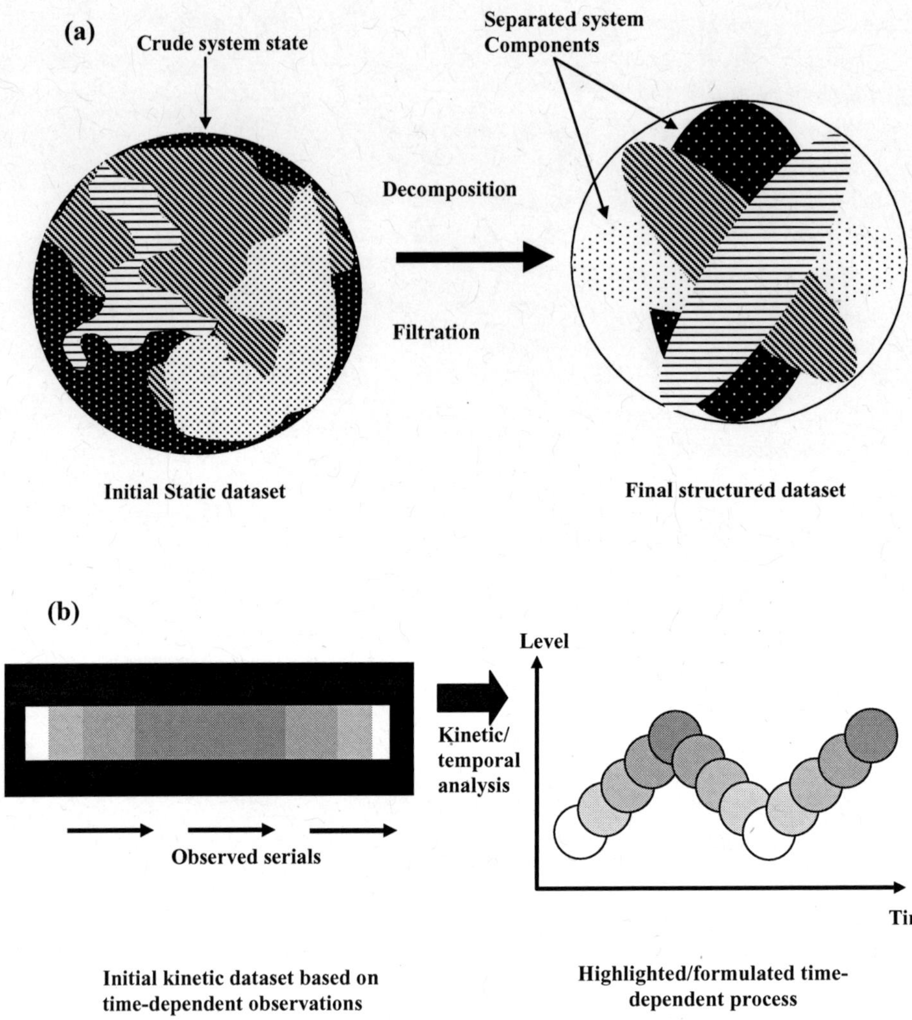

Figure 6. Schematic representation of static (a) and kinetic/temporal (b) analyses.

Metabolic systems are known to be complex networks in which many components/processes are interconnected. Representations of such inter-connections require matrix formulations which provide flexible tools to store multi-path information. On this basis, different metabolomic approaches can be considered by reference to the matrix tool used for metabolic system analysis. Matrix tools can be used to describe/treat distances, correlations, connectivity, transition, reactions, equilibrium, mixtures between different components of biological (metabolic) system (Figures 7-10, 13) (Crampin et al., 2004; Semmar et al., 2001, 2005b, 2008 ; Sumner et al., 2003; Gonzalez-Diaz et al., 2008; Gonzalez-Diaz, 2008; Kose et al., 2001; Llaneras and Picó, 2008; Steuer, 2007; Stelling, 2004).

This chapter will focus particularly on distance and correlation computation approaches used for static analysis of metabolic systems. Before the detailed sections on distance- and correlation-based approaches, a brief description of other matrix-based approaches will be presented in the following sections, particularly on the constraint and neighbouring notions.

III.2. Boolean Matrix Based Approaches

Connectivity between different (*p*) components of system can be codified by using a binary formalism (Boolean code) consisting of 1 if two components are connected and 0 if not (Estrada and Bodin, 2008; Estrada, 2006, 2007; Vilar et al., 2005; Kose et al., 2001; Janga and Babu, 2008) (Figure 7). For instance, the value 1 can be attributed for two neighbour or two linked metabolites (e.g. precursor-product) in the metabolic system. The resulting adjacency matrix can be graphically represented by a multigraph containing *p* nodes or vertices (corresponding to the *p* system components) which are connected by edges.

Metabolites	Unchanged states	Transformation reactions	
M1	M1 → M1	M1 → M2	M1 → M4
M2	M2 → M2	M2 → M3	M2 → M4
M3	M3 → M3	M3 → M4	M3 → M5
M4	M4 → M4	M4 → M5	
M5	M5 → M5		

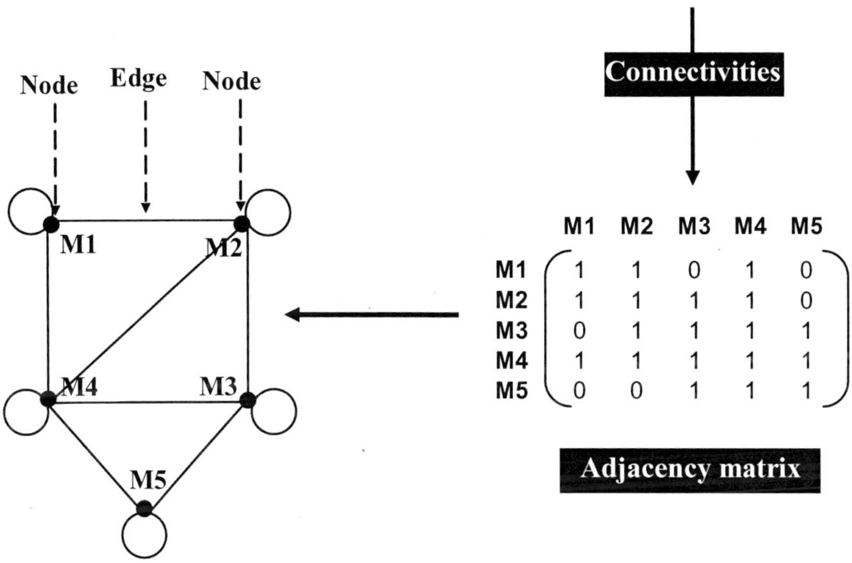

Figure 7. Boolean formalism of connectivities between metabolites in a metabolic system and corresponding graphical representation.

III.3. Transition Matrix Based Approaches

The variation of a biological (metabolic) system in time can be described by a finite number of successive states (Guttorp, 1995; Tamir, 1998). For example, at a given time, a metabolic system can be described by the set of the metabolites present in the network. Between two successive times *t* and *t*+1, each molecule of metabolite *j* can be subjected to different exclusive processes: it can remain unchanged, or be transformed to another metabolite among different possible ones. The exclusivity between the different metabolic processes makes possible to analyse the evolution of the metabolic system on the basis of probabilities of metabolites to transit between different successive states. These probabilities (0 ≤ ≤ 1) are stored into a transition matrix the rows and columns of which represent the initial (e.g. precursor) and final (e.g. product) elements (Figure 8).

(a)

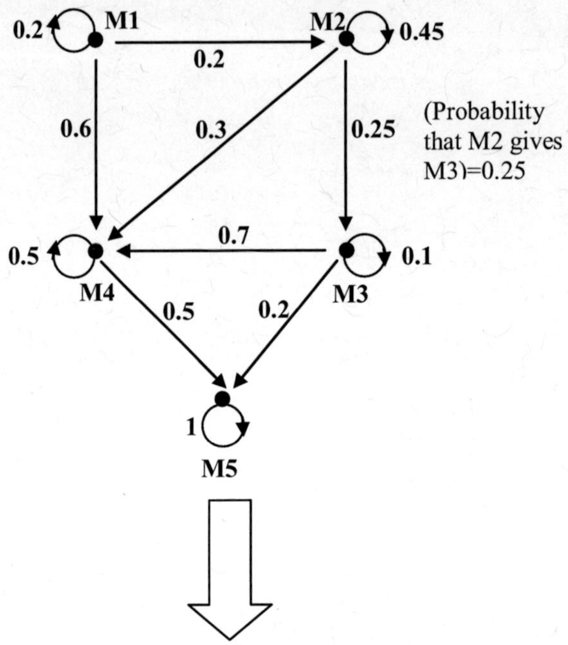

(b)

Final metabolite

Initial metabolite		M1	M2	M3	M4	M5
	M1	0.2	0.2	0	0.6	0
	M2	0	0.45	0.25	0.3	0
	M3	0	0	0.1	0.7	0.2
	M4	0	0	0	0.5	0.5
	M5	0	0	0	0	1

Tansition probability matrix

Figure 8. Basic example representing a transition probability matrix (b) and its graphical representation (a).

III.4. Stoichiometric Matrix Based Approaches

When all the metabolic reactions of a metabolic network are known, it is possible to translate the transformation processes between precursors and products in terms of stoichiometric coefficients. Such algebraic coefficients take positive or negative values for appearing and disappearing metabolites, respectively. The absolute value of a stoichiometric coefficient indicates the number of molecules implied in an elementary reaction. The set of coefficient is stored into a stoichiometric matrix the rows and columns of which represent the metabolites and the reactions, respectively (Figure 9).

Metabolites	Transformation reactions R_k	
M1	R_1: M1 → M2	R_2: M1 → M4
M2	R_3: M2 → M3	R_4: M2 → M4
M3	R_5: M3 → M4	R_6: M3 → M5
M4	R_7: M4 → M5	
M5	-	-

Reactions

Metabolites	R1	R2	R3	R4	R5	R6	R7
M1	-1	-1	0	0	0	0	0
M2	+1	0	-1	-1	0	0	0
M3	0	0	+1	0	-1	-1	0
M4	0	+1	0	+1	+1	0	-1
M5	0	0	0	0	0	+1	+1

Stoichiometric matrix

Figure. 9. Translation of a metabolic process network into a stoichiometric matrix based on the stoichiometric coefficients of the different metabolites for the different chemical reactions.

Stoichiometric approaches represent powerful tools for metabolic modelling when time measurements are not available. They make possible to exploit the knowledge about the cell metabolism structure, without considering the intracellular kinetic processes (complex and still not well understood). Stoichiometric models have been used to (Llaneras and Picó, 2008; Morgan and Rhodes 2002; Stelling, 2004):

- estimate the metabolic flux distribution under given circumstances in the cell at some given moment (metabolic flux analysis) (Williams et al., 2008; Ettenhuber et al., 2005; Kruger et al. 2003),
- predict the metabolic flux distribution on the basis of some optimality hypotheses (flux balance analysis) (Schilling et al., 2001),
- analyse the structure of metabolism by providing information about systemic characteristics of the cell under investigation (pathway analysis) (Schilling et al., 2001).

Using stoichiometric matrix, the mass balance for each intracellular behavior is disregarded with the assumption of pseudosteady state for internal metabolites. Thereby, the mass balances can be described by a homogeneous system of linear equations. This system constraints the flux distribution that can be achieved by the metabolic network, but it does not predict the actual distribution. To this end, additional constraints, such as irreversibility or capacity constraints, can be incorporated in order to determine what functional states, i.e. flux distributions, can and cannot be achieved by a cell under certain conditions.

III.5. Jacobian Matrix Based Approach

Biological (metabolic) systems can be analysed on the basis of their ability to opposite or to be subjected to perturbations. This approach is known under the term of stability analysis (Steuer, 2007; Fall et al., 2005):

Stability analysis aims to examine the behaviour of a system around its equilibrium state. Equilibrium state of continue dynamical system can be represented by a stationary regimen. The question of stability can be asked in different manners:

- If the system is deviated from the equilibrium, does it return to this state?
- Does small perturbation, moving away the system from its stationary regimen, result in amplifications in time?

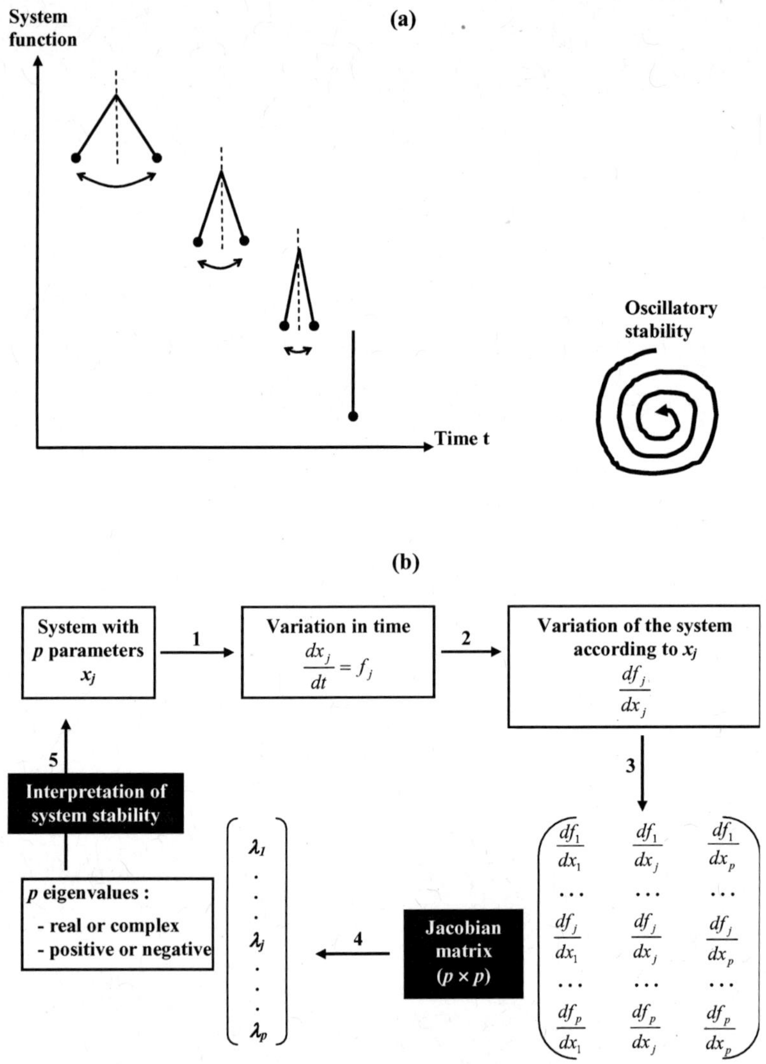

Figure 10. Basic concepts of stability analysis of dynamical systems; (a) basic example of a dynamic stability; (b) origin, form and usefulness of Jacobian matrix in stability analysis of dynamical system.

Such questions imply the analysis of all the possible perturbations of the system in relation to small variations of its variables in time (Figure 10a). In other words, we have to analyse the stability of a system in relation to its parameters x_j (e.g. metabolites' concentrations) varying in time (Figure 10b):

At equilibrium point, the derivatives of all the parameters x_j with respect to time are null: $f_j = \frac{dx_j}{dt} = 0$.

From the analytical form of f_j, the equilibrium point x_j^* for each parameter j will be calculated. With p parameters x_j (j=1 to p), one expects p values x_j^* to calculate from p derivative equations f_j=0. Moreover, the p functions f_j will be derived with respect to each x_j (one at once), resulting in ($p \times p$) partial derivatives. The set of all the partial derivatives $\frac{df_j}{dx_j}$ is called Jacobian matrix ($p \times p$) (Figure 10b3).

From the Jacobian matrix J, the stability of the system around the equilibrium point is analysed. For that, all its partial derivatives are calculated at the equilibrium values x_j^* to obtain the Jacobian matrix J^*. Therefore, stability analysis of the system consists in:

- Calculating the eigenvalues λ_j of J^* (there are as much eigenvalues as parameters) (Figure 10b4), and
- Interpreting their natures and their signs in terms of stability or non-stability of the system (Figure 10b5).

Eigenvalues of a biological (metabolic) system can be real or complex on the hand, and positive or negative on the other hand (Figure 11):

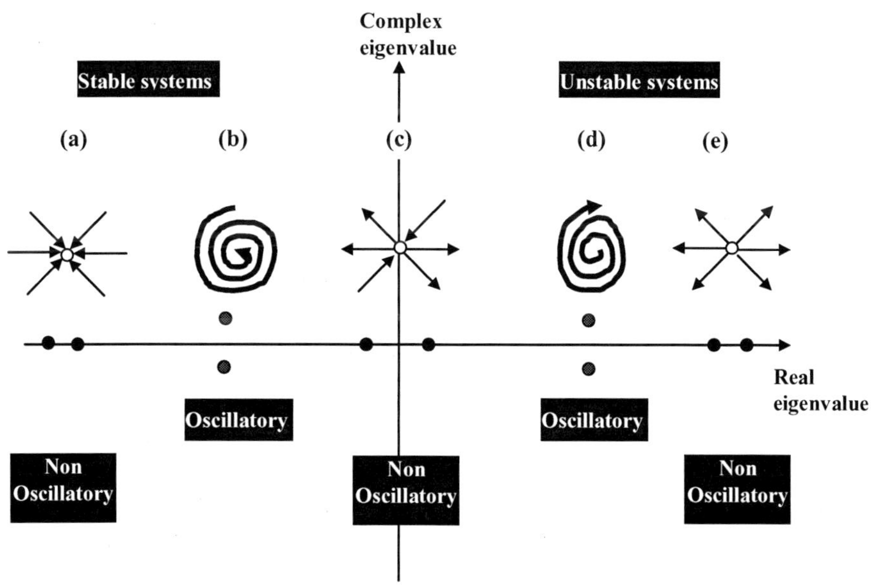

Figure 11. different equilibrium states of a dynamical system interpreted according to nature and sign of the eigenvalues of Jacobian matrix.

Complex eigenvalues indicate an oscillatory system (Figure 11b, d). Inversely, a system with only real eigenvalues is non-oscillatory (Figure 11a, c, e). Therefore, the sign of eigenvalue provides information on the convergence or divergence of the system, i.e. on its stability or non-stability, respectively: a negative real eigenvalue (or real part) indicates a stable system, i.e. a system which converges (returns) to steady state (equilibrium) (after disruption) (Figure 11a, b). A positive real eigenvalue (or real part) indicates an unstable solution which means that the system never converge to steady state (Figure 11d, e). When some eigenvalues are positive and others are negative, the system has a sell point, which represents a fragile equilibrium state leading the system to be unstable (Figure 11c).

III.6. Scheffe Matrix Based Approach

Metabolic system can be undertaken under a background consisting of different observed regulation patterns issued from a common metabolic backbone considered as a central black box. Such patterns represent extreme metabolic trends which are characterized by more or less high regulation ratios of some metabolites due to more or less high expressions of some metabolic pathways (Figure 12a). Therefore, any observed metabolic profile can be considered as more or less closer to one of these metabolic patterns. Statistically, any observed profile can be expressed by a particular combination of the extreme patterns affected by appropriate weights: the variation of the combined pattern weights leads to a set of combinations corresponding to different average patterns (Figure 12b); such mixture-resulting average patterns will be more or less close to the different observed profiles. Under a chemical aspect, the combination of different patterns can be assimilated to a concentration/dilution process where the more weighted patterns will be concentrated and the less ones will be diluted in the mixture.

After iterations of the complete set of combinations (Figure 12c), a response matrix of smoothed profiles is obtained by averaging the repeated average profiles' matrices (Figure 12d). Such a final smoothed data matrix is then used to analyze graphically the metabolic processes which would be responsible for the observed polymorphism (Figure 12e). More details are given in Figures 13 and 14.

The complete set of linear combinations of extreme states (or basic components) can be formalised by a mixture design represented by Scheffe matrix (Figure 13) (Sado and Sado, 1991; Scheffe, 1958, 1963; Duineveld et al., 1993). The total number N of combinations to carry out depends on two parameters: (i) the number of components (patterns) to combine and (ii) the number n (constant) of elements (e.g. metabolic profiles) to mix in each combination. An illustration of the Scheffe matrix is given for q=4 components and n=10 elements representing the q components in each mixture (Figure 13b). Each combination can be summarized by an average profile (Figure 14a). The mixture design is iterated several times to take into account the variability of the observed metabolic profiles (Figure 14b). From k iterations, a final response matrix containing a complete set of smoothed metabolic profiles is calculated by averaging all the k response matrices (Figure 14c). This smoothed final response matrix can be used to graphically analyse the variability between regulation ratios of different metabolites in order to understand metabolic processes responsible for the observed polymorphism (Figure 14d):

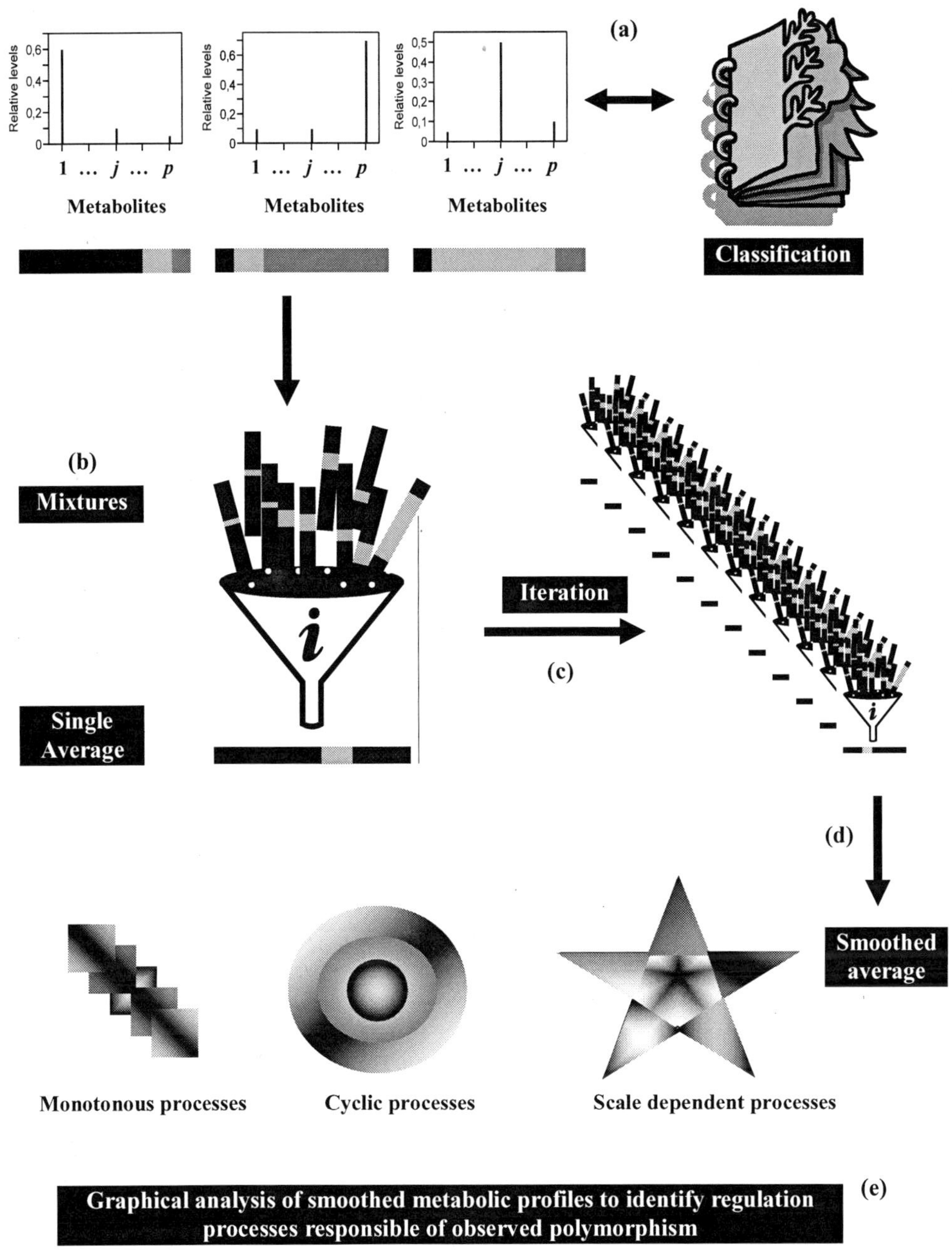

Figure 12. Schematic representation of the steps of metabolomic approach consisting in iteratively combining observed metabolic profiles representing different patterns to obtain a dataset of smoothed profiles helping to analyse graphically the regulation processes responsible of the observed chemical polymorphism (Semmar et al., 2007; Semmar, 2010).

As the observed patterns represent a background of the metabolic system, their iterative combinations can provide a way to access to a backbone of such common system. On the basis of this concept, a new metabolomic approach was developed from which the flexibility of metabolic regulations was graphically highlighted (Semmar et al., 2007; Semmar, 2010).

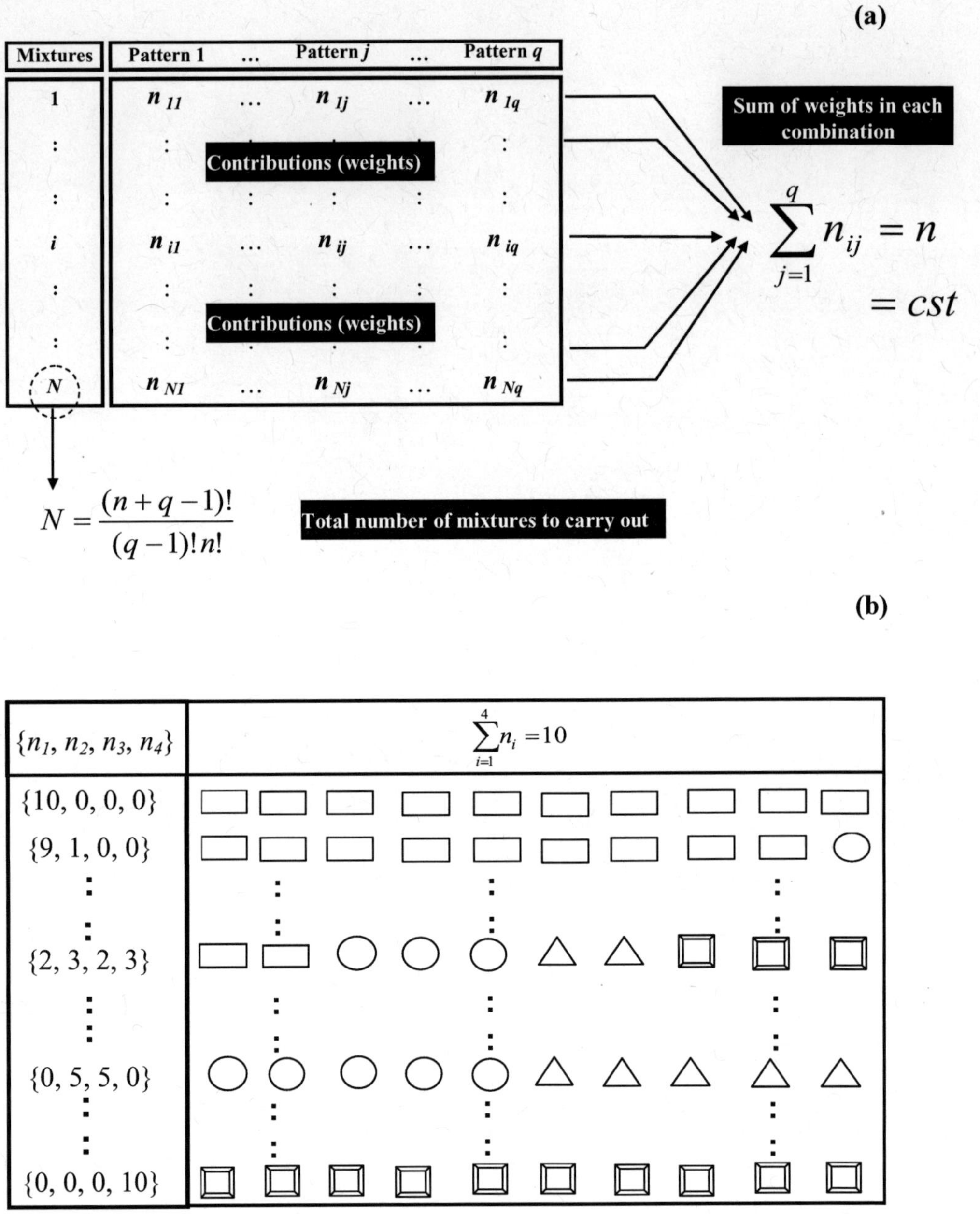

Figure 13. (a) General presentation of Scheffe mixture matrix and its parameters n (total number of mixed elements in each combination) and q (total number of components to combine); (b) illustrated example based on n=10 and q=4.

Such flexibility consisted of different scale- and/or phenotype-dependent processes constraining two given metabolites to have both positive and negative correlations according to the considered scale and/or phenotype (Figure 15). At local scale, two metabolites show systematic relationships consisting of a direct effect between them free from the effects of the other metabolites; such a systematic relationship can be affected (hidden or disturbed) at a higher scale from the development of global metabolic trend (a phenotype) resulting in a global relationship between the two considered metabolites. The correlation sign of such a global relationship depends on the effect of all the metabolites at the scale of the whole metabolic system. Thus, two metabolites can have a systematic affinity (positive local correlation) but will be constrained to be globally opposited (negative global correlation) under the development of a given metabolic trend, and vice versa.

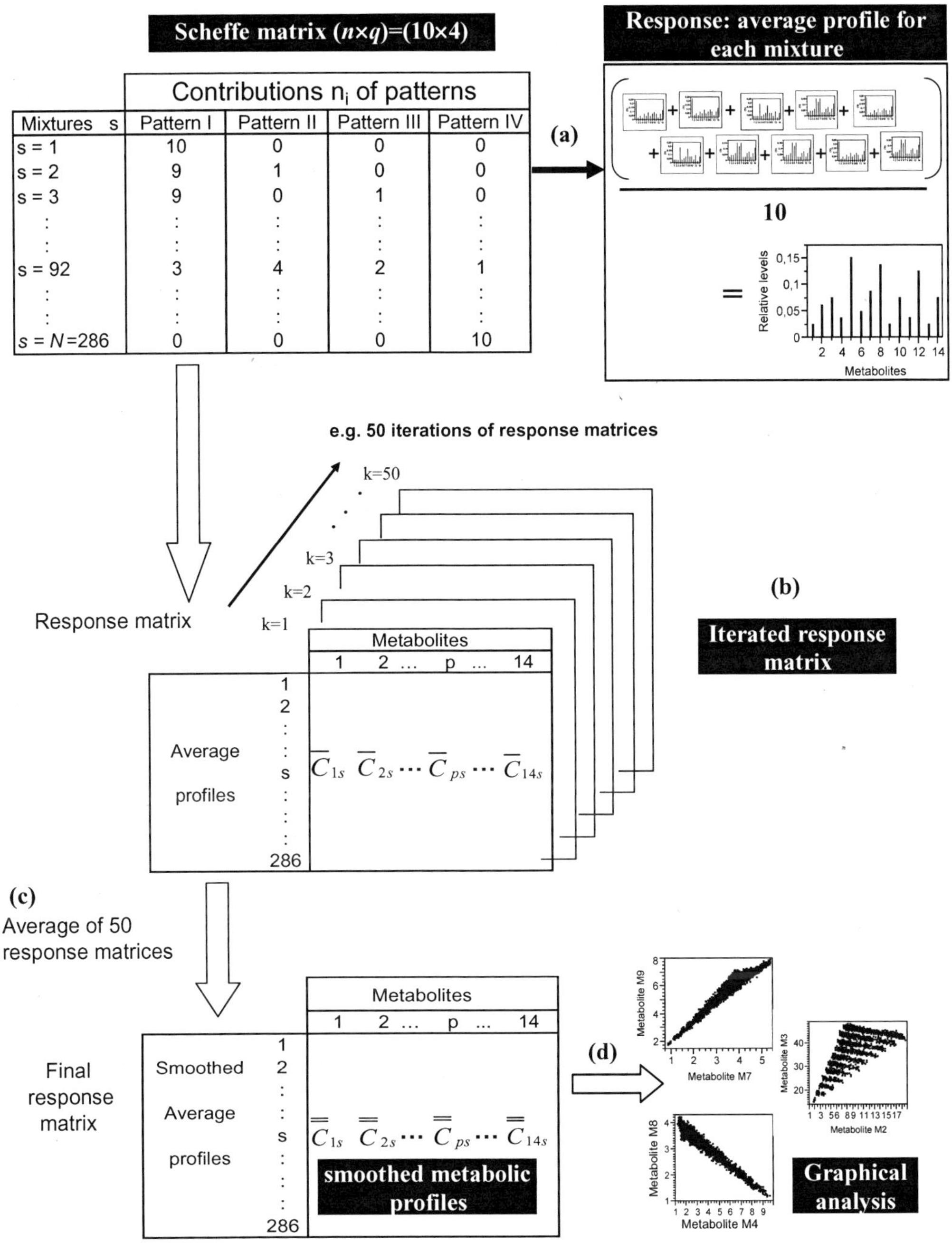

Figure 14. Metabolomic approach based on iterative Scheffe mixture design and leading to extract a set of smoothed profiles representing a backbone of metabolic system from combinations of observed profiles belonging to different patterns.

Figure 15a shows a relationship between two metabolites which is locally negative and globally positive. In terms of metabolic processes, this can concern two metabolites which are systematically competitive for a same precursor (negative local correlation) but which belong to a same metabolic pathway leading them to compete together against other competitive pathways (Fig 15b) (other metabolic trends) (Semmar et al., 2007).

Figure 15a shows that the cloud of points has the fingerprints of a triangular shape. This is due to the fact that the set of all the combinations of Scheffe matrix are contained within a simplex network with a vertices number equal to the number q of components to combine (e.g. q metabolic trends to combine) (Figure 16) (Eide I, 1996; Pattarino et al., 1993; Nyieredy et al., 1985; Glajch et al., 1982; Semmar, 2010). Iterations of the mixture design result in compressions and inclinations of the simplex space at degrees and under directions depending on the different relationships between metabolites.

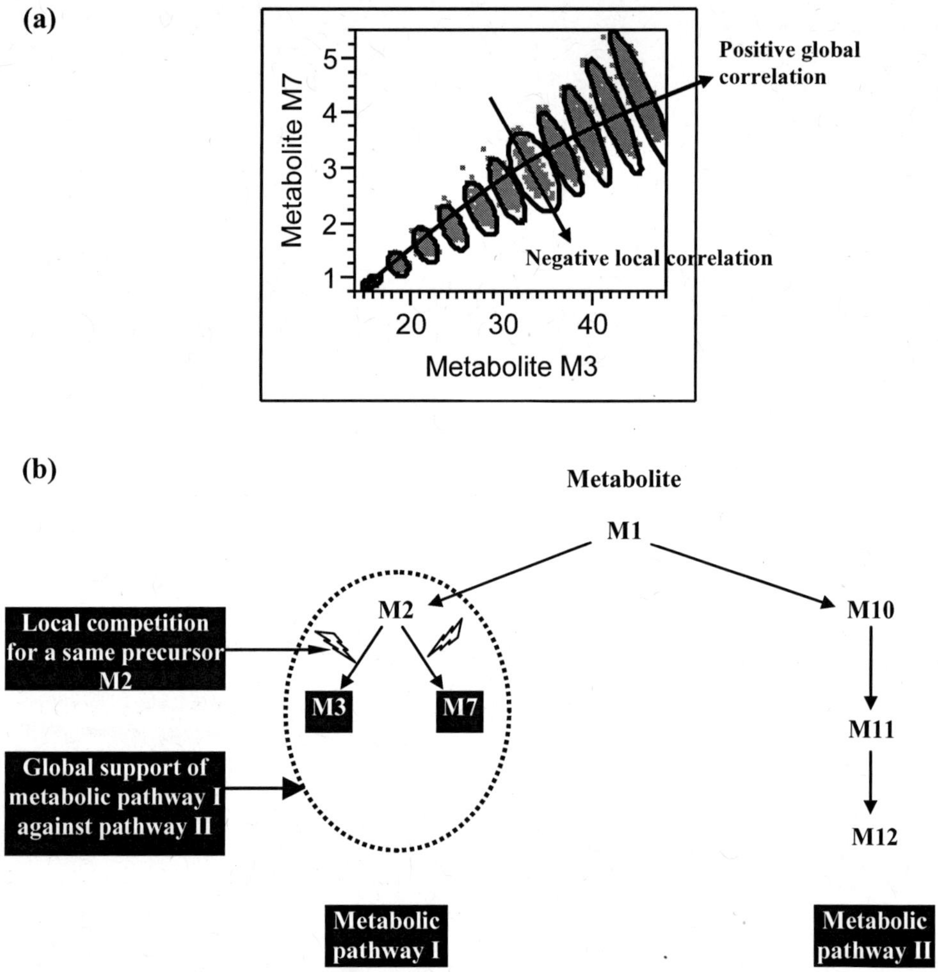

Figure 15. (a) Illustration of a correlation locally negative and globally positive; (b) Possible metabolic factor generating such scale dependent correlation, e.g. metabolites M3 and M7 compete each other in metabolic pathway I (negative local correlation) but sustain their common pathway I against the competitive pathway II (positive global correlation).

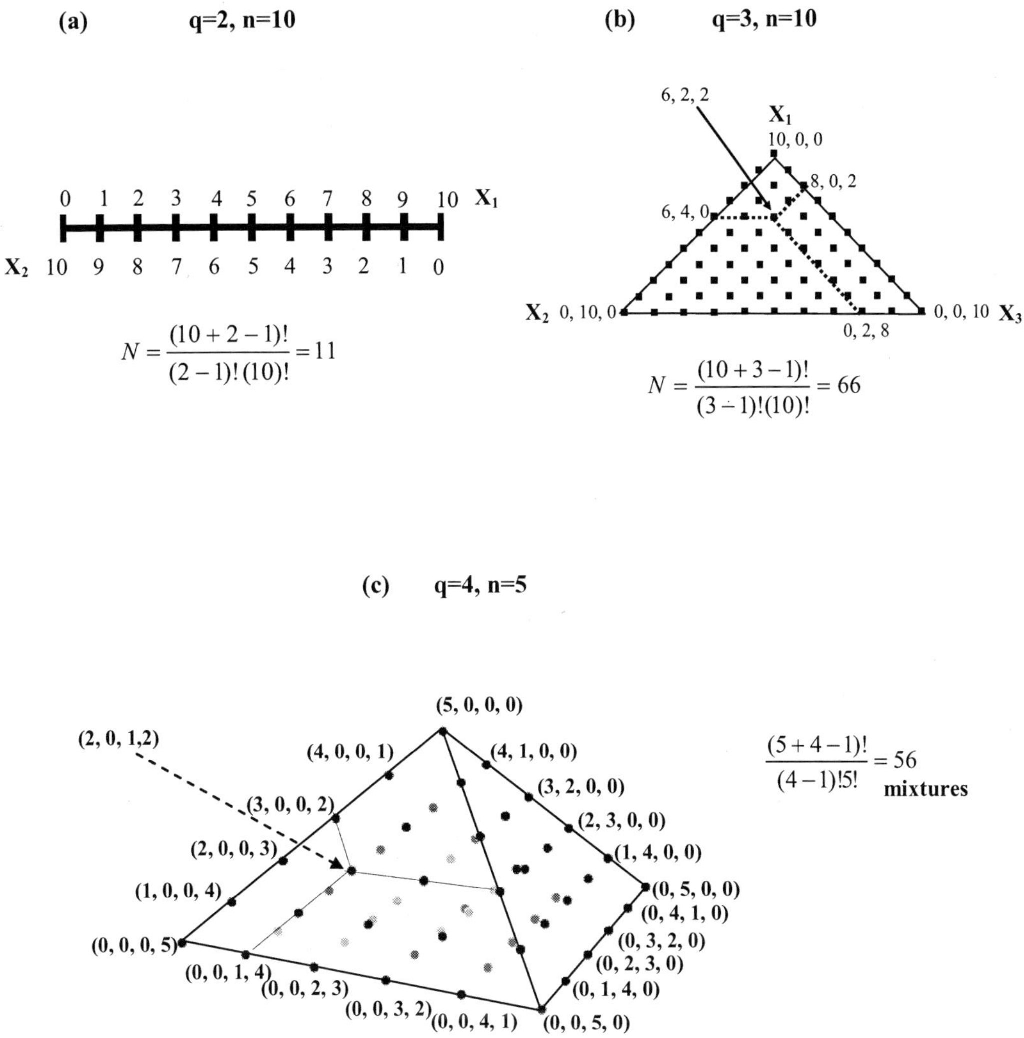

Figure 16. Different simplex representing different Scheffe mixture designs according to the number q of components to combine and the number n of elements representing the q components in each mixture.

IV. Metabolomic Approaches Based on Distance and Correlation Matrices

The variability of a metabolomic dataset (n rows × p columns) can be analysed under three aspects, viz. along rows, along columns, as well as through associations between rows and columns (Figure 17) (Lindon et al. 2007; Sumner et al., 2003):

Column analysis focuses on the relationships between variables (metabolites) in order to quantify and to fit the links between them. Such goals are provided by correlation analysis. Row analysis tends to screen the similarities and differences between individuals (e.g. metabolic profiles). This helps to classify the individuals into homogeneous groups that can be interpreted in terms of polymorphism poles within the studied population. Such fine segmentation of the dataset (population) can be reliably performed by means of cluster analysis. By considering both the rows and columns, extreme, atypical or original associations between individuals and variables can be identified in the dataset. This leads to analyse the heterogeneity or diversity degrees within the dataset and can be performed by different outlier diagnostic approaches (Figure 18).

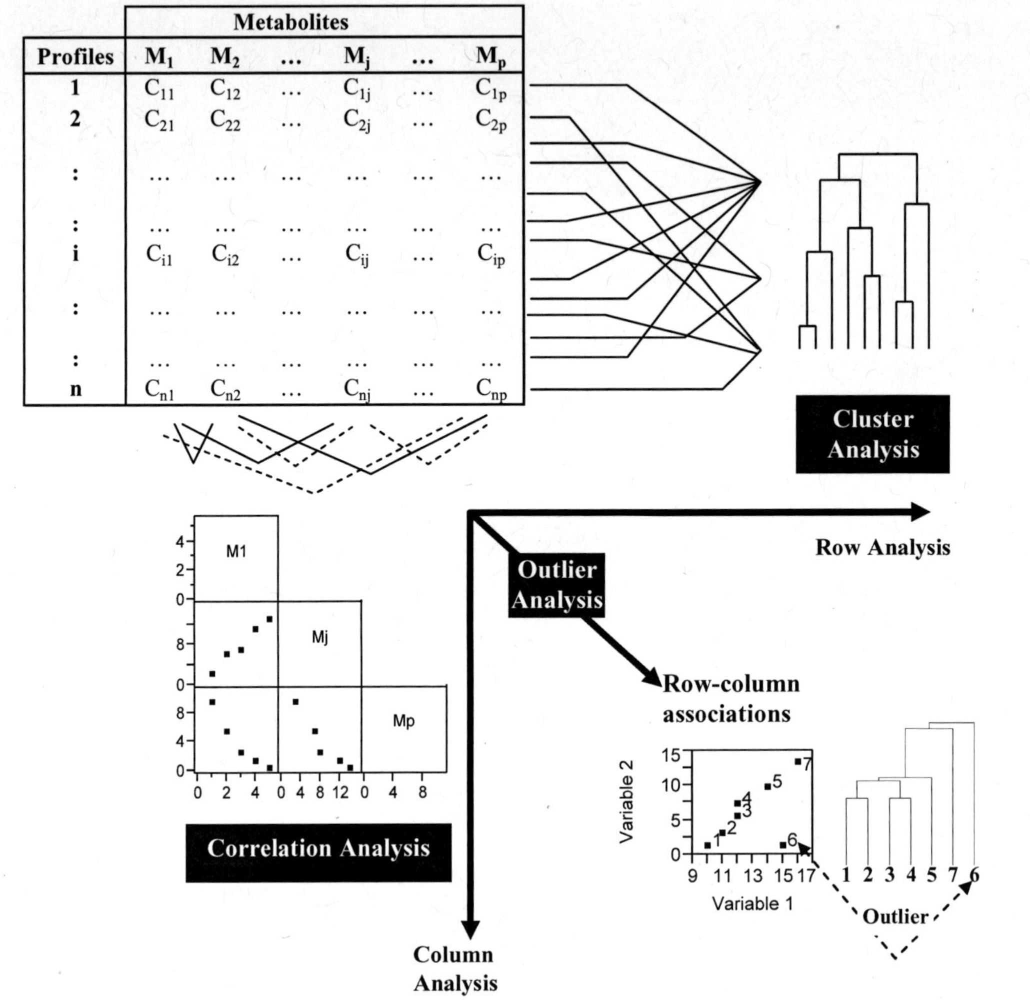

Figure 17. Different statistical approaches applied in metabolomics corresponding to horizontal or vertical data analysis.

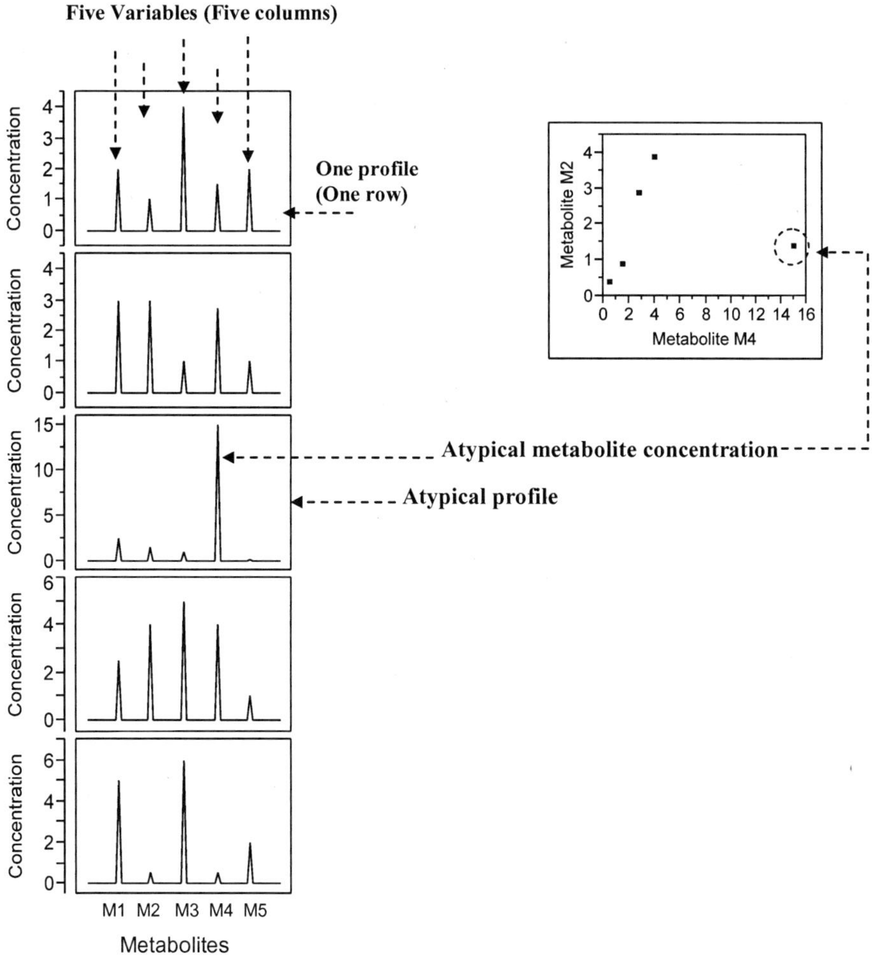

Figure 18. Simple illustration of identification of atypical profiles and concentration values based on profile (row) and variable (column) analyses, respectively.

IV.1. Correlation Based Approaches

Relationships between variables are subjected to correlation analysis which takes into account the dispersion, global inclination and shape of data. Correlation analysis leads to quantify the reciprocal effect of two variables each on the other. For that, different statistical parameters are calculated, viz. correlation coefficients, confidence ranges, slopes, etc. Correlation coefficient quantifies the monotony degree between variables, but it provides no information on the kind of their relationship. Correlation coefficient gives also qualitative information on the direction or inclination of the dataset through its sign: positive and negative signs indicate increasing and decreasing trends, respectively. The inclination of the cloud of points representing the dataset is quantified by the slope of the statistical model used to describe the data variability. The model is defined by an equation which is used to fit well the shape of the cloud of points. The most commonly used model is the linear model represented by the equation y=ax+b. Several other models can be used according to the shape of cloud of points (*y* vs *x*), viz. logarithmic (y=Ln(x)), square root (y=√x), inverse (y=1/x), exponential ($y=e^x$). These models are also applied in order to bring data linearization leading to benefit from computation and simplicity advantages of the linear model.

IV.1.1. Graphical Identification of Correlation Models

The first step in correlation analysis consists in visualising the bivariate data by means of naïve scatter plots. One obtains clouds of points from which the relationships between variables (metabolites) can be described on the basis of their dispersions, inclinations and shapes (Figure 19).

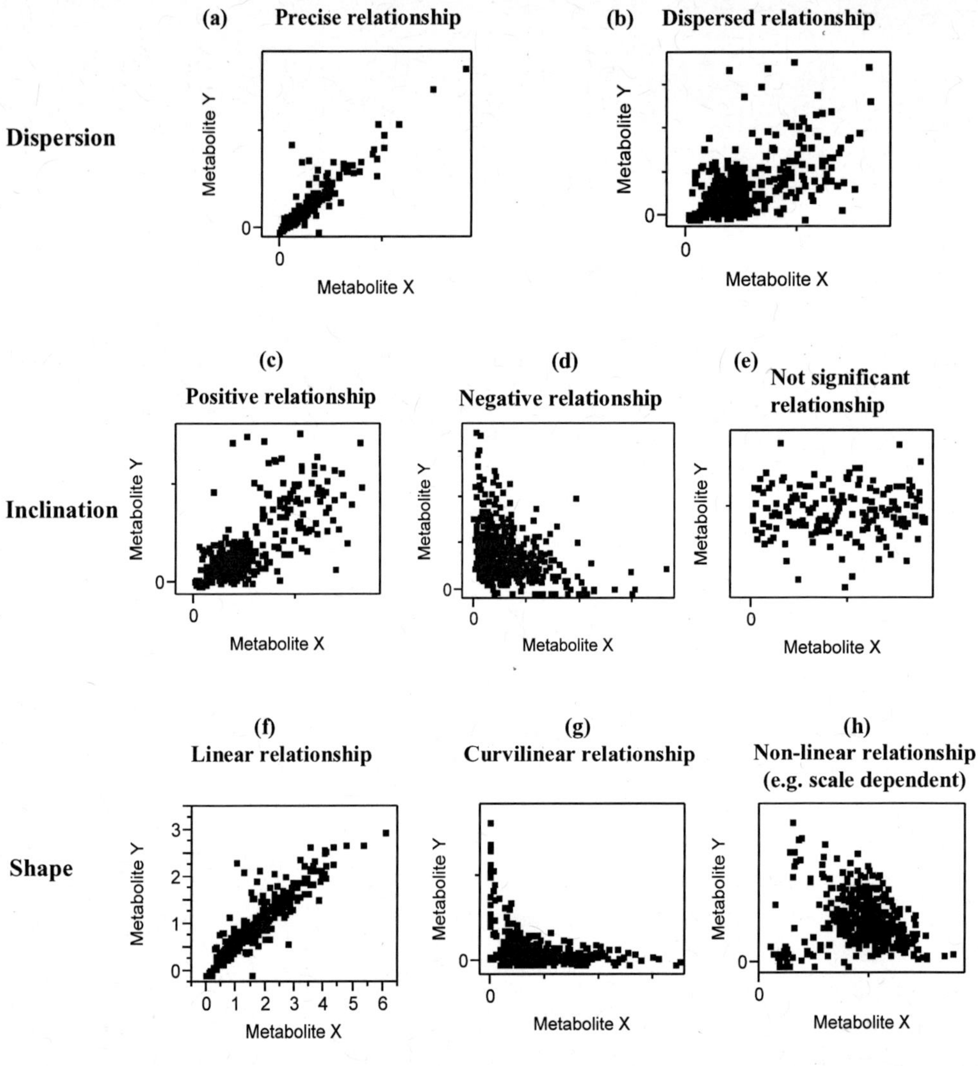

Figure 19. Different scatter plots showing different characteristics (dispersion, inclination, shape) from which statistical tools can be appropriately used to quantify and to fit relationships between variables (metabolites).

For thin or few dispersed clouds of points (Figure 19a, f), relationships between variables can be quantified by means of Pearson correlation coefficient. In the case of more dispersed data (Figure 19b, c, h), Spearman correlation coefficient can be used as robust statistic to detect trends between variables (metabolites). Positive (Figure 19a-c, f) and negative (Figure 19d, g) relationships will be indicated by positive and negative correlation coefficients, respectively.

Pearson correlation is sensitive to the non linearity of data (Figure 19d, g, h). In the case of curvilinear relationships, the use of Pearson coefficient can find application after data linearization using an appropriate transformation. Appropriate transformations provide symmetrical distributions (close to normal) of the data by reducing their dispersion, asymmetry and bias effects of isolated (extreme) points (Zar, 1999). Such transformations can be applied either on only one or on both variables of the pair (X, Y).

Moreover, such transformations are applied to stabilize the variances between several groups of the dataset, i.e. in the case of heteroscedastic data (non comparable variances between groups). Therefore, the resulting homoscedasticity will make possible the application of linear model.

IV.1.2. Data Transformation to Application of Linear Model

From a graphical visualisation, a curvilinear cloud of points (Y vs X) can be transformed into linear form by using an appropriate formula (Zar, 1999, Legendre and Legendre, 2000). Such a formula depends on the shape, intensity of curvature and number of inflexion point(s) of the cloud of points Y vs X (Figure 20).

Logarithmic transformations are appropriate to linearize curvature showing slow (i) or accelerated (ii) variations of Y vs X after an inflection (Figure 21). In the first case (i) (Figure 21a), linearization is obtained from Y vs Ln(X); in the second case (ii) (Figure 21b), linearization is obtained from Ln(Y) vs X. More precisely, the fonction $Y = a\ e^{bX}$ is linearized by taking the log of Y to give a straight-line equation with intercept Ln(a) and slope b, i.e. $\ln(Y) = \ln(a) + bX$. In the case where Y and X are linked by a power function $Y=a(X)^c$, such non-linear relationship can be linearized by taking the logarithms of both X and Y, giving linear equation $\ln(Y) = \text{Ln}(a) + c\ \ln(X)$ (Figure 21c). In general, from a curvilinear cloud of points, the appropriate model can be identified from the transformation by which the curve becomes aligned (Figure 21).

Taking into account the distribution of each variable, logarithmic transformation can be expected for a right asymmetric distribution, i.e. having a mode located at the left (a majority of low values). Therefore, logarithmic transformation results in more symmetrical distribution, i.e. a distribution which closer to normality conditions leading a possible application of the linear model (Figure 22).

Square root transformation can be applied to linearize parabolic cloud of points. Moreover, the square root can be preferred to the logarithm transformation (more generally used) in the case of small dataset (few number of observations). Graphically, models requiring square root transformation have more soft curvature than those requiring logarithmic transformation (Figure 20a).

Clouds of points can be also linearized by means of polynomial transformations. This is generally applied in the case where different inflection points are observed. Therefore, clouds with k inflexion points can be fitted by means of polynomes with degree k+1 (Figure 20d).

IV.1.3. Correlation Coefficient Computation

The correlation concept is used to measure the dependency degree between two variables (metabolites). Such dependency degree between variables is quantified by a correlation coefficient which can be characterised by two aspects: its absolute value and its sign. Absolute value of correlation coefficient varies between 0 and 1; higher value indicates a stronger dependency degree between the variables. All the same, small correlation values can be statistically significant because of a great number of points confirming it. This can be observed in large dataset containing many repeated experimental measurements. On the other hand, some high correlations can be not significant because they were calculated on few data.

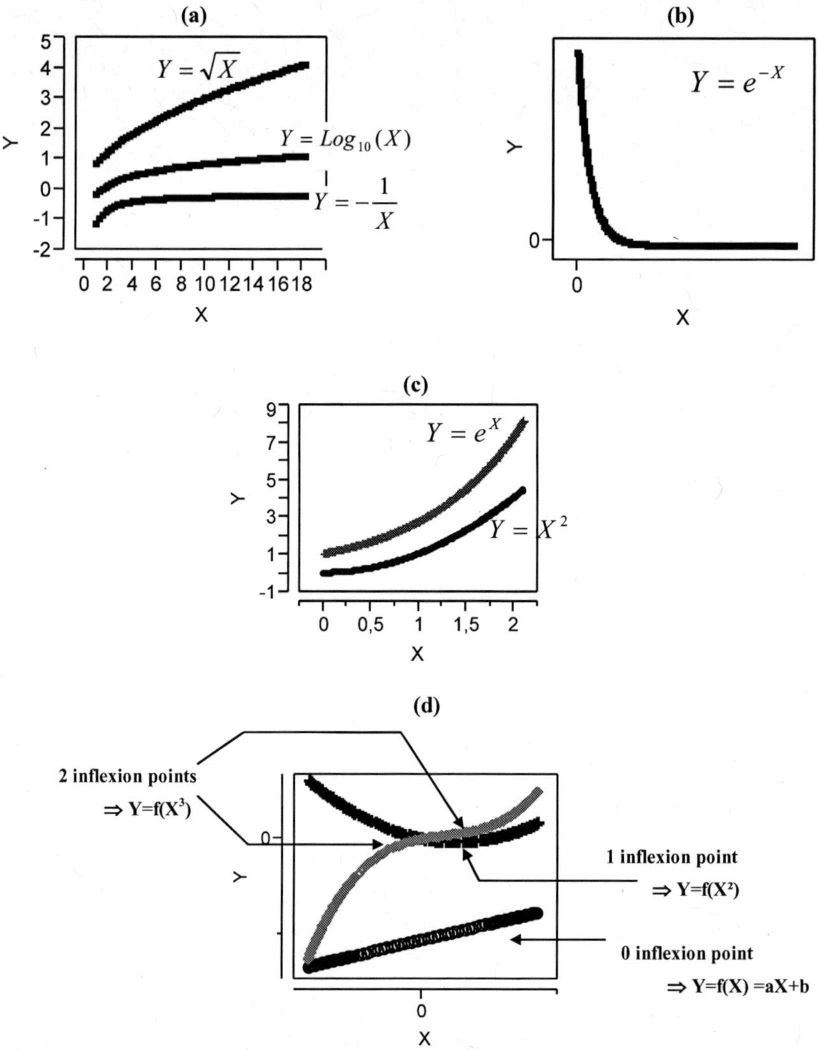

Figure 20. Linearization of different curvilinear relationships by using appropriate data transformations.

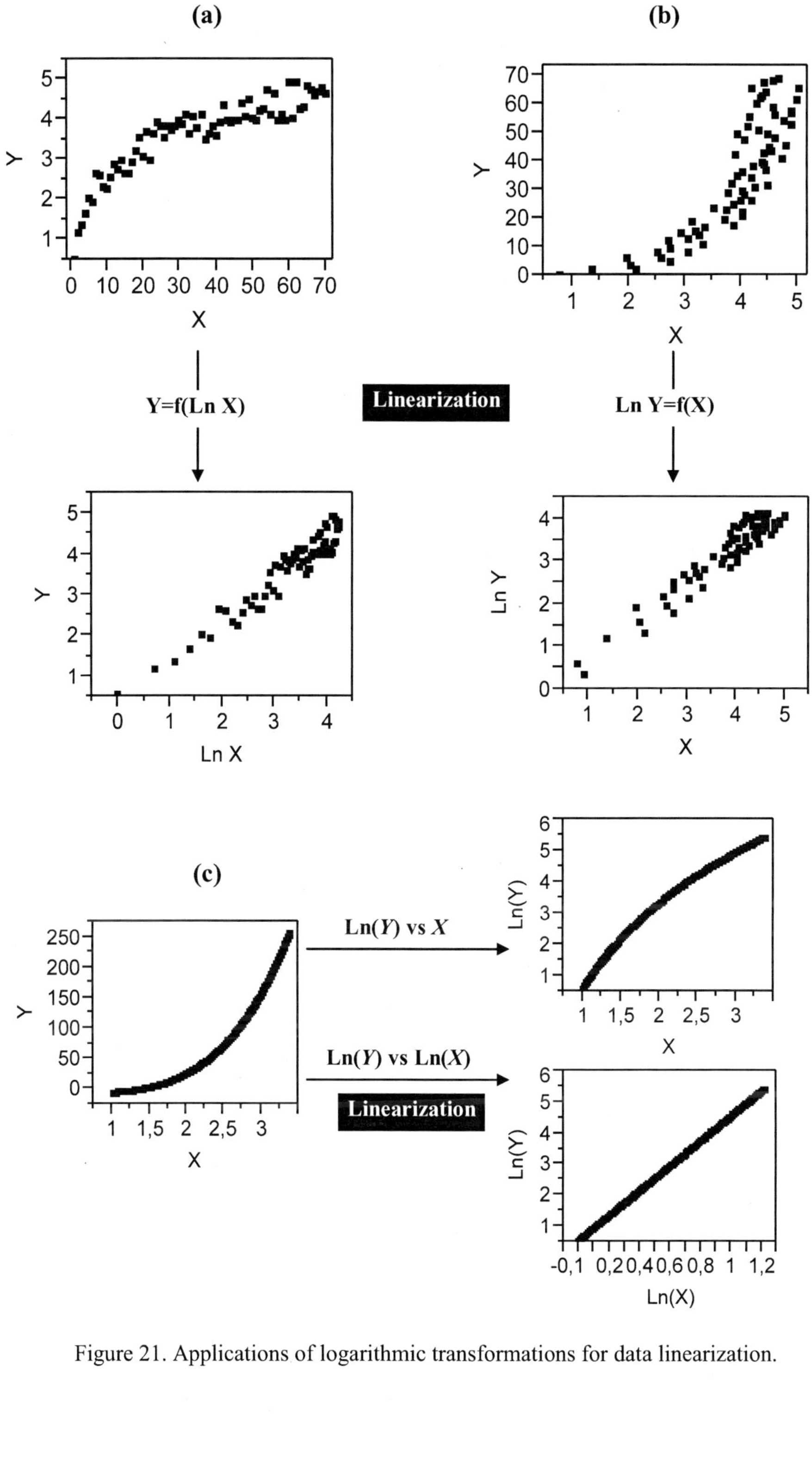

Figure 21. Applications of logarithmic transformations for data linearization.

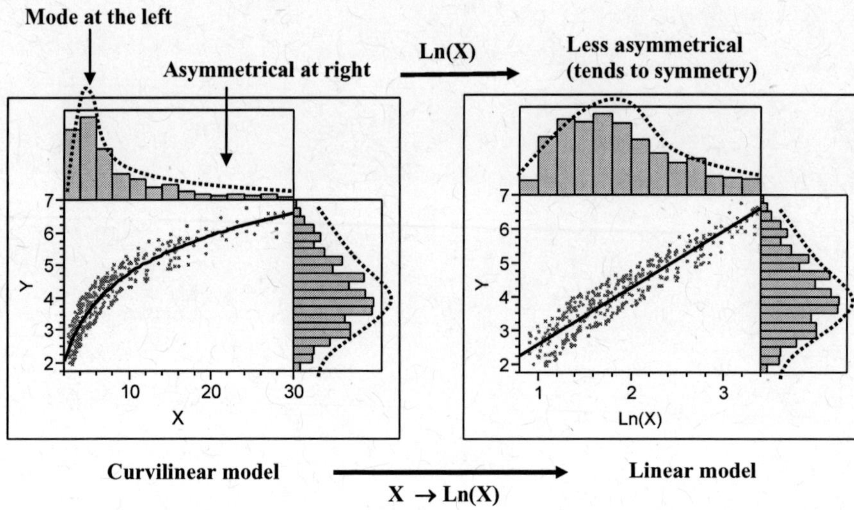

Figure 22. Logarithmic transformation leading to attenuate right asymmetric distribution to become close to normality conditions allowing linear model application.

IV.1.3.1. Pearson Correlation Computation

The Pearson correlation coefficient (r) between two variables x and y is calculated by using the following formula tacking into account their variances and covariance:

$$r = \frac{C_{xy}}{S_x \cdot S_y} = \frac{\dfrac{\sum_{i=1}^{n}(x_i - \bar{x})(y_i - \bar{y})}{n-1}}{\sqrt{\dfrac{\sum_{i=1}^{n}(x_i - \bar{x})^2}{n-1}} \cdot \sqrt{\dfrac{\sum_{i=1}^{n}(y_i - \bar{y})^2}{n-1}}} = \frac{\sum_{i=1}^{n}(x_i - \bar{x})(y_i - \bar{y})}{\sqrt{\sum_{i=1}^{n}(x_i - \bar{x})^2} \cdot \sqrt{\sum_{i=1}^{n}(y_i - \bar{y})^2}}$$

where:

C_{xy} is the covariance of the variables x and y
S_x and S_y: are the standard deviations of x and y
x_i and y_i are measured values (concentration values) of the variables x and y, respectively, in individual i
$\bar{x}$ and $\bar{y}$ are the means of the variables x and y, respectively.
n is the number of paired values (x_i, y_i) (total number of individuals or rows i in the dataset).

Let's give a numerical example to illustrate the calculus of Pearson correlation (Figure 23). Suppose we have a metabolic dataset (10 rows × 4 columns) describing 10 profiles by the concentrations of 4 metabolites:

i	PROFILES	METABOLITES M: M1	M2	M3	M4
1	P1	1.81	2.03	4.66	1.38
2	P2	1.54	3.91	4.3	6.5
3	P3	2.16	4.73	4.84	4.98
4	P4	2.68	5.02	3.82	10.13
5	P5	3.39	7	4.08	1.14
6	P6	3.83	7.11	4.23	0.61
7	P7	4.37	8.58	4	0.78
8	P8	5.47	9.95	3.66	3.49
9	P9	5.59	10.95	3.46	3.32
n=10	P10	6.65	12.84	2.56	6
	Means $\bar{x}$	3.75	7.21	3.96	3.83

$(x_i - \bar{x})$

M1	M2	M3	M4
-1.94	-5.18	0.7	-2.45
-2.21	-3.3	0.34	2.67
-1.59	-2.48	0.88	1.15
-1.07	-2.19	-0.14	6.3
-0.36	-0.21	0.12	-2.69
0.08	-0.1	0.27	-3.22
0.62	1.37	0.04	-3.05
1.72	2.74	-0.3	-0.34
1.84	3.74	-0.5	-0.51
2.9	5.63	-1.4	2.17

$(\ )^2$ → $(x_i - \bar{x})^2$

$(1.81 - 3.75)^2 = 3.76$

	M1	M2	M3	M4
	3.76	26.85	0.49	6.02
	4.88	10.9	0.11	7.11
	2.52	6.16	0.77	1.32
	1.14	4.8	0.02	39.65
	0.13	0.04	0.01	7.25
	0.01	0.01	0.07	10.39
	0.39	1.87	0,00	9.32
	2.96	7.5	0.09	0.12
	3.39	13.97	0.25	0.26
	8.42	31.67	1.96	4.7
Sum Σ	27.6	103.79	3.79	86.14

$(x_i - \bar{x})(y_i - \bar{y})$

$(-1.94 \times -5.18) = 10.05$

	(M1, M2)	(M1, M3)	(M1, M4)	(M2, M3)	(M2, M4)	(M3, M4)
	10.05	-1.36	4.75	-3.63	12.69	-1.72
	7.29	-0.75	-5.9	-1.12	-8.81	0.91
	3.94	-1.4	-1.83	-2.18	-2.85	1.01
	2.34	0.15	-6.74	0.31	-13.8	-0.88
	0.08	-0.04	0.97	-0.03	0.56	-0.32
	-0.01	0.02	-0.26	-0.03	0.32	-0.87
	0.85	0.02	-1.89	0.05	-4.18	-0.12
	4.71	-0.52	-0.58	-0.82	-0.93	0.1
	6.88	-0.92	-0.94	-1.87	-1.91	0.26
	16.33	-4.06	6.29	-7.88	12.22	-3.04
Sum Σ	52.46	-8.86	-6.13	-17.2	-6.69	-4.67

$$\frac{\sum_{i=1}^{n}(x_i - \bar{x})(y_i - \bar{y})}{\sqrt{\sum_{i=1}^{n}(x_i - \bar{x})^2 \sum_{i=1}^{n}(y_i - \bar{y})^2}}$$

$$\frac{52.46}{\sqrt{27.6 \times 103.79}}$$

Pearson correlations

	M1	M2	M3
M2	0.98		
M3	-0.87	-0.87	
M4	-0.13	-0.07	-0.26

Figure 23. Computation of Pearson correlations between four variables (metabolite concentrations) M1, M2, M3, M4 from a dataset of 10 individuals (10 metabolic profiles).

One obtains six correlation values ($0 \le r \le 1$) varying by their absolute values and their signs. The highest correlation value concerns the pair of metabolites (M1, M2) (+0.98), whereas the lowest concerns (M2, M4) (-0.07). Metabolite M4 appears to be the less correlated to all the others. From the signs of correlations, metabolite M3 appears to be strongly negatively correlated to M1 and M2 ($r_{1,3}$=-0.87 and $r_{2,3}$=-0.87, respectively). Parallely to its quantification by Pearson coefficient, the correlation can be qualitatively analyzed by graphical visualisation of the clouds of points (Figure 24).

The scatter plot matrix shows that the strong positive correlation between M1 and M2 was due to a thin (few dispersed) cloud of points. The absolute values of correlations decrease with the dispersion of the cloud of points; such dispersion can be showed by the confidence ellipse thickness. This can be further illustrated by the lowest correlations between M4 and the other metabolites which don't correspond to elliptic shapes, but rather to spherical shapes; such spherical shapes can be interpreted by absences of linearity.

After correlation computations, conclusions will be finally established by testing the significance of each correlation. Two variables (metabolites) will be concluded to be linked if their correlation coefficient is significant. Pearson correlations are tested by using Student t statistics by reference to the value zero: r=0 represents absence of correlation, and therefore the test will respond to the question: does the tested correlation r is significantly different from 0 or no?. The Student test consists in calculating a standardized value t of r:

$$t = \frac{|r - 0|}{s_r}$$

The standard deviation s_r of the correlation coefficient is calculated by the following formula:

$$s_r = \sqrt{\frac{1 - r^2}{n - 2}}$$

where n is the number of measurements (rows, individuals, profiles). Therefore the formula of t can be written:

$$t = \left| \frac{r\sqrt{n - 2}}{\sqrt{1 - r^2}} \right|$$

The calculated t value will be compared to a cut-off value given by the Student table for a low risk α (e.g. 0.05 = 5%) and a degree of freedom (n-2) (Figure 25).

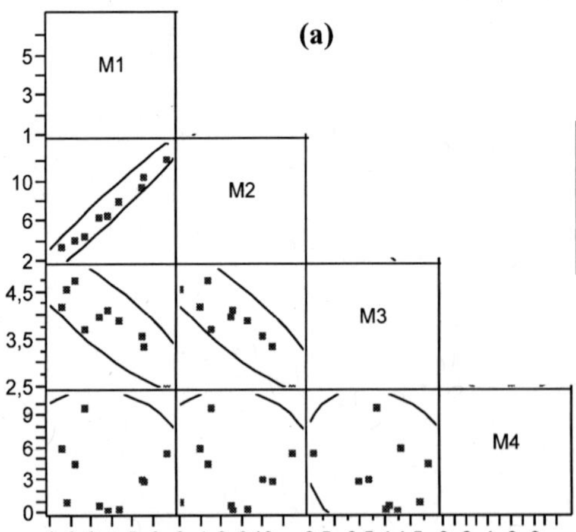

(b)

M2	0.98		
M3	-0.87	-0.87	
M4	-0.13	-0.07	-0.26
	M1	M2	M3

Figure 24. (a) Scatter plot matrix visualizing the variations between different variables (metabolite concentrations); (b) Correlation matrix corresponding to the scatter plot matrix.

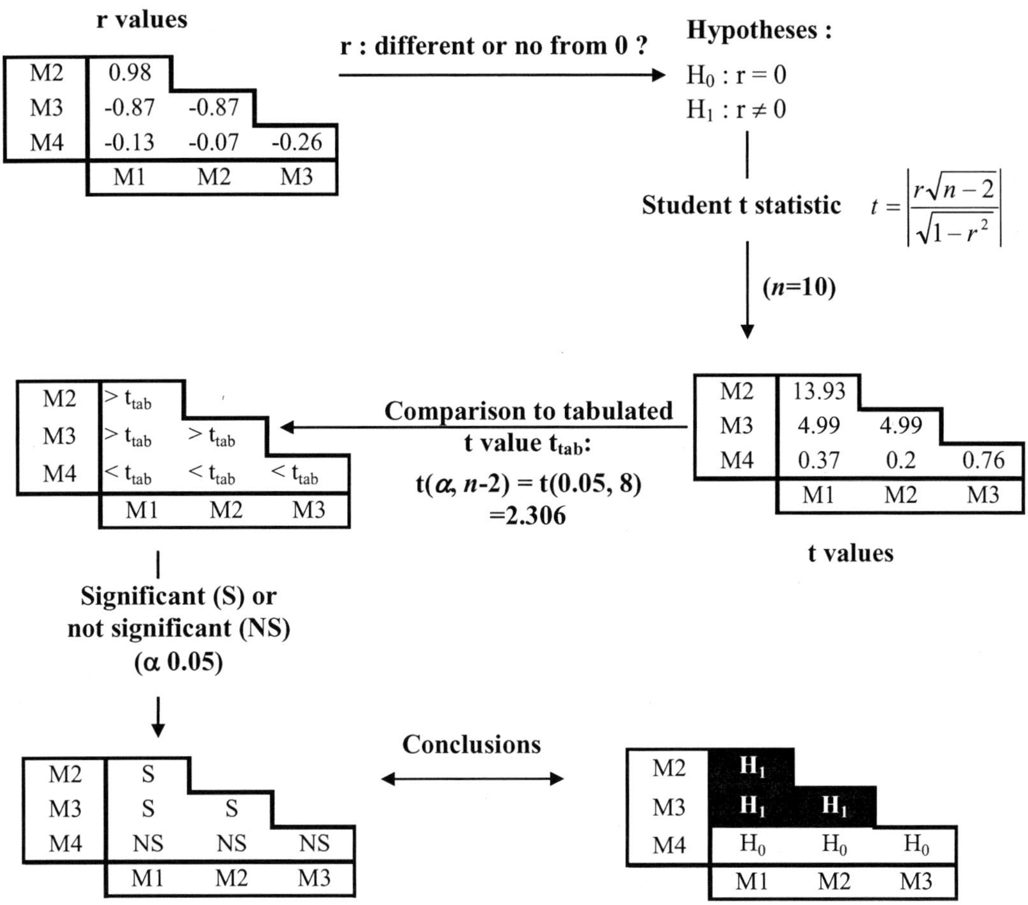

Figure 25. Student t statistics calculated to test the significance of correlation coefficients.

The results show that the correlation correlations are significantly different from 0 with α risk ≤ 5% for the pairs (M1, M2), (M1, M3) and (M2, M3). However, the correlations between M4 and M1, M2, M3 were not significantly different from 0 at the α level = 5%.

IV.1.3.2. Matrix Correlation Computation

Generally, experimental datasets (e.g. metabolomic datasets) contain more variables than the previous simple illustrative example. Therefore, it becomes necessary to handle information and to carry out calculus directly by means of matricial formulation leading to avoid time-consuming repeated calculus. Pearson correlation matrix of a dataset (n rows × p columns) is calculated by a single product between the standardized data matrix S and its transposed S' (S'S), divided by the degree of freedom (n-1) (Figure 26) (Legendre and Legendre, 2000). A numerical example is given in Figure 27.

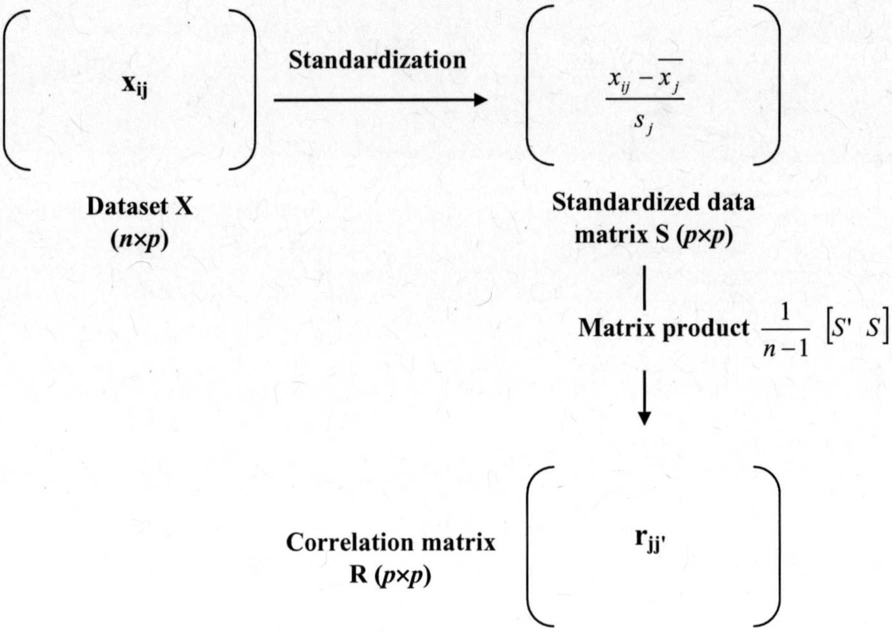

Figure 26. Principle of correlation matrix computation.

IV.1.3.3. Spearman Correlation Calculation

Spearman coefficient are non parametric correlations which require less conditions than parametric Pearson correlations. They can be calculated without to have to check or to assume normality, homoscedasticity of variable, and linearity between variables. However, the number n of paired measures must be higher to 10 in order to be able to test the significance of Spearman correlation. In other words, the use of Spearman correlation is advised for datasets with great number of measures. This is all the more since such datasets have generally high dispersions from which significant trends can be reliably extracted by Spearman correlation. If either Spearman or Pearson correlation analysis is applicable (checked application conditions), the former is $9/\pi^2 = 0.91$ as powerful as the later (Daniel, 1978; Hotelling and Pabst, 1936). The significance of calculated Spearman rank correlations are accessed by consulting statistical tables giving critical values in relation to the number of measurements n and α level.

The calculation of Spearman correlation requires the values x_i, y_i (of the variables x, y) to be ranked (not sorted). Each variable is ranked with reference to itself only: individual values are replaced by a number which gives the ranked position of that value; the association degree between the ranks of the two variables is then quantified by using the Spearman correlation coefficient ρ (Zar, 1999):

$$\rho = 1 - \frac{6\sum_{i=1}^{n} d_i^2}{n^3 - n}$$

Where :

d_i is the difference between the ranks of x_i and y_i values.
n is the number of paired values.

The computation of Spearman correlations (ρ) is illustrated by a numerical example consisting of a dataset of 12 rows (n>10) and 4 columns (Figure 28). We suppose we have a concentration dataset of 4 metabolites analysed in 12 individuals to obtain 12 concentration profiles (in arbitrary unit).

Dataset X = (x_{ij}) —— **Standardization** ——→ **S = $(x_{ij} - \overline{x}_j)/s_j$**

i \ j	1	2	3	4
1	1.81	2.03	4.66	1.38
2	1.54	3.91	4.3	6.5
3	2.16	4.73	4.84	4.98
4	2.68	5.02	3.82	10.13
5	3.39	7,00	4.08	1.14
6	3.83	7.11	4.23	0.61
7	4.37	8.58	4,00	0.78
8	5.47	9.95	3.66	3.49
9	5.59	10.95	3.46	3.32
10	6.65	12.84	2.56	6,00
Mean $\overline{x}_j$	3.75	7.21	3.96	3.83
Standard deviation s_j	1.75	3.4	0.65	3.09

1	2	3	4
-1.11	-1.52	1.08	-0.79
-1.26	-0.97	0.52	0.86
-0.91	-0.73	1.35	0.37
-0.61	-0.64	-0.22	2.04
-0.21	-0.06	0.18	-0.87
0.05	-0.03	0.42	-1.04
0.35	0.4	0.06	-0.99
0.98	0.81	-0.46	-0.11
1.05	1.1	-0.77	-0.17
1.66	1.66	-2.15	0.7

Transposition S'

S' =

j \ i	1	2	3	4	5	6	7	8	9	10
1	-1.11	-1.26	-0.91	-0.61	-0.21	0.05	0.35	0.98	1.05	1.66
2	-1.52	-0.97	-0.73	-0.64	-0.06	-0.03	0.4	0.81	1.1	1.66
3	1.08	0.52	1.35	-0.22	0.18	0.42	0.06	-0.46	-0.77	-2.15
4	-0.79	0.86	0.37	2.04	-0.87	-1.04	-0.99	-0.11	-0.17	0.7

Product S'S

1.11×-1.52 - 1.26×-0.97 - 0.91×-0.73 - 0.61×-0.64 - 0.21×-0.06
+ 0.05×-0.03 + 0.35×0.4 + 0.98×0.81 + 1.05×1.1 + 1.66×1.66
= 8.82

j \ j	1	2	3	4
1	9.00	8.82	-7.79	-1.13
2	8.82	9.00	-7.8	-0.64
3	-7.79	-7.8	9.00	-2.33
4	-1.13	-0.64	-2.33	9.00

× 1/(n-1) ←

j \ j	1	2	3	4
1	1.00	0.98	-0.87	-0.13
2	0.98	1.00	-0.87	-0.07
3	-0.87	-0.87	1.00	-0.26
4	-0.13	-0.07	-0.26	1.00

Correlation matrix R (4×4)

Figure 27. Numerical example illustrating the computation of correlation matrix from a standardized dataset.

Concentration dataset (n=12 × p=4) (i =1 to n=12; j =1 to 4)

PROFILES		METABOLITES			
		M1	M2	M3	M4
	P1	1	2	5	1.59
	P2	2.25	4	6	2.12
P	P3	4	7.5	7	1.7
R	P4	5	8.5	10	0.9
O	P5	1.5	2.5	6.5	1.29
F	P6	0.75	1	3.5	1.83
I	P7	0.5	1.2	3.3	2.08
L	P8	2.5	5	6.75	1.75
E	P9	4.5	7.9	8.5	1.37
S	P10	1.2	2.2	5.8	1.58
	P11	2	4.5	6.2	2.5
	P12	4.8	8	9	1.5

Ranks (1 to 12) →

Rank matrix

	M1	M2	M3	M4
P1	3	3	3	6
P2	7	6	5	11
P3	9	9	9	7
P4	12	12	12	1
P5	5	5	7	2
P6	2	1	2	9
P7	1	2	1	10
P8	8	8	8	8
P9	10	10	10	3
P10	4	4	4	5
P11	6	7	6	12
P12	11	11	11	4

$$d_i^2 = [\text{Rank}(x_i) - \text{Rank}(y_i)]^2$$

	M1M2	M1M3	M1M4	M2M3	M2M4	M3M4
	0	0	9	0	9	9
	1	4	16	1	25	36
	0	0	4	0	4	4
	0	0	121	0	121	121
	0	4	9	4	9	25
	1	0	49	1	64	49
	1	0	81	1	64	81
	0	0	0	0	0	0
	0	0	49	0	49	49
	0	0	1	0	1	1
	1	0	36	1	25	36
	0	0	49	0	49	49
Sum $\sum d_i^2$	4	8	424	8	420	460

$$\rho = 1 - \frac{6\sum_{i=1}^{n} d_i^2}{n^3 - n}$$

Correlation matrix

	M1	M2	M3
M2	0.99		
M3	0.97	0.97	
M4	-0.48	-0.47	-0.61

Figure 28. Numerical example illustrating the computation of Spearman correlations (ρ) between paired variables.

The calculated ρ values showed positive correlations between metabolites M1, M2 and M3, and negative correlations between these three metabolites and M4. A statistical table gives for α=0.05 and n=12, a tabulated value ρ_{tab}=0.587, leading to conclude that there are four significant correlations with α risk ≤5% (M1-M2; M1-M3; M2-M3; M3-M4), against two not significant at α level = 5% (M1-M4; M2-M4) (from ρ absolute values). From the scatter plot matrix (Figure 29a), the significant correlations correspond to thin and sharply inclined clouds of points, whereas the not significant ones correspond to weakly inclined clouds of points (nearly horizontal; Figure 19e). Note that the significant negative correlation between M3 and M4 corresponds also to a weakly inclined cloud, but which is less dispersed (thin confidence ellipse) than the pairs (M1, M4) and (M2, M4). This shows that a correlation coefficient takes into account both the covariance (inclination) and the variance (dispersion) of the variables.

As the correlations were calculated on concentrations, they have to be interpreted in terms of biosynthesis or availability processes because the concentration is all the more high since the biosynthesis or absorption process are important. On this basis, significantly positive correlations between M1, M2 and M3 can be indicative of common factors favouring the biosynthesis of such metabolites (common metabolic pathways, common resources, sensitivity toward same stimulus factors, same cell transport paths, etc.). Concerning the pair (M3, M4), its significantly negative correlation can be originated from different situations e.g. metabolites which have opposite or not shared characteristics (e.g. biosynthesis and elimination which are rapid for one metabolite and slow for the other), which belong to two alternative/successive metabolic pathways, which are stimulated by different factors, etc. . Finally, the not significant correlations of M4 toward M1, M2 indicate that there are not sufficient oriented factors/characteristics to group or to opposite the concerned metabolites.

(a)

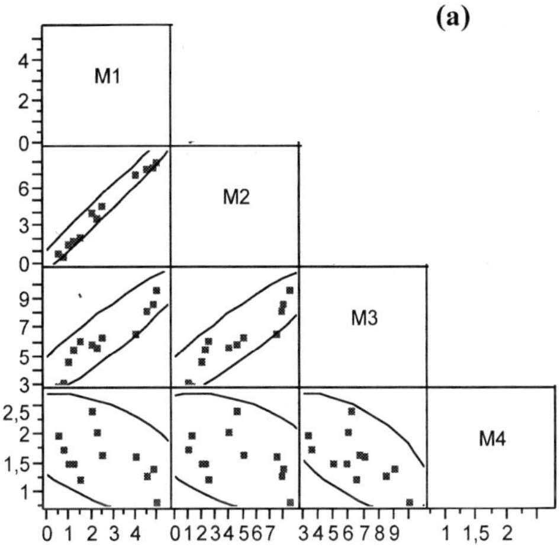

M2	0.99		
M3	0.97	0.97	
M4	-0.48	-0.47	-0.61
	M1	**M2**	**M3**

(b)

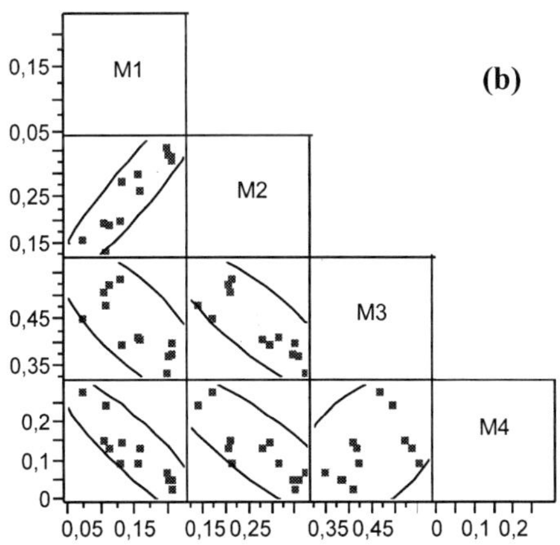

M2	0.87		
M3	-0.75	-0.83	
M4	-0.9	-0.86	0.55
	M1	**M2**	**M3**

Figure 29. Scatter plot matrix providing a visualization of relationships between concentration (a) and relative levels (b) of different variables, and corresponding correlation matrices.

Apart from the concentration variables which are directly interpretable in terms of synthesis or availability, metabolomic focuses on the analysis of the relative levels of such concentrations which are interpretable in terms of metabolic regulation ratios. Regulation ratios of different metabolites provide information on the internal structure/organization of their metabolic systems, whereas concentrations are particularly appropriate to analyse the metabolic machine in relation to external conditions.

Spearman statistic can be applied on relative level data to calculate correlations between regulation ratios of different metabolites. Such a computation is illustrated from the previous numerical example (Figure 30) (Figure 29b).

Five among the six correlation values are significant with $\alpha \leq 5\%$, because they are higher than the cut off tabulated value $\rho_{tab}=0.587$ ($\alpha=0.05$ and $n=12$). Although at α level of 5%, the positive correlation 0.55 is not significant, it is enough high to be considered as significant with α risk $\leq 10\%$ ($\rho_{tab}(\alpha=10\%, n=12)=0.503$).

Relative levels' matrix

	M1	M2	M3	M4	Sum
P1	0.1	0.21	0.52	0.17	1
P2	0.16	0.28	0.42	0.15	1
P3	0.2	0.37	0.35	0.08	1
P4	0.2	0.35	0.41	0.04	1
P5	0.13	0.21	0.55	0.11	1
P6	0.11	0.14	0.49	0.26	1
P7	0.07	0.17	0.47	0.29	1
P8	0.16	0.31	0.42	0.11	1
P9	0.2	0.35	0.38	0.06	1
P10	0.11	0.2	0.54	0.15	1
P11	0.13	0.3	0.41	0.16	1
P12	0.21	0.34	0.39	0.06	1

Ranks →

Rank matrix

M1	M2	M3	M4
2	4	10	10
8	6	6	8
9	12	1	4
11	10	5	1
5	5	12	6
3	1	9	11
1	2	8	12
7	8	7	5
10	11	2	2
4	3	11	7
6	7	4	9
12	9	3	3

$$d_i^2 = [\text{Rank}(x_i) - \text{Rank}(y_i)]^2$$

M1M2	M1M3	M1M4	M2M3	M2M4	M3M4
4	64	64	36	36	0
4	4	0	0	4	4
9	64	25	121	64	9
1	36	100	25	81	16
0	49	1	49	1	36
4	36	64	64	100	4
1	49	121	36	100	16
1	0	4	1	9	4
1	64	64	81	81	0
1	49	9	64	16	16
1	4	9	9	4	25
9	81	81	36	36	0

Sum Σd_i^2	36	500	542	522	532	130

$$\rho = 1 - \frac{6\sum_{i=1}^{n} d_i^2}{n^3 - n}$$

Correlation matrix

M2	0.87		
M3	-0.75	-0.83	
M4	-0.9	-0.86	0.55
	M1	**M2**	**M3**

Figure 30. Numerical example illustrating the computation of Spearman correlations (ρ) between regulation ratio variables.

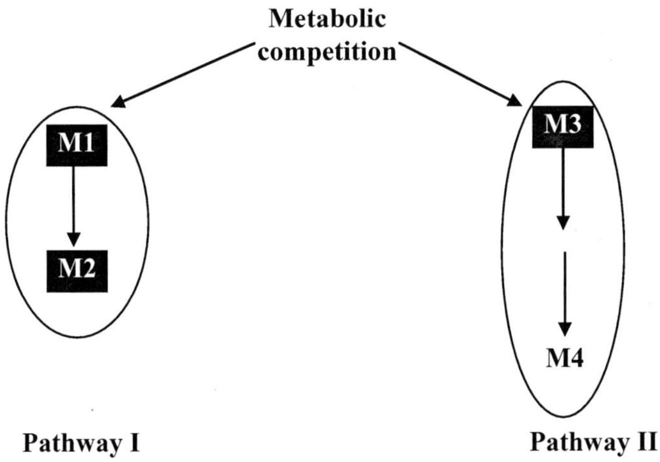

Figure 31. Hypothetic scheme on the global organisation of metabolic system interpreted from Spearman correlations between relative levels of metabolites (M1, M2, M3, M4). Black squares (M1-M3) indicate metabolites sharing some factors favouring their biosynthesis, and interpreted from correlations between their concentrations (rather than relative levels). Double arrow between M3 and M4 is indicative of a lesser neighbouring between them, interpreted from a lower absolute value of correlation between their relative levels.

From positive and negative correlations, the four compounds are organized into two subsets each one containing positively correlated metabolites: M1, M2 on the hand, and M3, M4 on the other hand. The compounds of each subset are negatively correlated to those of the other subset. The negative correlations can be indicative of the presence of two competitive metabolic pathways (M1, M2) against (M3, M4). In other words, the metabolic regulations of M1, M2 occur at the expense of M3, M4, and vice versa. From the positive correlations, the value of the pair (M1, M2) which is higher (and more significant) than that of (M3, M4) can be indicative of more shared factors (metabolic processes, chemical structure similarities, etc) between M1 and M2 than between M3 and M4. A hypothetical organization of metabolic system from these correlations is presented in Figure 31.

Interestingly, some positive correlations observed between concentrations corresponded to negative ones between relative levels; this concerns the pairs (M1, M3) and (M2, M3). Moreover, the negative correlation previously observed between concentrations of M3 and M4 showed a positive value when calculated on relative levels. By combining the negative and positive correlations observed with relative levels and concentrations, respectively, metabolite M3 can be considered as belonging to a different pathway but sharing some biosynthetic factors with M1 and M2 (Figure 31).

More details on the origins of correlations in metabolomic datasets will be presented in the next section.

IV.1.4. Origins and Interpretation of Correlations in Metabolic Systems

A high correlation between two metabolites can be originated from several mechanisms (Camacho et al. 2005):

1) Chemical equilibrium
2) Mass conservation
3) Assymetric control
4) Unusually high variance in the expression of a single gene

IV.1.4.1. Chemical Equilibrium

Two metabolites near chemical equilibrium will show a high positive correlation, with their concentration ratio approximating the equilibrium constant. As a consequence, metabolites with negative correlation are *not* in equilibrium. Positive correlation can be observed between a precursor and its product which have synchronous metabolic variations (Figure 32a).

IV.1.4.2. Mass Conservation

Within a moiety-conserved cycle, at least one member should have a negative correlation with another member of the conserved group. This may be the case of two metabolites competing for a same substrate (precursor) representing a limited source which has to be shared (Figure 32b-c).

IV.1.4.3. Assymetric Control

Most high correlations may be due (a) to either strong mutual control by a single enzyme (Figure 32b), or (b) to variation of a single enzyme level much above others (Figure 32c). This may result from a metabolic pathway effect (Figure 32d): the variation of a single enzyme level within a metabolic pathway will have direct or indirect repercussions on metabolites of such a pathway leading to their positive correlation(s). In the case where two metabolites are controlled by a same enzyme, the activity of such enzyme in favour to the first path (or subpath) will be at the expense of the second one; this contributes to negative correlation between metabolites of the two paths (e.g. M1, M5) or subpaths (e.g. M7, M8). In more general terms, if one parameter dominates the concentration of two metabolites, intrinsic fluctuations of this parameter result in a high correlation between them.

Assymetric control can be graphically analysed by a log-log scatter plot between metabolites' concentrations (Camacho et al., 2005). From such graphic, change in correlation reflects change in the co-response of the metabolites in relation to the dominant parameter (Figure 33).

IV.1.4.4. Unusually High Variance in the Expression of a Single Gene

This is similar to the previous situation but the resulting correlation is not due to a high sensitivity toward a particular parameter, but due to an unusually high variance of this parameter. In particular, a single enzyme that carries a high variance will induce negative correlations between its substrate and product metabolites (Steuer, 2006).

IV.1.5. Scale-Dependent Interpretations of Correlations

The analysis of correlations exploits the intrinsic variability of a metabolic system to obtain additional features of the state of the system. The set of all the correlations (given by the correlation matrix) is a global property of the metabolic system, i.e. whether two metabolites are correlated or not does not depend solely on the reactions they participate in, but on the combined result of all the reactions and regulatory interactions present in the system. In this sense, the pattern of correlations can be interpreted as a global fingerprint of the underlying system integrating environmental conditions, physiological states, etc., at a given time.

Apart from the temporal, physiological and environmental factors, the correlation between two metabolites can show a scale-dependent variation within a same metabolic system; this provides evidence on the flexibility of metabolic processes and on the complexity of metabolic network:

At a local scale, two metabolites are closely considered the one toward the other without consideration of the other metabolites. For example, two metabolites can be competitive for a same enzyme (Figure 32b) or a same precursor (Figure 32c) within a common metabolic pathway leading to a locally negative correlation between them. However, when they are considered together into their common pathway in presence of other competitive pathways, these two metabolites can manifest a positive correlation at the global scale (Figure 32d: Metabolites M7, M8).

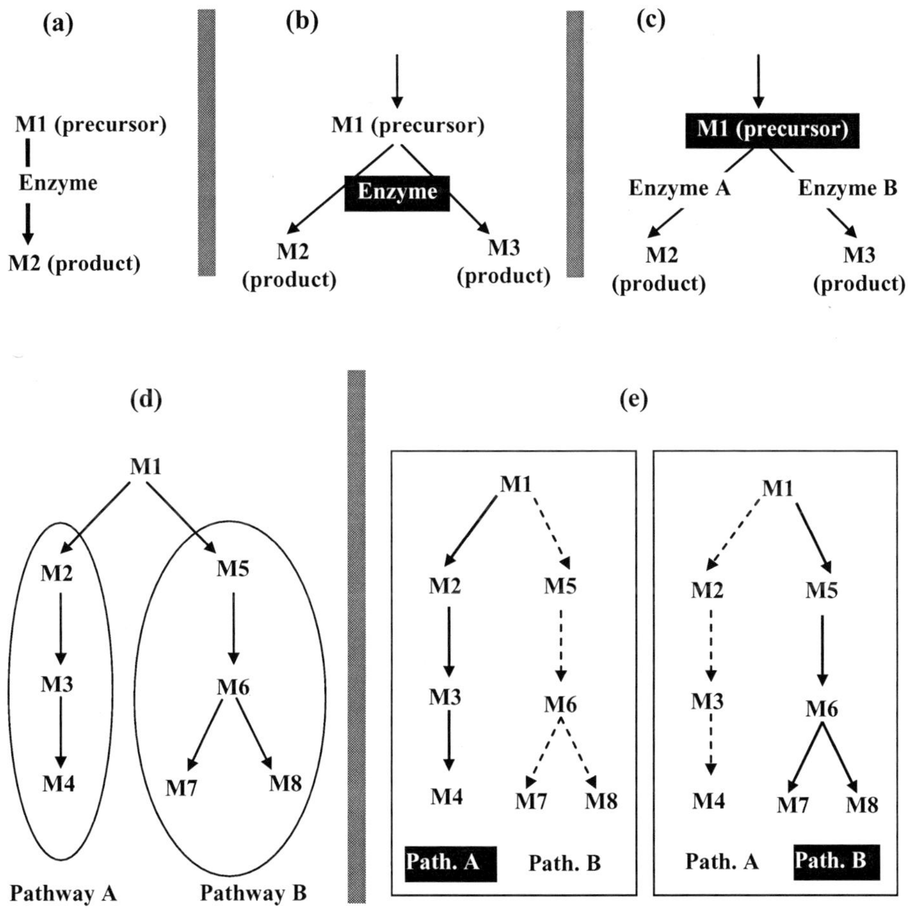

Figure 32. Different scales at which correlation between metabolites can be interpreted: metabolite scale (a-c); metabolic pathway scale (d); Network (physiological) scale (e).

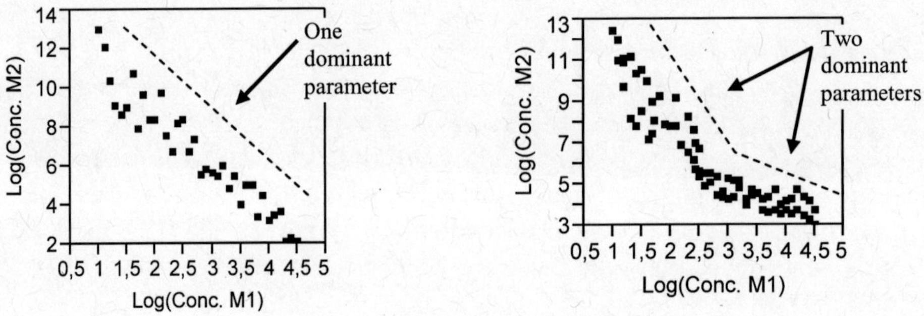

Figure 33. Some examples of Log-Log scatter plots used to detect co-response of two metabolites under the effect of some dominant parameter(s).

At a global scale, several metabolites can be biosynthesized within a same metabolic pathway in which they share a serial of regulation enzymes, by competting other metabolites belonging to other metabolic pathways (Figure 32d).

At a higher scale, diminutive fluctuations within the metabolic system or in the environment conditions induce correlations which will propagate through the system to give rise to a specific pattern of correlations depending on the physiological state of the system (Camacho et al., 2005; Steuer et al., 2003a, b; Morgenthal et al., 2006) (Figure 32e).

A transition from a physiological state to another may not only involve changes in the average levels of the measured metabolites but additionally may also involve changes in their correlations.

There are many pairs of metabolites that are neighbours in the metabolic map but which have low correlations, and others that are not neighbours but have high correlations. This is due to the fact that the correlations are shaped by both stoichiometric and kinetic effects (Steuer et al., 2003a, b).

IV.1.6. Multidimensional Correlation Screening by Means of Principle Component Analysis

IV.1.6.1. Aim

Principle component analysis (PCA) is a multivariate analysis which uses the linear algebra rules to provide graphical representations where the n rows and p columns of a dataset will be restricted to n and p points, respectively, on a single axis or in a plan (Waite, 2000). PCA aims to represent the complexity of relationships between variables in the minimum number of dimensions. The relative positions of row- and column-points given by PCA are interpretable in terms of affinities, oppositions or independences between them; this helps to understand:

- specific characteristics of individuals (e.g. metabolic profiles),
- relative behaviours of variables (e.g. metabolites),
- associations between individuals and variables.

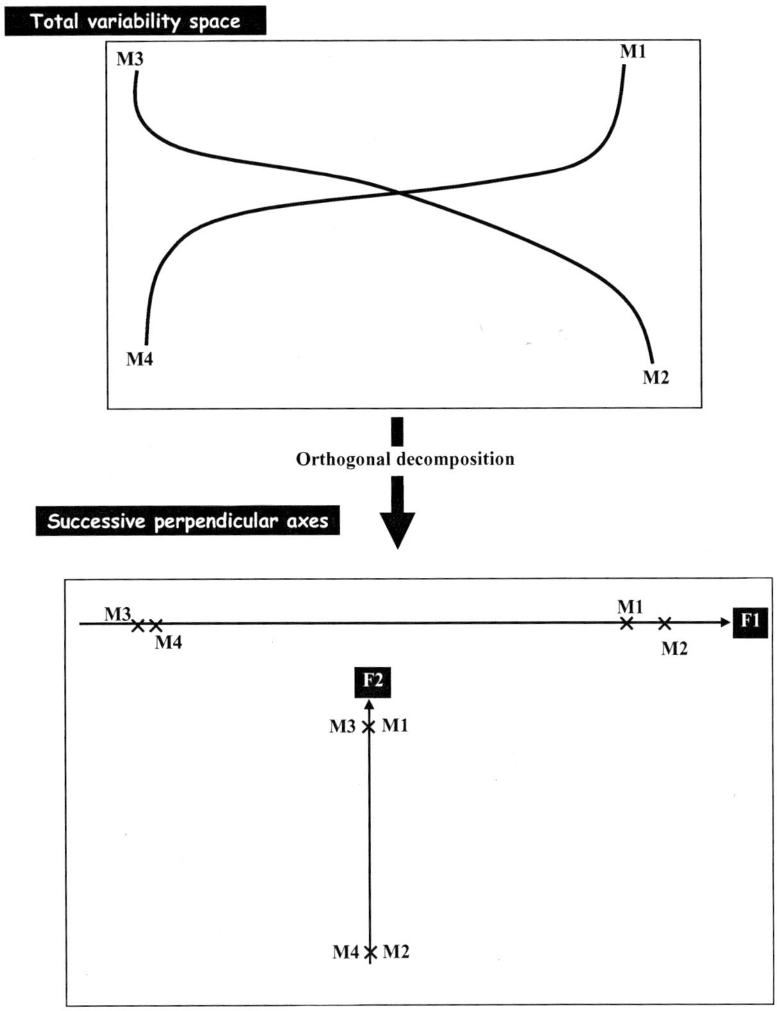

Figure 34. Simplistic illustration of decomposition of the total variability into additive (complementary) parts along perpendicular axes.

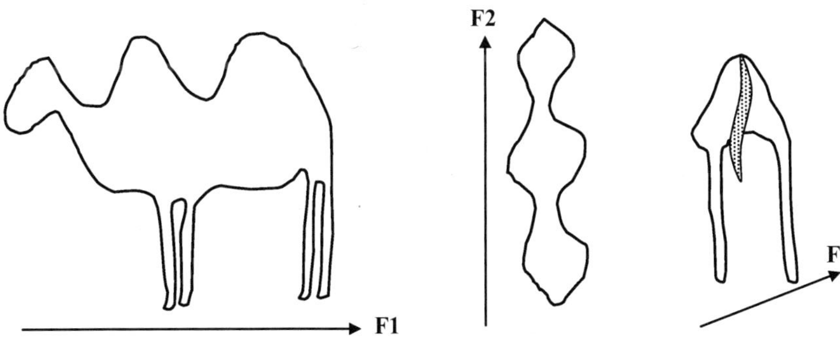

Figure 35. Intuitive illustration of the usefulness of orthogonal decomposition to describe a complex variability according to decreasing complementary parts (Fj).

In the plan, row-points can show grouping into different "constellations" indicating the presence of different trends or sub-populations in the dataset.

For that, PCA decomposes the variability space of a dataset into a succession of orthogonal axes representing decreasing and complementary parts of the total variability (Figure 34). From the simplistic illustration, decomposition of the total variability into two orthogonal directions F1 and F2 highlights clearly some similar and opposite behaviours of the different variables M_j: along F1, the variables M1 and M2 show a certain affinity and seem to be opposite to the variables M3 and M4 (projected on the other extremity of F1). Such information is completed by that along F2 where M1 and M3 share a similar behaviour opposite to that of the variables M2 and M4.

This illustrates the aim of PCA consisting in handling the complex variability under successive complementary view angles.

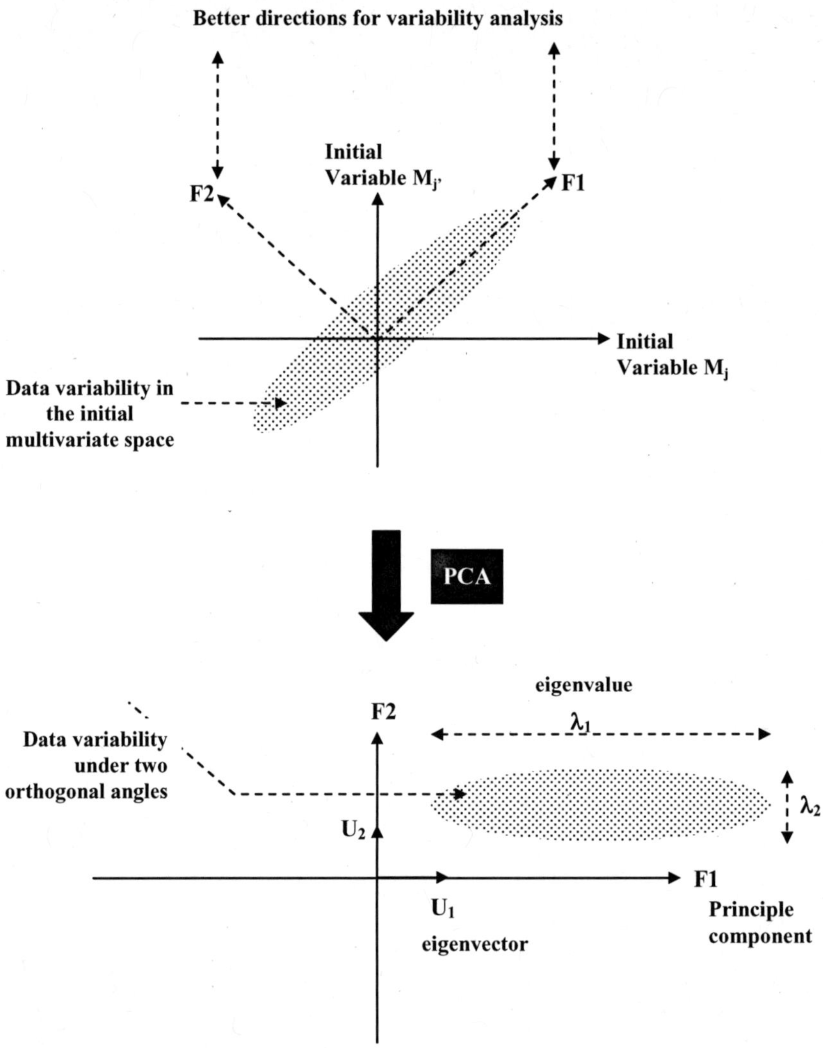

Figure 36. Graphical illustration of principle of PCA based on calculation of eigenvalues λ_k, eigenvectors U_k and principle components F_k

IV.1.6.2. General Principle of PCA

PCA is a decomposition approach based on the extraction of the eigenvalues and eigenvectors of a dataset. The eigenvectors give orthogonal directions called the principle components (F_j) which describe complementary and decreasing parts of the total variability (Figure 35).

The decrease in explained variability is closely linked to the eigenvalues sorted by decreasing order. To each eigenvalue λ_j of the dataset corresponds an eigenvector U_j which gives the direction of principle component F_j; the variability explained along F_j is equal to λ_j and it can be expressed in terms of relative part by $\lambda_j/\Sigma(\lambda_j)$ (Figure 36) (Waite, 2000).

IV.1.6.3. Computation of Eigenvalues, Eigenvectors and Principle Components

Eigenvalues and eigenvectors are calculated for a square ($p \times p$) and invertible (i.e. not null determinant) matrix A. Therefore, any square matrix A ($p \times p$) can be decomposed into p directions F_k defined by p eigenvectors U_k and weighted by p eigenvalues λ_k. From an experimental dataset X, a square matrix A can be directly obtained by the product A= X'X; therefore, the eigenvalues and eigenvectors are calculated from A.

The eigenvalues λ_k and their corresponding eigenvectors U_k are calculated for a square matrix A ($p \times p$) by solving the following matricial equation:

$$A.U = \lambda.U \Leftrightarrow A.U - \lambda.U = 0 \Leftrightarrow (A - \lambda.I).\ U = 0 \Leftrightarrow (A - \lambda.I) = 0$$

where I is a ($p \times p$) identity matrix:

$$I = \begin{pmatrix} 1 & 0 & \dots & 0 & 0 \\ 0 & 1 & \dots & 0 & 0 \\ 0 & 0 & 1 & 0 & 0 \\ 0 & 0 & \dots & 1 & 0 \\ 0 & 0 & \dots & 0 & 1 \end{pmatrix} \begin{matrix} 1 \\ . \\ . \\ . \\ p \end{matrix}$$

$$\begin{matrix} 1 & \dots & \dots & \dots & p \end{matrix}$$

This matricial equation is solved by setting its determinant to zero: det(A - λ.I) = 0, leading to solve a p equation system with p unknown λ_k. After computation of the eigenvalues λ_k, the corresponding eigenvectors U_k are calculated from the initial equation A.U = λ.U.

Finally, from the eigenvectors U_k, the initial variables M_j of the dataset X are replaced by "synthetic" variables F_k (called principle components) obtained by linear combinations of the p initial variables M_j affected by the coordinates of the corresponding eigenvectors U_k:

$$F_{ik} = \sum_{j=1}^{p} X_{ij} U_{jk} = x_{i1}.u_{1k} + x_{i2}.u_{2k} + x_{i3}.u_{3k} + \dots + x_{ij}.u_{jk} + \dots + x_{ip}.u_{pk}$$

In other words, from the p coordinates x_{ij} of a row i corresponding to the p columns j, one new coordinate F_{ik} is calculated to represent the new position of row i along the principle component F_k (Figure 37). The new coordinates, called factorial coordinates, are more appropriate to associate behaviours of different individuals i to some levels of variables M_j, leading to understand the variability structure of the initial dataset X.

To understand more the calculation and the interpretation of eigenvalues, eigenvectors and factorial coordinates in PCA, let's give a simplistic numerical example based on a square matrix A (2 × 2).

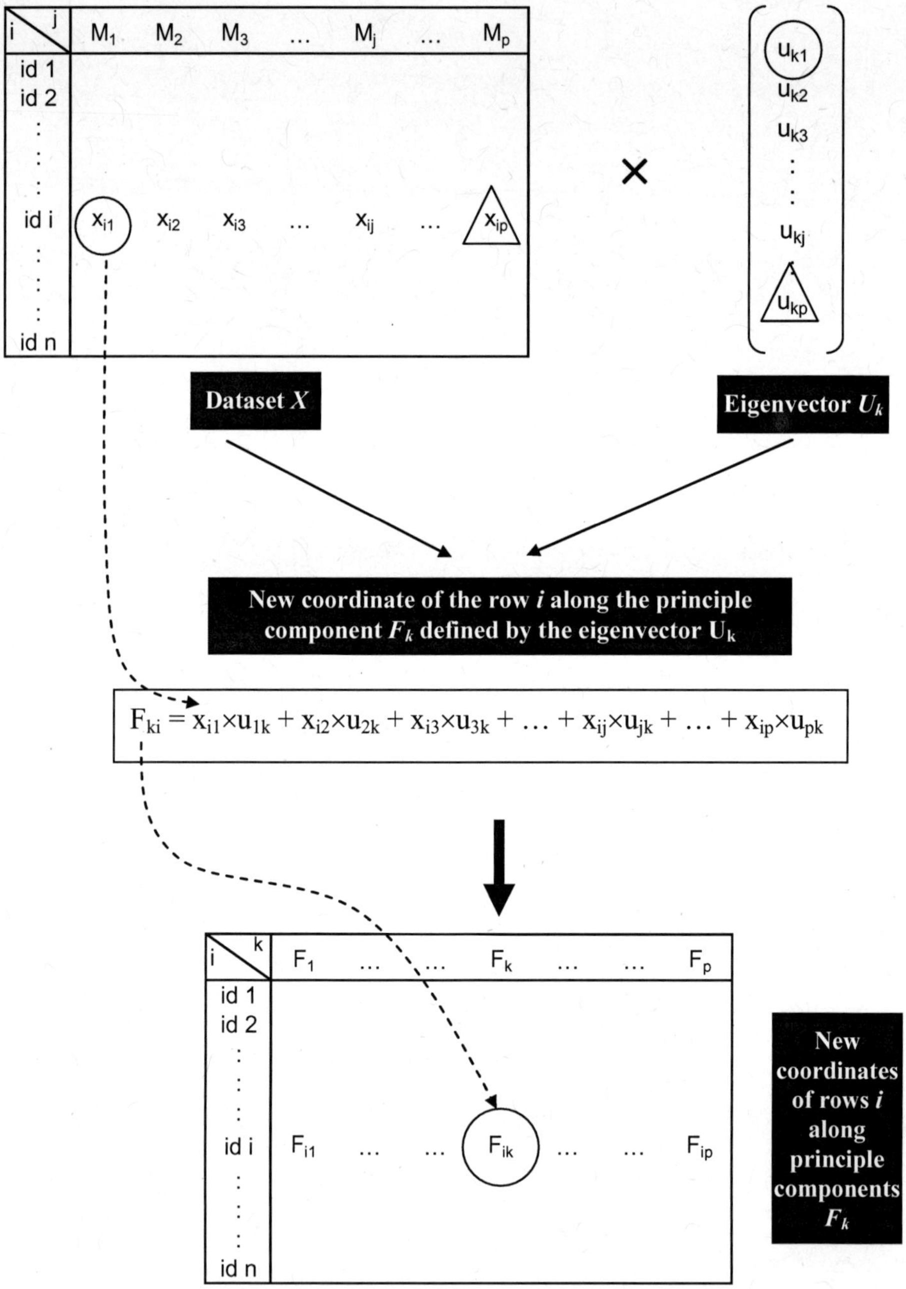

Figure 37. Computation of new coordinates (factorial coordinates) of an individuals i along a principle component F_k by a linear combination of its initial coordinates x_{ij} affected by the coordinates of the eigenvector U_k.

$$A = \begin{pmatrix} 2 & 3 \\ 3 & -6 \end{pmatrix}$$

$$A - \lambda.I = \begin{pmatrix} 2 & 3 \\ 3 & -6 \end{pmatrix} - \lambda \begin{pmatrix} 1 & 0 \\ 0 & 1 \end{pmatrix} = \begin{pmatrix} 2 & 3 \\ 3 & -6 \end{pmatrix} \begin{pmatrix} \lambda & 0 \\ 0 & \lambda \end{pmatrix} = \begin{pmatrix} 2-\lambda & 3 \\ 3 & -6-\lambda \end{pmatrix}$$

$$\det(A - \lambda.I) = \det \begin{pmatrix} 2-\lambda & 3 \\ 3 & -6-\lambda \end{pmatrix} \longrightarrow \det \begin{pmatrix} a & c \\ b & d \end{pmatrix} = ad - bc$$

$$\det \begin{pmatrix} 2-\lambda & 3 \\ 3 & -6-\lambda \end{pmatrix} = [(2-\lambda)(-6-\lambda) - 9] = \lambda^2 + 4\lambda - 21$$

Setting $\lambda^2 + 4\lambda - 21$ to 0 leads to the equivalent form: $(\lambda - 3)(\lambda + 7) = 0$, so the eigenvalues λ_k of A are 3 and -7. After sorting these two λ_k by decreasing absolute value, we have $\lambda_1 = -7$ and $\lambda_2 = 3$.

For each eigenvalue λ_k, the corresponding eigenvector U_k is calculated by solving the matricial equation $(A - \lambda.I).U = 0$:

For $\lambda_1 = -7$, the matricial equation will be:

$$\left(\begin{pmatrix} 2 & 3 \\ 3 & -6 \end{pmatrix} - (-7)\begin{pmatrix} 1 & 0 \\ 0 & 1 \end{pmatrix}\right)\begin{pmatrix} u_{11} \\ u_{21} \end{pmatrix} = 0 \quad \Leftrightarrow \quad \left(\begin{pmatrix} 2 & 3 \\ 3 & -6 \end{pmatrix} + \begin{pmatrix} 7 & 0 \\ 0 & 7 \end{pmatrix}\right)\begin{pmatrix} u_{11} \\ u_{21} \end{pmatrix} = 0$$

$$\Leftrightarrow \begin{pmatrix} 9 & 3 \\ 3 & 1 \end{pmatrix}\begin{pmatrix} u_{11} \\ u_{21} \end{pmatrix} = 0$$

This leads to the following equation system:

$$9u_{11} + 3u_{21} = 0 \quad \Leftrightarrow \quad 9u_{11} = -3u_{21}$$
$$3u_{11} + u_{21} = 0 \quad \Leftrightarrow \quad 3u_{11} = -u_{21}$$

For $u_{11} = 1$, we have $u_{21} = -3$. Therefore, $U_1 = (1, -3)$ is the first eigenvector of A.

Note that due to the fact that the equation system is reduced to one equation with two unknown, results in the existence of infinity of eigenvectors proportional to U_1.

For $\lambda_2 = 3$, the matricial equation will be:

$$\left(\begin{pmatrix} 2 & 3 \\ 3 & -6 \end{pmatrix} - (3)\begin{pmatrix} 1 & 0 \\ 0 & 1 \end{pmatrix}\right)\begin{pmatrix} u_{12} \\ u_{22} \end{pmatrix} = 0 \quad \Leftrightarrow \quad \left(\begin{pmatrix} 2 & 3 \\ 3 & -6 \end{pmatrix} - \begin{pmatrix} 3 & 0 \\ 0 & 3 \end{pmatrix}\right)\begin{pmatrix} u_{12} \\ u_{22} \end{pmatrix} = 0$$

$$\Leftrightarrow \begin{pmatrix} -1 & 3 \\ 3 & -9 \end{pmatrix}\begin{pmatrix} u_{12} \\ u_{22} \end{pmatrix} = 0$$

This leads to the following equation system:

$$-u_{12} + 3u_{22} = 0 \Leftrightarrow u_{12} = 3u_{22}$$

$$3u_{12} - 9u_{22} = 0 \Leftrightarrow 3u_{12} = 9u_{22}$$

For $u_{22} = 1$, we have $u_{12}=3$. Therefore, $U_2 = (3, 1)$ is the second eigenvector of A.

Also, the fact that the equation system is reduced to one equation with two unknown results in the existence of infinity of eigenvectors proportional to U_2.

The two calculated eigenvectors U1 and U2 define a new basis of orthogonal directions along which the row and column variability of the dataset A can be topologically analysed (Figure 38).

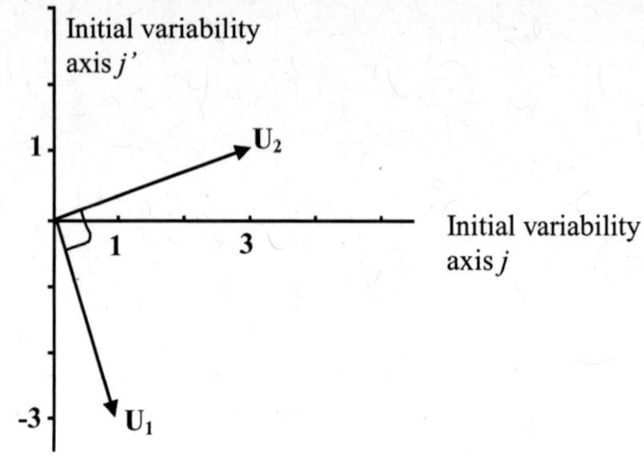

Figure 38. Illustration of the orthogonality between the eigenvectors of a matrix.

After calculation of the eigenvectors U_1 and U_2, the new coordinates F_{ik} of the rows *i* along the principal components *k* (*k*=1 to 2) can be calculated by the scalar product $A.U_k$. Thus, along the principle component F1 defined by the direction of U1, the two rows of the matrix A will be represented by two coordinates given by:

$A.U_1 = \begin{pmatrix} 2 & 3 \\ 3 & -6 \end{pmatrix} \begin{pmatrix} 1 \\ -3 \end{pmatrix} = \begin{pmatrix} -7 \\ 21 \end{pmatrix}$; this result is also obtained by the product $\lambda_1.U_1$.

Along the second principle component F2, each row of the matrix A will have a new coordinate given by:

$A.U_2 = \begin{pmatrix} 2 & 3 \\ 3 & -6 \end{pmatrix} \begin{pmatrix} 3 \\ 1 \end{pmatrix} = \begin{pmatrix} 9 \\ 3 \end{pmatrix}$; this result is also obtained by the product $\lambda_2.U_2$.

Finally, the dataset A can be replaced by the new matrix F giving the factorial coordinates of the rows (individuals) *i* along each principle component F_k (*k*=1-2):

$F = \begin{pmatrix} -7 & 9 \\ 21 & 3 \end{pmatrix}$; from F, the individuals (the rows) of the dataset A can be projected on the plane F1F2 for a topological analysis of their variability (Figure 39). To link the variability of individuals to that of variables, a variable plot can be obtained from the coordinates of the eigenvectors by which the initial variables were weighted (Figure 39). According to their absolute values, such coordinates attribute more or less importance to the initial variables Mj in the new (factorial) coordinates of individuals i. For example, the individual id1 has a factorial coordinate equal to -7 on F1; this value was calculated by the following linear combination:

$$-7 = (\text{id1}).U1 = (2\ 3)\begin{pmatrix} 1 \\ -3 \end{pmatrix} = (2 \times 1) + (3 \times \mathbf{-3})$$

In this linear combination, the second variable M2 is affected by an eigenvector score equal to -3 the absolute value of which (Abs(-3)=3) is higher than the coordinate=1 by which is affected the first variable M1. This remark concerning the role of M2 on F1 can be generalised for all the factorial coordinates along F1. This helps to conclude that the variability of all the individuals on F1 is mainly due to the variable M2. Graphically, this can be showed by a projection of M2 both at extremity and close to the axis F1 (Figure 39).

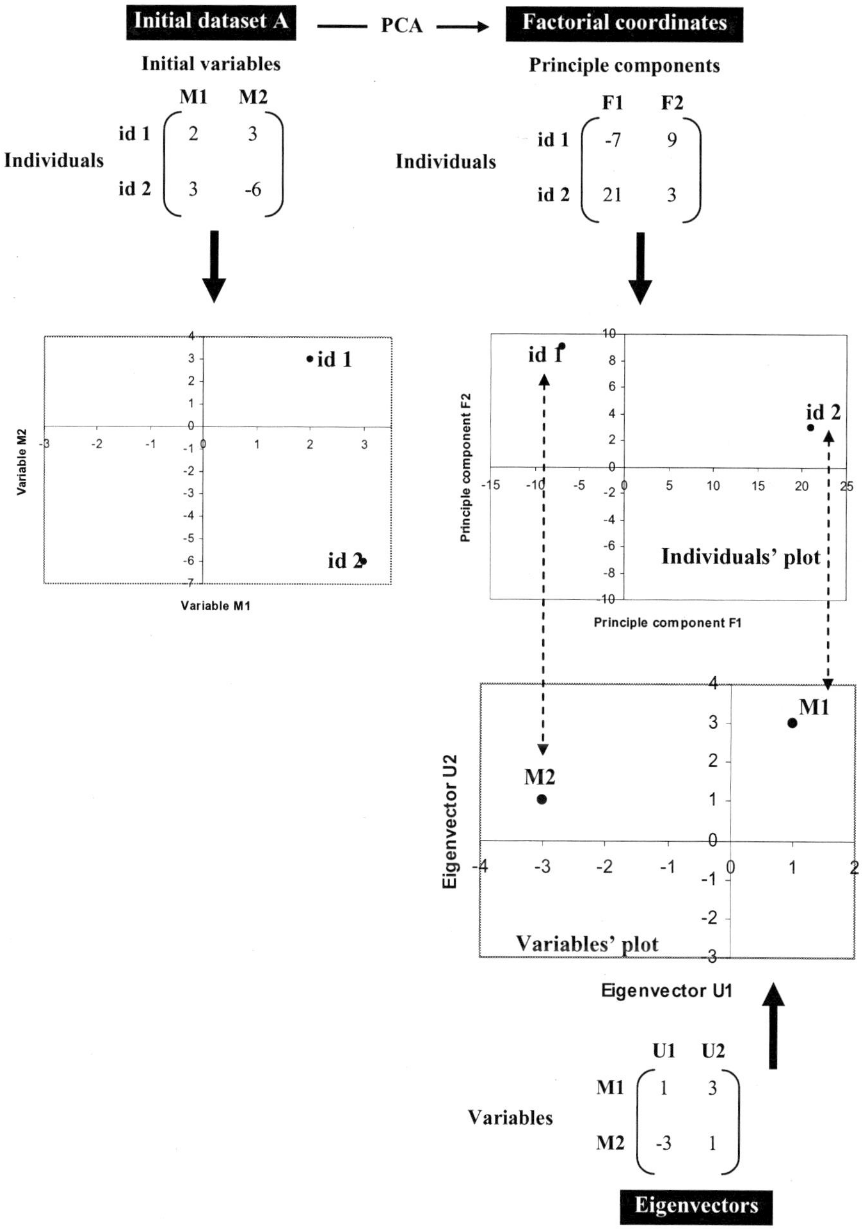

Figure 39. Graphical analysis of links between the variability of individuals and that of variables by means of PCA.

IV.1.6.4. Graphical Interpretation of Factorial Plans

According to the factorial plan F1F2 of individuals (Figure 39), id1 and id2 show opposition along F1. According to the variable plot, the variables M1 and M2 seem to be opposite, and projected on the same sides than id2 and id1, respectively. Taking into account the importance of variable M2 on F1, and the graphical proximity between M2 and id1, the opposition of id1 to id2 can be explained by a high value of M2 in id1 and a low one in id2. In fact, the initial dataset A shows values of 3 and -6 for M2 in id1 and id2, respectively. Thus, the PCA helped to identify that the highest variability source in the dataset A consisted of an important opposition between id1 and id2 for variable M2. In metabolomic terms, this can correspond to a situation where some individuals are productive of a metabolite M2 whereas others are relatively deficient in M2.

For F2, the highest coordinate of corresponding eigenvector U2 concerns variable M1, leading to deduce that the role of M1 on F2 is relatively more important than that of M2. Graphically, the individual id2 projects closer to M1 than it is id1. This translates a higher value of M1 in id2 than in id ; this can be checked in the initial dataset A. From this simplistic example, variable M2 appears to play a separation role between individuals (profiles), whereas the variable M1 seems to group the individuals according to a more or less affinity. The fact that id1 and id2 are bot opposite alonf F2 can be attributed to their relatively close positive values (2 and 3, respectively).

Apart from the dual analysis between rows (individuals) and columns (variables), the interpretations in PCA can be focused on the variability of variables and individuals, separately: on the plan F1F2 (Figure 39), the variables M1 and M2 seem to have mainly opposite behaviours from their projections in two different parts of the plan. This opposition is observed for individuals, and seems to indicate the presence of two trends in the initial dataset A.

IV.1.6.5. Different Types of PCA

The variability of a dataset X ($n \times p$) can be analysed by PCA on the basis of different criteria by considering (Figure 40):

- The crude effects of variables leading to give more importance to the most dispersed variables from the axes' origin.
- The variations of data around their mean vector (centered PCA) leading to analyse the variability of the dataset around its gravity centre GC.
- Standardized data obtained by homogenizing the variation scales of all the variables through their weighting by their variances. This leads to analyse the variability of the dataset around the gravity centre and within a unity scale space.
- Ranked data consisting in using the ranks of data rather than their values.
- These different PCA are performed from different square matrices ($p \times p$):
- PCA on crude data is performed on the square matrix X'X.
- Centred PCA is performed on the square matrix C'C, with $C = X - \overline{X}$, and where $\overline{X}$ is the mean vector of the different variables.

- Standardized PCA is applied from the square matrix Z'Z, with $Z = \frac{X - \overline{X}}{SD}$, and where $\overline{X}$ and SD are the mean and standard deviation of each corresponding variable, respectively.
- Rank-based PCA is applied on the square matrix K'K, where K is the rank matrix representing the ranked data for each variable of dataset X.

The applications of these different kinds of PCA require some conditions and have different interests:

Centred PCA application is applied when all the variables have the same unit (e.g. µg/mL). Its interest consists in highlighting the effect of the most dispersed variables on the structure of the dataset. Thus, the most dispersed variables can be considered as more rich in information than the less dispersed ones. Centred PCA helps to identify how the individuals (profiles) are separated the ones from the others under the dispersion effect of some variables. Moreover, such a multivariate analysis allows classification of the different variables according to their variation scales and directions (i.e. according to their covariances). In centred PCA, the sum of the eigenvalues is equal to the total variance of the dataset.

Standardized PCA is required when the dataset consists of heterogeneous variables expressed with different measure units (µg, mL, °c, etc.). Also, it is required when the variables have different variation scales due to incomparable variances. In these cases, the values of each variable X_j are standardized by subtracting the mean $\overline{X}_j$ and by dividing by the standard deviation SD_j. Graphically, the set of standardizations attributes to the variables different relative positions which are interpretable in terms of Pearson correlations: the co-response of two variables will be highlighted by two vectors which will be projected along a same direction in the multivariate space. If two variables are positively correlated, their corresponding vectors will have a very sharp angle ($0 \leq \leq \pi/4$); in the case of negatively correlated variables, the corresponding vectors will be opposite, i.e. their angle will be strongly obtuse ($3\pi/4 \leq \leq \pi$). In the case of low correlations, the two vectors corresponding to the paired variables will have almost perpendicular directions. In standardized PCA, the sum of the eigenvalues is equal to the number (p) of variables.

Rank-based PCA finds an exclusive application on ordinal qualitative dataset where the variables are not measured but consist of different classification modalities of the individuals (e.g. modalities low, intermediate, high levels). After substitution of the ordinal data by their ranks, a standardized PCA can be applied to analyse correlations between the qualitative variables on the basis of Spearman statistics. Rank-based PCA finds also application on heterogeneous datasets because of different variable units or because of imbalanced variation ranges of the variables.

IV.1.6.6. Numerical Application and Interpretation of Standardized PCA

The application of standardized PCA will be illustrated by a numerical example based on a dataset of n=9 rows and p=5 columns (Figure 41). Under a metabolomic aspect, let's consider the rows as metabolic profiles, the columns as metabolites and the data as concentrations.

The PCA gives two principle components F1 and F2 represented by two eigenvalues λ_1=3.74 and λ_2=1.20. Such eigenvalues correspond to 75% (3.74/*p*) and 24% (1.20/*p*) of the total variability extracted by F1 and F2, respectively.

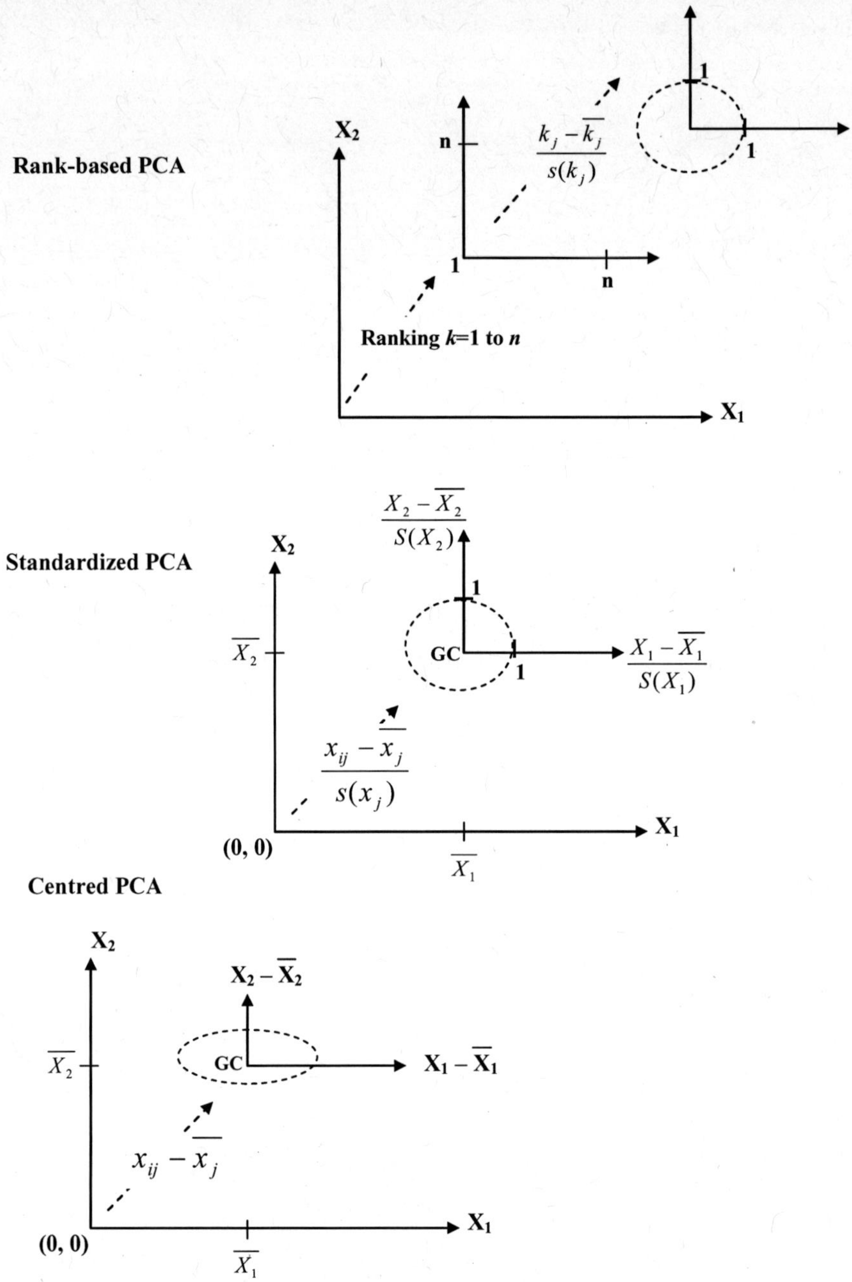

Figure 40. Illustration of different numerical transformations in PCA.

Initial dataset

	M1	M2	M3	M4	M5
id1	1.80	3.88	10.10	1.89	2.33
id2	2.21	3.58	11.25	1.96	2.74
id3	2.72	4.51	11.28	2.17	3.97
id4	9.03	4.23	3.35	10.83	10.82
id5	9.84	5.43	3.64	10.87	10.55
id6	10.4	5.18	4.44	11.42	11.59
id7	1.55	2.26	3.32	4.83	5.19
id8	1.81	2.83	3.81	4.88	6.12
id9	2.70	3.00	4.14	5.72	6.71

Standardized PCA

Individual factorial coordinates

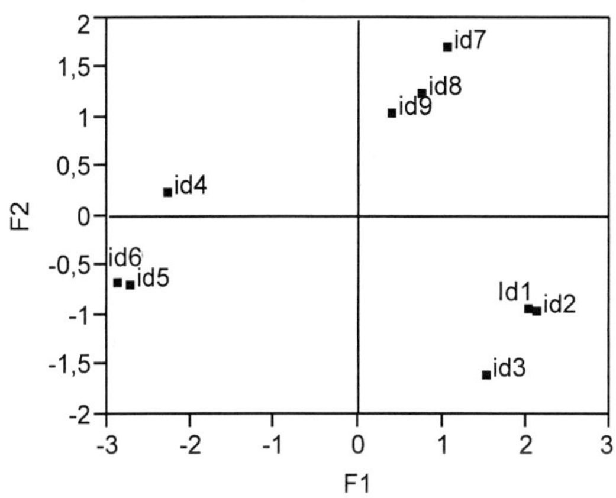

Correlation circle

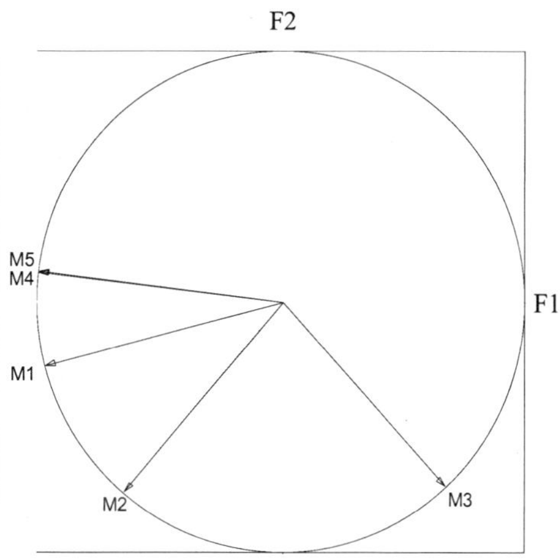

Figure 41. Graphical representations of a standardized PCA based on the factorial coordinates' plot of individuals and correlation circle of variables.

From the plot of individuals, the nine individuals are projected according to three trends (Figure 41): id1, 2, 3 (group G1), id4, 5, 6 (group G2) and id7, 8, 9 (group G3). Groups G1 and G2 are opposite along the first component F1; this means that they have opposite characteristics: according to the correlation circle, the variable M3 projects closely to the individuals of G1, meaning that its values are high in these individuals. On this same basis, the graphical proximity between variables M1, M4, M5 and individuals id4, 5, 6 leads to conclude that the group G2 is characterized by high values for these variables. Finally, the variable M2 projects in a part where no individual is concerned. However, it appears to be opposite to G1 along F1 and to G3 (particularly) along F2. This means that the variable M2 is an opposition variable characterizing individuals by its low values: in fact the individuals id1-id3 and id7-id9 have relatively low values for M2.

From the correlation circle, affinity and opposition between the variables can be highlighted from sharp or obtuse angles between corresponding vectors: thus, the vectors M4, M5 and M1 show very sharp angles between them meaning positive correlations between corresponding variables (Figure 42). On the other hand, the vector of M3 seems to be particularly opposite to those of M4, M5 meaning negative correlations between their corresponding variables. M1 and M3 have almost perpendicular obtuse vectors (Figure 41) meaning a low or not significant correlation between them (Figure 42). The vectors M2 and M3 are closer to orthogonality than M1, M3, and represent a stronger independence state between corresponding variables. Finally, the vector M2 shares a sharp angle with M1 and in a lesser measure with M4 and M5. This means a positive correlation of variable M2 toward M1, which is higher than those toward M4 and M5.

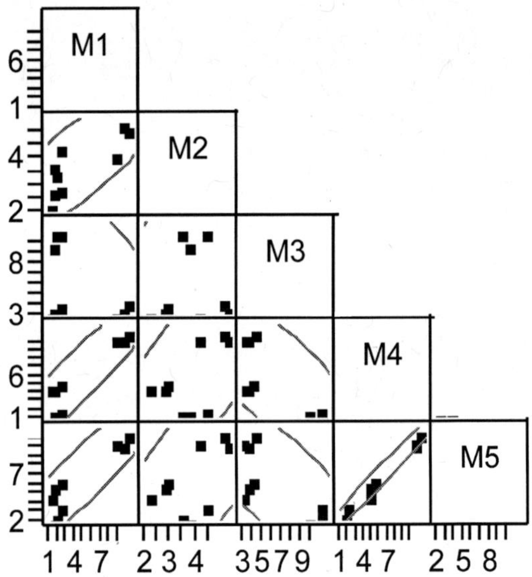

Figure 42. Scatter plot matrix showing the correlations between different variables M1-M5 of the dataset of figure 41. High correlations are indicated by thin confidence ellipses.

IV.2. Distance Matrix-Based Approach: Cluster Analysis

IV.2.1. Introduction

Population analysis is closely linked to the variability and diversity concepts. A population consists of a great number of individuals that are more or less similar/different. To understand better the complex structures of a population, it is helpful to classify it into complementary and homogeneous subsets (Maharjan and Ferenci, 2005; Semmar et al., 2005; Everitt et al., 2001; Gordon, 1999; Dimitriadou et al., 2004; Jain et al., 1999; Milligan and Cooper, 1987).

When the individuals are characterized by several variables, it becomes difficult to separate them easily into homogeneous groups because their similarity/dissimilarity must be evaluated by considering all the variables at once. Such high-dimension problem can be overcame by means of multivariate analyses: cluster analysis is particularly appropriate to classify populations by different manners based on different techniques leading to different classification patterns.

Cluster analysis (CA) is performed into two steps: (a) computation of distances between all the individual pairs to quantify the closeness/farthness degree between individual cases; (b) grouping the most similar (the less distant) cases into homogeneous subsets (clusters) according to a certain criterion (Figure 43). Different classification patterns can be obtained by using different distance kinds and different aggregation criteria; this allows to analyse what approach gives the best interpretable classification by reference to the biological (metabolic) context.

There are two main clustering methods: hierarchical and non-hierarchical clustering. This chapter will focus on hierarchical clustering.

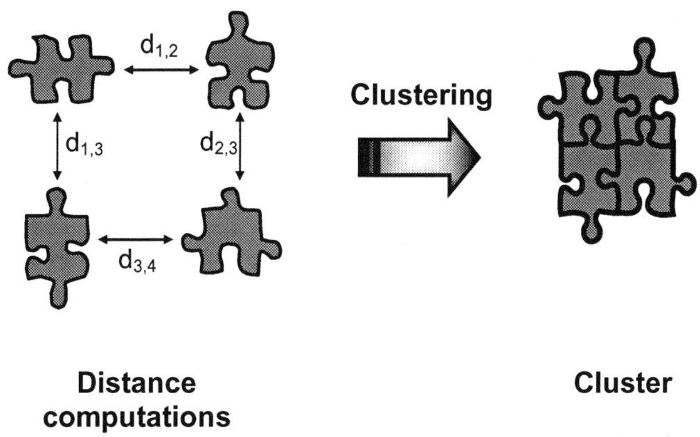

Figure 43. Intuitive presentation of the two main steps in cluster analysis _ distance computations and clustering _.

In metabolomics, the classification can play important role in the analysis of the complex variability of a metabolic dataset. This is all the more important since the metabolic profiles in a dataset can vary gradually by slight fluctuations in the relative levels of metabolites, leading to the absence of frank borders between profiles.

IV.2.2 Goal of Cluster Analysis

Cluster analysis, also called data segmentation aims to partition a set of experimental units (e.g. metabolic profiles) into two or more subsets called clusters. More precisely, it is a classification method for grouping individuals or objects into clusters so that the objects in the same cluster are more similar to one another than to objects in other clusters.

IV.2.3. General Protocols in Hierarchical Cluster Analysis (HCA)

The hierarchical classification structure given by HCA is graphically represented by a tree of clusters, also known as a dendrogram. The cluster protocols can be subdivided into divisive (top-down) and agglomerative (bottom-up) methods (Figure 44) (Lance and Williams, 1967):

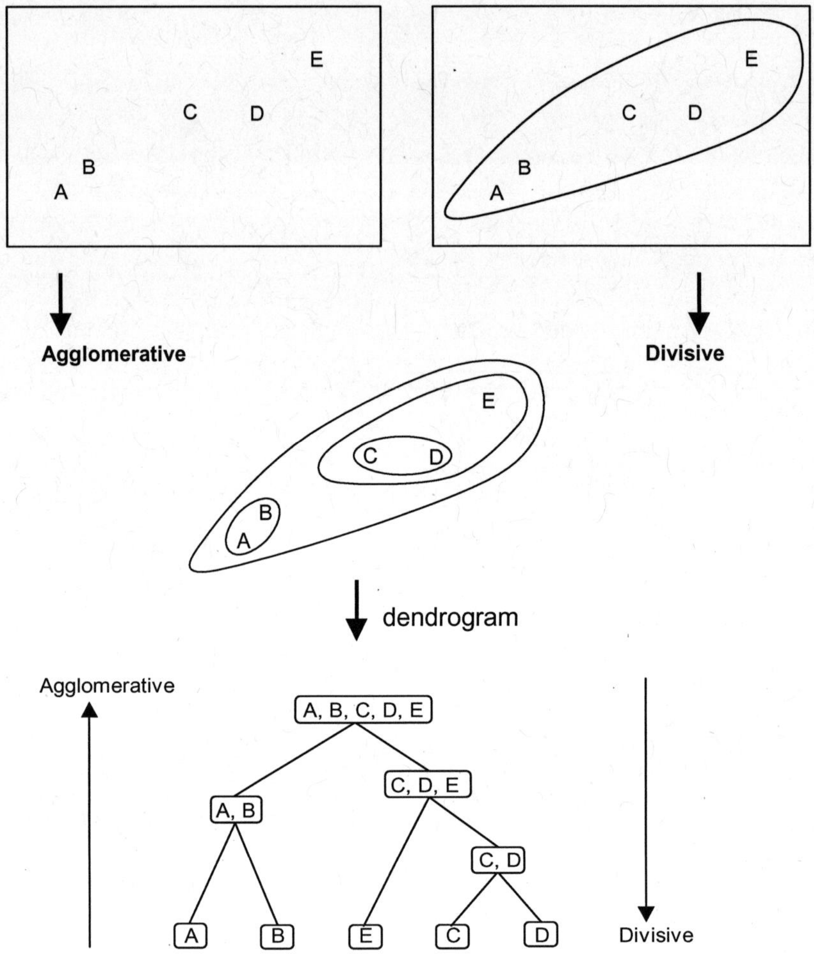

Figure 44. Two tree-building protocols in hierarchical cluster analysis (HCA) consisting in grouping (agglomerative) or separating (divisive) progressively the individuals.

The divisive method, less common, starts with a single cluster containing all objects and then successively splits resulting clusters until only clusters of individual objects remain. Although some divisive techniques attempt to minimize the within-cluster error sum of squares, they face problems of computational complexity that are not easily overcome (Milligan and Cooper, 1987).

The agglomerative method starts with every single object in a single cluster. Then, in a series of successive iterations, it agglomerates (merges) the closest pair of clusters by satisfying some similarity criteria, until all of the data is in one cluster. The agglomerative method is the one especially described in this chapter.

The complete process of agglomerative hierarchical clustering requires defining an inter-individual distance and an inter-cluster linkage criterion, which can be represented by two iterative steps:

1. Calculate the (dis)similarities or distances between all individual cases;

2. Fuse the most appropriate (close, similar) clusters by using a clustering algorithm, and then recalculate the distances. This step is repeated until all cases are in one cluster.

IV.2.4. Dissimilarity Measures

Dissimilarities are calculated in order to quantify the degree of separation between points. On continuous data, distances are calculated to evaluate dissimilarities between individuals. However, on qualitative data (binary, counts), the dissimilarities are indirectly evaluated from similarity indices (SI) which can be transformed into dissimilarities by single operations, e.g. (1 – SI). A part from distances and SI, there are many ways to measure a dissimilarity/similarity according to circumstances and data type: correlation coefficient, non metric coefficient, cosine, information-gain or entropy-loss (Everitt, et al., 2001; Gordon, 1999; Arabie et al., 1996; Lance and Williams, 1967; Shannon, 1948).

IV.2.4.1. Continuous Data and Distance Computation

IV.2.4.1.1. Euclidean Distance

Euclidean distance is appropriately calculated between profiles containing continuous data. It is a particular case of Minkowski metric:

$$dist(x_i, x_k) = \left[\sum_{j=1}^{p} \left| x_{ij} - x_{kj} \right|^r \right]^{1/r}$$

where:

- r is an exponent parameter defining a distance type (=1 for Manhattan distance, =2 for Euclidean distance, etc.);
- x_{ij}, x_{kj} are values of variable j for the objects i and k respectively;
- p is the total number of variables describing the profiles x_i, x_k.

Let's give a numerical example of three concentration profiles containing three metabolites:

	Metabolites		
Profiles	**M1**	**M2**	**M3**
X1	10	6	4
X2	10	4	3
X3	5	3	2

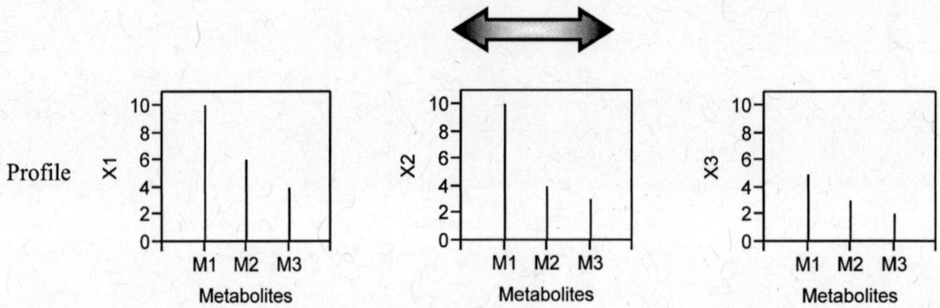

By applying the Euclidean distance, one would know which profiles are the closest the one the other?

We have to calculate three distances between profiles: X1-X2, X1-X3 and X2-X3.

	Metabolites				Euclidean distances d
Profiles	M1	M2	M3	Sum	d=√Sum
$(X1-X2)^2$	0	4	1	5	2.24
$(X1-X3)^2$	25	9	4	38	6.16
$(X2-X3)^2$	25	1	1	27	5.20

From the lowest Euclidean distance, one can deduce that profiles X1 and X2 are the closest between them, whereas X1 and X3 and the farthest.

The distance can be calculated either on crude data or after data transformation. Using crude data is appropriate when the variables have comparable variances or when one would attribute domination to higher variance variable. In the second case, data transformation can be used to gives to the variables comparable scales and equal influence in cluster analysis. The most common transformation (standardization) consists of the conversion of crude data into standard scores (*z*-scores) by subtracting the mean and dividing by the standard deviation of each variable.

Many other distance measures are appropriate according to the data types: Mahalanobis, Hellinger, Chi-square distance, etc. (Blackwood et al., 2003; Gibbons and Roth, 2002).

IV.2.4.1.2. Chi-Square Distance

Chi-square distance is applied on dataset the values of which are additive both on rows and columns. This is the case for concentration datasets which are common in metabolomics. This distance can be calculated according to the formula:

$$\chi^2(X1,X2)=\sum_{j=1}^{p}\frac{Sum_{tot}}{Sum_j}\left(\frac{X1_j}{Sum_{X1}}-\frac{X2_j}{Sum_{X2}}\right)^2$$

where :

X1, *X2* denotes individual profiles (e.g. metabolic profile)
j: index of column or variable *j* (e.g. metabolite *j*)
$X1_j$, $X2_j$: values of variables j in the profiles X1 and X2, respectively
Sum_{X1}, Sum_{X2} are the sums of values in each individual *X1* and *X2*, respectively

Sum_j is the sum of the values of variable *j* (e.g. sum of concentrations of metabolite *j*)

Sum_{tot} is the sum of all the values of the whole dataset

According to the χ^2 distance, two individuals are all the more close since their relative profiles are similar. This can be checked when the values of a given profile are multiple of the values in another one.

Let's calculate the χ^2 distances between the three profiles X1, X2, X3 (Figure 45).

Initial dataset: (3 profiles × 3 metabolites)

Profiles	**Metabolites** M1	M2	M3	**Sum row** Sum_{Xi}
X1	10	6	4	20
X2	10	4	3	17
X3	5	3	2	10
Sum col. Sum_j	25	13	9	$Sum_{tot} = 47$

↓

Profiles	$\frac{X_{ij}}{Sum_{Xi}}$ M1	M2	M3
X1	0.500	0.300	0.200
X2	0.588	0.235	0.176
X3	0.500	0.300	0.200

↓

Pairs	$\left(\frac{X_{ij}}{Sum_{Xi}} - \frac{X_{i'j}}{Sum_{Xi'}}\right)^2$ **M1**	**M2**	**M3**
(X1, X2)	0.0078	0.0042	0.0006
(X1, X3)	0	0	0
(X2, X3)	0.0078	0.0042	0.0006

↓

Pairs	$\left(\frac{Sum_{tot}}{Sum_j}\right) * \left(\frac{X_{ij}}{Sum_{Xi}} - \frac{X_{i'j}}{Sum_{xi'}}\right)^2$ **M1**	**M2**	**M3**
(X1, X2)	0.0147	0.0152	0.0031
(X1, X3)	0	0	0
(X2, X3)	0.0147	0.0152	0.0031

Sum →

Chi2 distances

Chi2
0.033
0
0.033

Figure 45. Numerical example illustrating the computation of Chi2 (or χ^2) distances between three pairs of profiles.

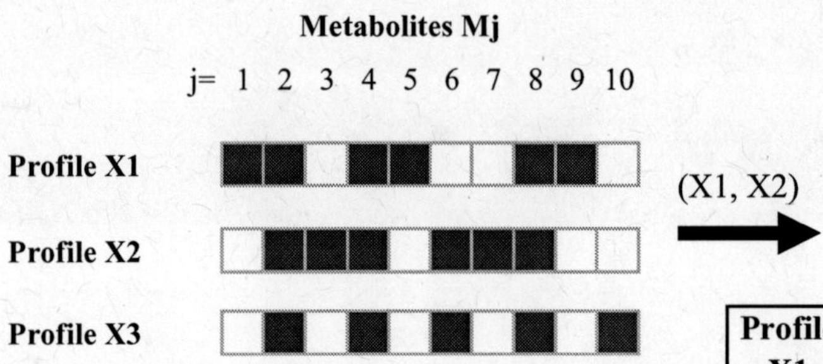

		Profile X2	
		Present	Absent
Profile X1	Present	a = 3	b = 3
	Absent	c = 3	d = 1

Similarity indices	Formula	Result
Kulizinsky	$\frac{a}{b+c}$	0.5
Jaccard	$\frac{a}{a+b+c}$	0.33
Russel-Rao	$\frac{a}{a+b+c+d}$	0.3
Dice	$\frac{2a}{2a+b+c}$	0.5
Sokal-Michener	$\frac{a+d}{a+b+c+d}$	0.4
Roger-Tanimoto	$\frac{a+d}{a+2b+2c+d}$	0.25
Sokal-Sneath	$\frac{a}{a+2(b+c)}$	0.2
Yule	$\frac{ad-bc}{ad+bc}$	-0.5
Correlation	$\frac{\lvert ad-bc \rvert}{\sqrt{(a+b)\cdot(a+c)\cdot(b+d)\cdot(c+d)}}$	0.33

Figure 46. Calculus of similarity between two profiles according to different similarity indices.

The computations show that the minimal χ^2 distance concerns the pairs (X1, X3) by opposition to the Euclidean distance. This χ^2 is minimal, indeed null, because the absolute profiles X1 (10, 6, 4) and X3 (5, 3, 2) correspond to the same relative profile (0.5, 0.3, 0.2).

IV.2.4.2. Qualitative Variables and Similarity Indices

For qualitative data (binary, counting), many similarity indices (SI) could be used as intuitive measures of the closeness between individuals: Jaccard, Sorensen-Dice, Tanimoto, Sokal-Michener indices, etc. (Jaccard , 1912; Duatre et al., 1999; Rouvray, 1992). The similarity indices are less sensitive to the null values of the variables, and thus they are useful in the case of sparse data. To evaluate similarity between two individuals *X1* and *X2*, we need three or four essential elements: a = number of shared characterisrics; b = number of characteristics present in *X1* and absent in *X2*; c = number of characteristics present in *X2* and absent in *X1*; d = number of characteristics absent both in *X1* and *X2* (required for some SI).

The different *SI* can be converted into dissimilarity *D* according to the formula:

- $D = 1 - SI$ if SI $\in$ [0, 1]
- $D = \dfrac{1 - SI}{2}$ if SI $\in$ [-1, 1]

To illustrate the concept of similarity index, let's give a numerical example concerning three metabolic profiles characterized by 10 metabolites the concentration of which are not known (Figure 46). In such case, quantitative data (concentrations) are not available, and consequently, distances can't be computed. However, information on presence/absence of metabolites j in the different profiles X_i can be used to calculate SI between the profiles.

IV.2.5. Clustering Techniques

After computation of distances or dissimilarities between all the individuals of the dataset (e.g. metabolic profiles), it becomes possible to merge them into homogeneous and well separated groups by using an aggregation algorithm: initially, the most close (the less distant) individuals are merged to give a group. After the apparition of some small groups, the immediate next step consists in merging the most similar groups into larger groups by reference to a certain homogeneity criterion (aggregation rule). Such procedure is iteratively applied until all the individuals/groups are merged into one entity; the most separated (dissimilar) groups will be merged at the final step of the clustering procedure. This leads to a hierarchical stratification of the whole population into well homogeneous and separated groups (called clusters).

For the clustering procedure, there are several aggregation algorithms which are based on different homogeneity criteria. Two clustering principles will be illustrated here: distance-based (a) and variance-based (b) clustering. The distance-based clustering will be illustrated by four algorithms (single, average, centroid and complete links) (Figure 48), whereas the variance-based clustering will be illustrated by one method (Ward method or second order moment algorithm) (Figure 47) (Ward, 1963; Everitt, 2001; Gordon, 1999; Arabie, 1996).

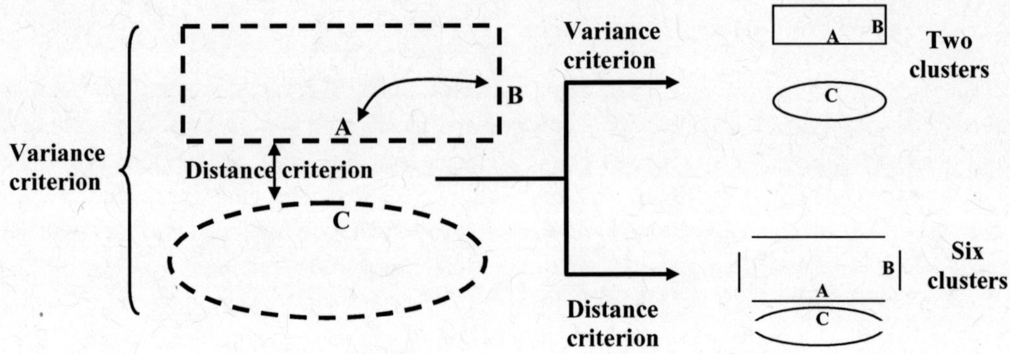

Figure 47. Intuitive representation of clustering based on distance and on variance criteria.

Using the distance criterion, let :

- r and s be two clusters with n_r and n_s elements respectively,
- x_{ri} and x_{sk} the ith and k^{th} elements in clusters r and s, respectively,
- $D(r, s)$ the inter-cluster distance.

It is assumed that $D(r, s)$ is the smallest measure remaining to be considered in the system, so that r and s fuse to form a new cluster t with n_t $(=n_r+n_s)$ elements:

IV.2.5.1. Single Link-Based Clustering

In single-link, two clusters are merged if they have the two closest objects (nearest neighbors) (Figure 48).

Single-link rule strings objects together to form clusters, and consequently it tends to give elongated chain clusters. This elongation is due to the tendency to incorporate intermediate objects into an existing cluster rather than to form a new one. A single linkage algorithm would perform well when clusters are naturally elongated. It is often used in numerical taxonomy.

IV.2.5.2. Complete Link-Based Clustering

In complete-link, two clusters are merged if their farthest objects are separated by a minimal distance by comparison with all other distances between the farthest neighbors of all the clusters (Figure 48). This rule leads to minimize the distance between the most distant objects in the new cluster.

Complete-link rule results in dilatation and may produce many clusters. This algorithm is known to give well compact clusters and usually performs well when the objects form naturally distinct "clumps", or when one wishes to emphasize discontinuities (Jain et al., 1999; Milligan and Cooper, 1987). Moreover, if unequal size clusters are present in the data, complete-link gives superior recovery than other algorithms (Milligan and Cooper, 1987). Complete-link, however, suffers from the opposite defect of single-link: it tends "to break" groups presenting a certain lengthening in space, so as to provide rather spherical classes.

IV.2.5.3. Centroid Link-Based Clustering

In centroid-link, a cluster is represented by its mean position (i.e. centroid). The joining between clusters will be based on the smallest distance between their centroids (Figure 48). This method is a compromise between single and complete linkages.

The centroid method is more robust to outliers than most other hierarchical methods, but in other respects, this method can produce a cluster tree that is not monotonic. This occurs when the distance from the union of two clusters, r and s, to a third cluster u is less than the distance from either r or s to u. In this case, sections of the dendrogram change direction. This change is an indication that one should use another method.

IV.2.5.4. Average Link-Based Clustering

In average-link algorithm, the closest clusters are those having the minimal average distance calculated between all their point pairs. The basic assumption regarding this rule is that all the elements in a cluster contribute to the inter-cluster similarity.

Average linkage is also as interesting compromise between the nearest and the farthest neighbor methods. Average linkage tends to join clusters with small variances; it is slightly biased toward producing clusters with the "same" variance. The agglomeration levels can be difficult to interpret with this algorithm.

IV.2.5.5. Variance Criterion Clustering: Ward Method

Ward's method (also called incremental sum of squares method) is distinct from all other methods because it uses an analysis of variance to evaluate the distances between centroids of clusters; it builds clusters by maximizing the ratio of between- on within-cluster variances. Under the criterion of minimization of the within-cluster variance, two clusters are merged if they result in the smallest increase in variance within the new single cluster (Duatre et al., 1999) (Figure 47). In other words, the Ward algorithm compares all the pairs of clusters before any aggregation, and selects the pair (r, s) with the minimum value of $D(r, s)$:

$$D(r,s) = \frac{d^2\left(\overline{x_r}, \overline{x_s}\right)}{\left(\frac{1}{n_r} + \frac{1}{n_s}\right)} = \frac{1}{\left(\frac{1}{n_r} + \frac{1}{n_s}\right)} (\overline{x_r} - \overline{x_s})' \ (\overline{x_r} - \overline{x_s})$$

where:

n_r, n_s : total numbers of objects into clusters r and s respectively ;

$D(r, s)$: second order moment of clusters r and s;

$\overline{x_r}$, $\overline{x_s}$: coordinates of centroids of clusters r and s respectively;

$d(\overline{x_r}, \overline{x_s})$: distance between centroids of clusters r and s .

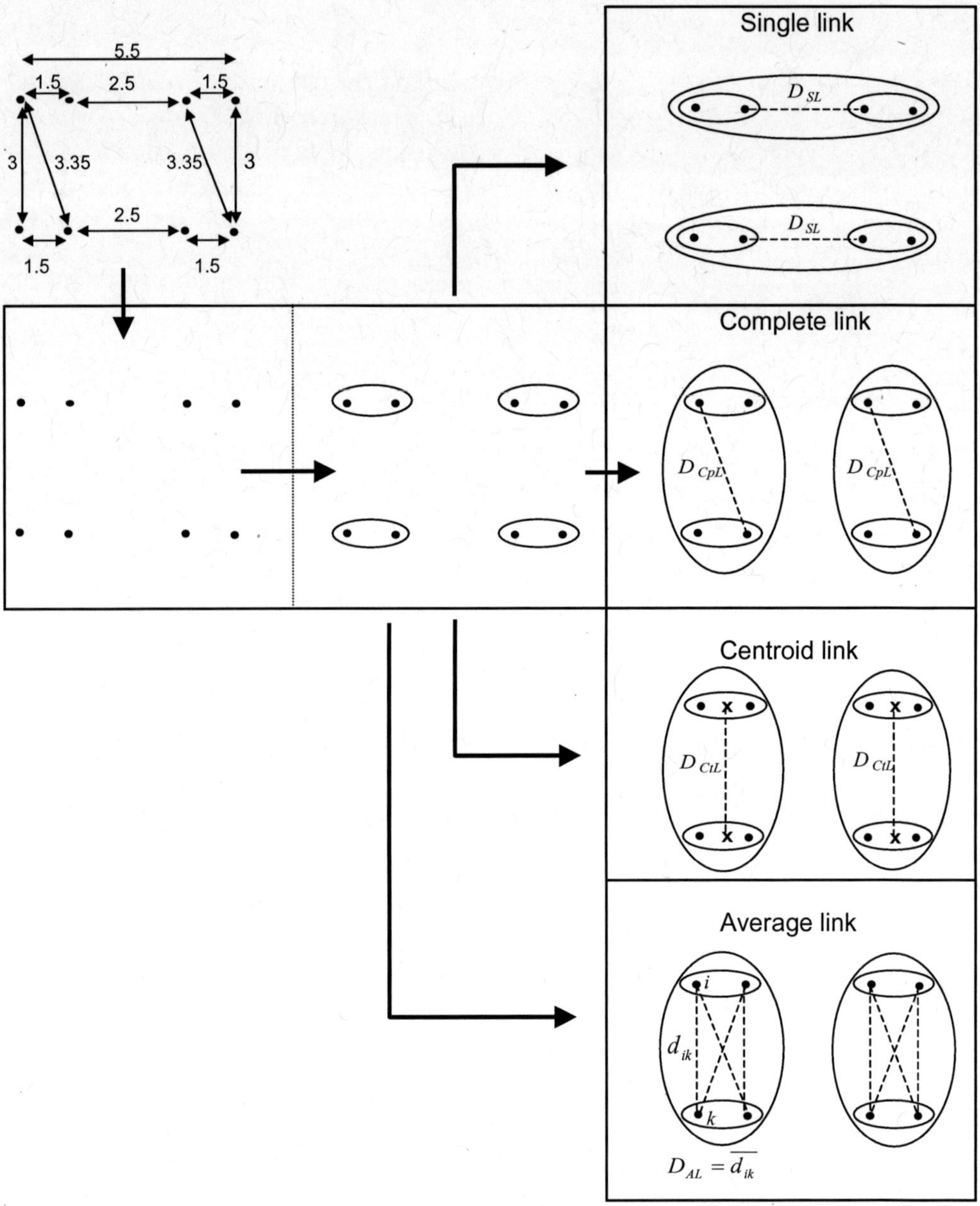

Figure 48. Schematic representations of different clustering rules in agglomerative cluster analysis. D_{SL}, D_{CpL}, D_{CtL}, D_{AL}: distances used in single, complete, centroid and average link, respectively. d_{ik}: distance between elements i and k belonging to two different clusters.

Ward's method is regarded as very efficient and makes the agglomeration levels clear to interpret. However, it tends to give balanced clusters of small size, and it is sensitive to outliers (Milligan, 1980).

IV.2.6. Identification and Interpretation of Clusters from Dendrogram

After clustering of all individuals according to a given criterion, HCA provides a dendrogram which is a tree-like diagram informing about the classification structure of the population (Figure 49). In the dendrogram, a certain number of clusters (groups) can be retained on the basis of high homogeneity and separation levels. For each cluster, the homogeneity and separation levels can be graphically evaluated on the dendrogram from its compactness and distinctness, respectively:

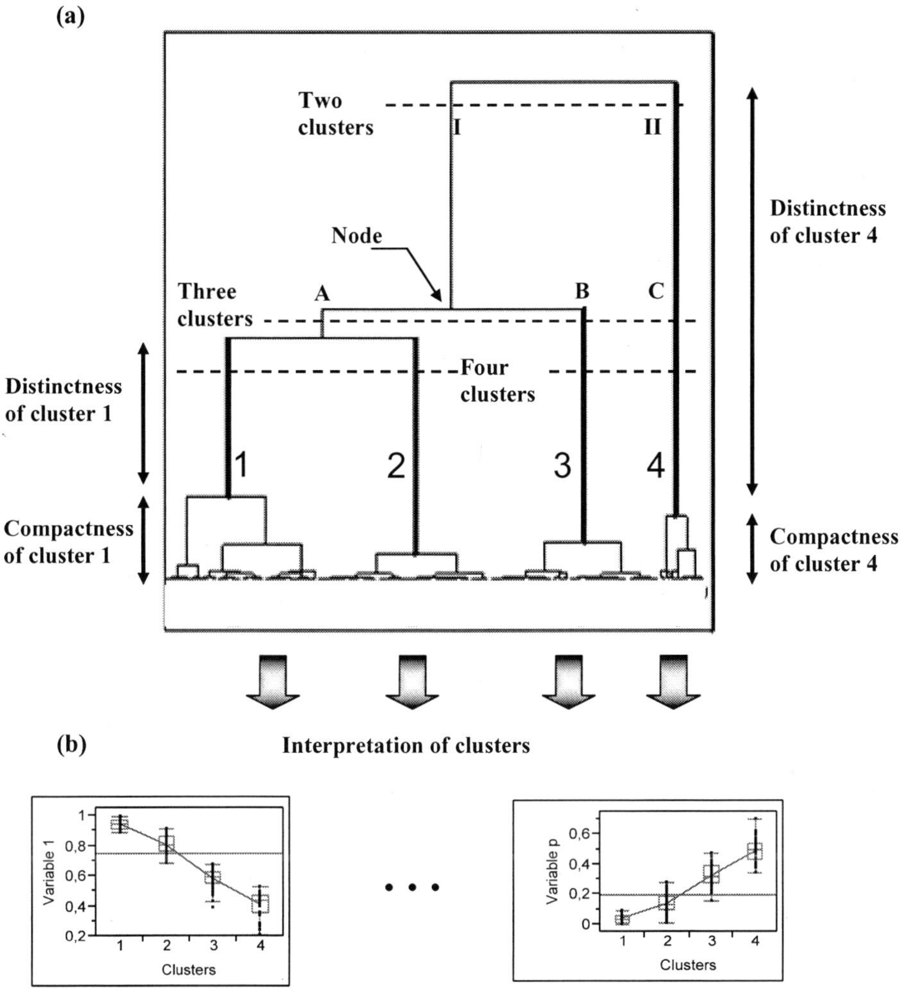

Figure 49. Illustration of the different parameters required for the identification and interpretation of clusters in a dendrogram.

In a dendrogram (Figure 49a), the number of clusters increases from the top to the bottom. This number is often empirically determined by how many vertical lines are cut by a horizontal line. Validation depends on whether the resulting clusters have a clear biological (clinical) meaning or not. Raising or lowering the horizontal line varies the number of vertical lines cut, i.e. the number clusters resulting from the subdivision of the population.

The dissimilarity level or distance between two clusters or two subunits is determined from the height of the node that joins them. This height represents also the compactness of the parent cluster formed by the merging of the two children clusters. In other words, the compactness of a cluster represents the minimum distance at which the cluster comes into existence (Figure 49a). At the lowest levels, the subunits are individuals.

When the classification is well structured, each cluster contains individuals which are similar between them and dissimilar with regard ti the individuals of other clusters. It results in clusters with low compactness and long distinct branches (high distinctness). The distinctness of a cluster is the distance from the point (node) at which it comes into existence to the point at which it is aggregated into a larger cluster.

The interpretation of distinct clusters can be easily guided by box-plots highlighting the dispersions of the *p* initial variables (e.g. the *p* metabolites) in the different identified clusters (Figure 48b). These graphics help to detect which variable(s) significantly influences the distinction between clusters. This step serves to determine the meaning of each cluster.

V. Outlier Analyses

V.1. Introduction

Biological populations can be characterized by a high variability consisting of more or less similar/dissimilar individuals. Beyond of such a diversity concept, it is important to identify the eventual occurrence of atypical individuals which can be considered as potential sources of heterogeneity. Detection of such individual cases is interesting to avoid to work on heterogeneous dataset on the hand, and to detect original/rare information which needs some particular consideration on the other hand (Figure 50). From these two cases, outliers can be either suspect values or represent interesting points which provide evidence of new phenomena or new populations. In all cases, a dataset needs to be treated with and without its detected outliers; then comparisons will help to conclude on the diversity or heterogeneity of the studied population.

For example in metabolomics, some individuals can have atypical biosynthesis, secretion, storage or transformation (elimination) of certain metabolites compared to the whole population. In clinics, such cases need to be identified in order to optimize their treatments. Moreover in statistical analysis of biological populations, identification and removing of outliers allow to extract more reliable information on the studied population, because atypically high or atypically located values of outliers can be responsible for bias in the results: for instance, the mean of the population can be significantly shifted to higher values under the effect of some outliers.

Figure 50. Intuitive examples illustrating two meaning of outliers; outliers can be suspect points resulting in biased results (a), or can provide original information on extreme states in the population or on new populations (b).

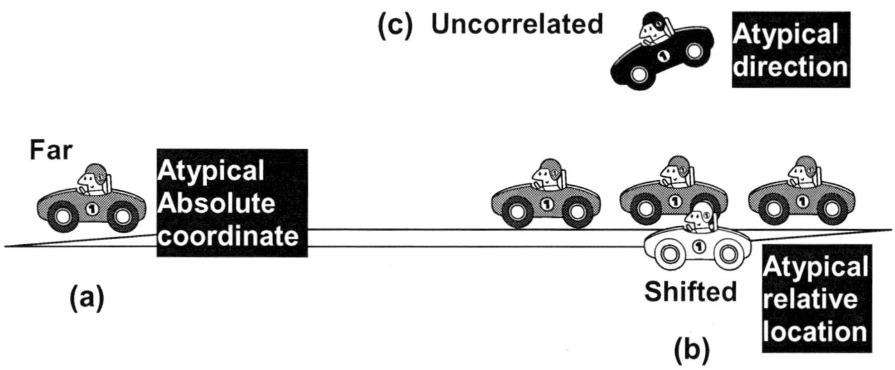

Figure 51. Intuitive representation of different types of outliers.

V.2. Different Types of Outliers

Outliers can be defined according to three criteria: remoteness, gap, deflection (Figure 51).

- Remoteness concerns individuals (e.g. metabolic profiles) that are atypically far from the whole population because of atypically high or low coordinates (Figure 51a).
- Gap concerns individuals that are shifted within the population because of discordance in their coordinates (Figure 51b).
- Deflection concerns individuals that are not oriented along the global direction of the whole population (Figure 51c).

V.3. Statistical Criteria for Identification of Outliers

Identification of outliers is closely linked to the criterion under which the differences between individuals are evaluated. The greatest dissimilarities can help to detect the most atypical/original individuals. By reference to the three types of outliers, differences can be described on the basis of three criteria (Figure 52):

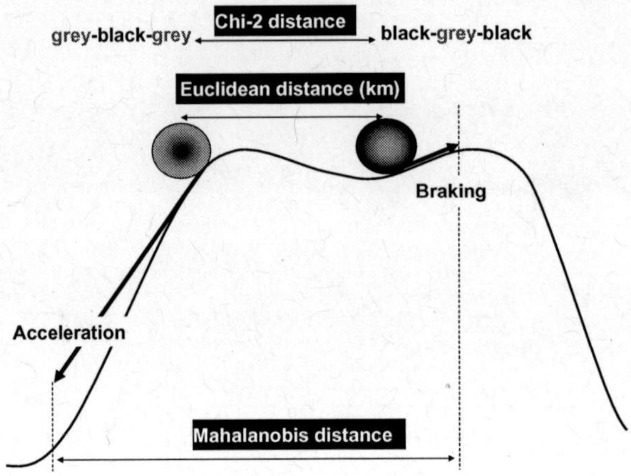

Figure 52. Illustration of three distance criteria to evaluate the outlier/non-outlier states of individuals within a population.

- Differences can be undertaken on the basis of measurable data (continue variables). Classic example is given by kilometric measurements leading to conclude about the remoteness of individuals to a reference point. Such remoteness is evaluated by means of Euclidean distance.
- Differences between individuals can be described on the basis of presence-absence for qualitative characteristics, or relative values for quantitative measures. In a given individual, the number of presences and absences of characteristics are compared to the corresponding total numbers in the population. Rarely present or absent characteristics in a given individual lead to consider such individual as atypical. The evaluation of atypical individuals on the basis of such relative states can be performed by means of the Chi-2 distance.
- Atypical individuals can be identified on the basis of their role to stretch and/or disturb the global shape of a population. For that, the variance-covariance matrix of the whole population is considered as a metric on the basis of which atypical variations in the coordinates of some individuals can be reliably identified. The distance calculated taking into account the variances-covariances corresponds to the Mahalanobis distance.

The three different criteria presented above show that the outlier concept is closely linked to the used metric distance.

V.4. Graphical Identification of Univariate Outliers

The simplest outlier identification method consists in analyzing the values of all the individuals for a given variable. In such case, the atypical individuals correspond only to range outliers because of their atypically high or low values of the considered variable (Figure 51a). Graphically, such outliers can be identified by means of box-plots as points located beyond the cut-off values corresponding to the extremities of the whiskers (Figure 53)

(Hawkins, 1980; Filzmoser et al., 2005). These two extremities are calculated by adding and subtracting (1.5*inter-quartile range) to third and first quartiles, respectively.

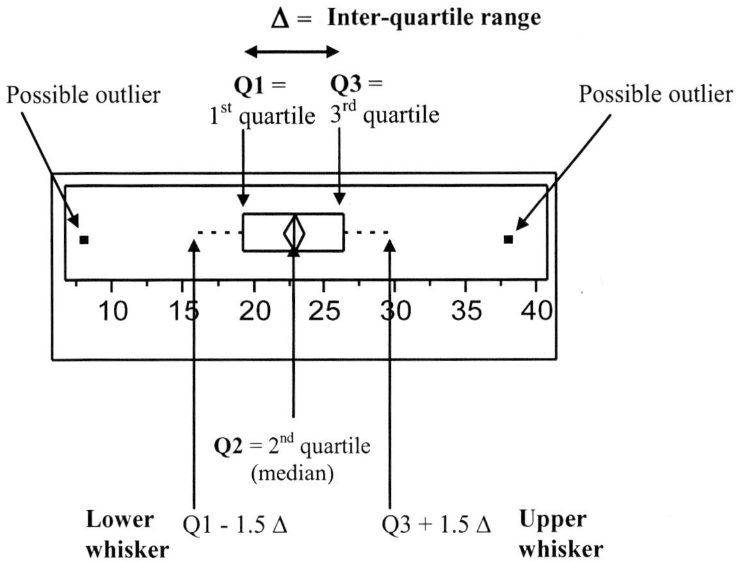

Figure 53. Tuckey Box-plot showing univariate outlier detection from the upper and/or lower limits of whiskers.

V.5. Graphical Identification of Bivariate Outliers

When two variables X, Y are considered, the dataset can be represented graphically by using a scatter plot Y versus X. In the case of linear model, three kinds of outliers can be detected on the scatter plot viz., range (a), spatial (b) and relationship (c) outliers (Rousseeuw and Leroy, 1987; Cerioli and Riani, 1999; Robinson, 2005) (Figure 54):

For (a), the high coordinates (x,y) of the point will inflate variances of both variables, but will have little effect on the correlation; in this case, the point (x, y) is a univariate outlier according to each variable X, Y, separately.

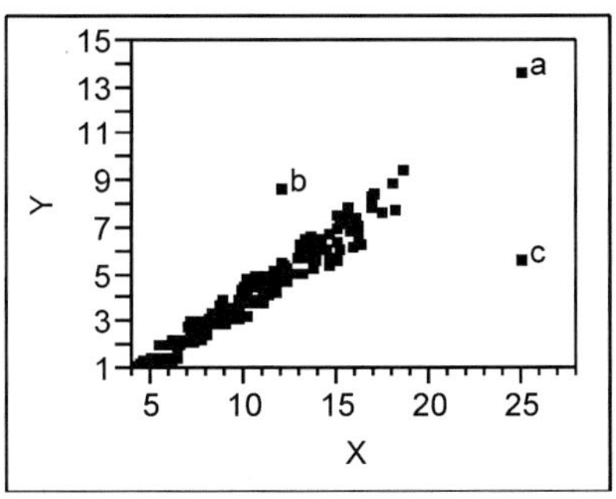

Figure 54. Graphical illustration of different types of oultiers that can be detected from a scatter plot of two variables Y vs X.

Observation (b) is extreme with respect to its neighboring values. It will have little effect on variances but will reduce the correlation.

For (c), outlier can be defined as an observation that falls outside of the expected area; it has a high moment (leverage point) through which it will reduce the correlation and inflate the variance of X, but will have little effect on the variance of Y.

V.6. Identification of Multivariate Outliers Based on Distance Computations

When more than two variables are considered, the identification of outliers requires more sophisticated tools and computations on the multivariate matrix X consisting of (n rows × p columns) and where each element x_{ij} represents the value of the variable j for the case i :

$$\overset{j\,(1 \text{ to } p)}{X = \begin{pmatrix} x_{11} & x_{12} & \cdots & x_{1j} & \cdots & x_{1p} \\ x_{21} & x_{22} & \cdots & x_{2j} & \cdots & x_{2p} \\ \cdots & \cdots & \cdots & \cdots & \cdots & \cdots \\ x_{i1} & x_{i2} & \cdots & x_{ij} & \cdots & x_{ip} \\ \cdots & \cdots & \cdots & \cdots & \cdots & \cdots \\ x_{n1} & x_{n2} & \cdots & x_{nj} & \cdots & x_{np} \end{pmatrix}} \quad i\,(1 \text{ to } n)$$

For that, appropriate metric distances have to be computed by combining all the variables X_j describing individuals i. In metabolomics, such matrix can be represented by a dataset describing n metabolic profiles i by p metabolites j.

The calculated distance from a neutral state representing the population will be used to visualize the relative state of the corresponding individual within the population. Three multivariate outlier cases can be detected by three types of distances viz., Euclidean, Chi-2 and Mahalanobis distances.

These distances are computed between individuals X_i and a reference individual X_0 by using three parameters: the coordinates x_{ij} and x_{0j} of the observed and reference individual X_i and X_0, and a metric matrix Γ (Gnanadesikan and Kettenring, 1972; Barnett, 1976; Barnett and Lewis, 1994):

$$d^2(x_i, x_0) = \sum_{j=1}^{p} \left(x_{ij} - x_{0j}\right)^t \Gamma^{-1} \left(x_{ij} - x_{0j}\right)$$

The kind of distance depends on the matrix Γ:

- If Γ=identity matrix, d corresponds to the Euclidean distance;
- If Γ= matrix of the products (sum of lines × sum of columns), d corresponds to the Chi-2 distance;
- Γ=variance-covariance matrix, d corresponds to the Mahalanobis distance.

The three approaches based on the three kinds of distance are: Andrews curves (Andrews, 1972; Barnett, 1976; Everitt and Dunn, 1992), correspondence analysis (CA) (Greenacre, 1984, 1993; Mortier and Bar-Hen, 2004) and Jackknifed Mahalanobis distance (Swaroop and Winter, 1971; Robinson, 2005), respectively. These different methods provide complementary diagnostics of the states of individuals in a dataset, leading to extract a diversity of outliers under different criteria: among all the extracted outliers, the most marked can be identified as points confirmed by the three diagnostics (Semmar et al., 2008).

Another approach used in multivariate data, consists in performing multiple regression analysis between a depend variable Y and several explanative ones X_j, then a scatter plot can be visualized between observed and predicted Y (Y_{obs} vs Y_{pred}) (Figure 54). However, this approach has the disadvantage to be model-dependent by opposition to the three distance-based approaches which advantageously extract independent-model outliers.

V.6.1. Standard Mahalanobis Distance Computation

This section presents the basic concepts of the Mahalanobis distance (MD) computation; it will be followed by a presentation (**V.6.2**) of the Jackknifed technique which is mainly used to calculate robust MD. The two techniques (ordinary and Jackknifed) will be illustrated by a numerical example.

The Mahalanobis distance provides a multivariate measure of how much a multivariate point is far from the centroid (average vector) of the whole database. Using Mahalanobis distance, we can assess how similar/dissimilar each profile x_i is to a typical (average) profile $\bar{x}$.

The Mahalanobis distance takes into account the correlation structure of the data, and it is independent of the scales of the descriptor variables. It is computed as (Rousseeuw and Leroy, 1987):

$$MD_i^2 = (x_i - \bar{x})C^{-1}(x_i - \bar{x})^t, \qquad \text{(eq. 1)}$$

Where:

MD_i^2 is the squared Mahalanobis distance of the subject i from the average vector (or centroid) $\bar{x}(\bar{x}_1,...,\bar{x}_p)$,

x_i: a p-row vector ($x_{i1}, x_{i2},...,x_{ip}$) representing subject i (e.g. patient i) characterized by p variables (e.g. p concentration values measured at p successive times).

$\bar{x}$: vector of the arithmetic means of the p variables

$$\bar{x} = \frac{1}{n}\sum_{i=1}^{n} x_i \text{ (with } n \text{ : total number of individuals)} \qquad \textbf{(eq. 2)}$$

C: the covariance matrix of the p variables

$$C = \frac{1}{n-1}\sum_{i=1}^{n}(x_i - \bar{x})^t(x_i - \bar{x}) \qquad \text{(eq. 3)}$$

The Mahalanobis distance measures how far is each profile x_i from the average profile $\overline{x}$ in the metric defined by *C*. It is the Euclidean distance if the covariance matrix is replaced by the identity matrix. The purpose of these MD_i^2 is to detect observations for which the explanatory part lies far from that of the bulk of the data: according to Mahalanobis criteria, a subject *i* described by *p* variables *j* tends to be outlier if its coordinates x_{ij} increase the variance of the variable *j* by comparison with all other coordinates x_{kj} ($k \neq i$). This situation can be due to:

- a great difference of x_{ij} to the mean $\overline{x}_j$ (high numerator) (eq. 1).
- a weak variance s_j^2 of the variable *j*, i.e. when the set of values x_{kj} ($k \neq i$) represents a homogenous group (weak denominator) (eq. 1).

Let's illustrate the Mahalanobis calculus by a numerical example (Figure 55):

X =

i = 1 to *n*=5 individuals	*j* = 1 to p=3 metabolites M1	M2	M3
X1	1	2	20
X2	1	2	2
X3	2	1	3
X4	4	4	4
X5	0	7	0
Average $\overline{X}$	1.6	3.2	5.8

$X - \overline{X} =$

j = 1 to *p*=3 metabolites M1	M2	M3
-0.6	-1.2	14.2
-0.6	-1.2	-3.8
0.4	-2.2	-2.8
2.4	0.8	-1.8
-1.6	3.8	-5.8

$$\frac{(X-\overline{X})'(X-\overline{X})}{n-1}$$

$(X-\overline{X})^t$

	X1	X2	X3	X4	X5
M1	-0.6	-0.6	0.4	2.4	-1.6
M2	-1.2	-1.2	-2.2	0.8	3.8
M3	14.2	-3.8	-2.8	-1.8	-5.8

$(X-\overline{X})$

	M1	M2	M3
X1	-0.6	-1.2	14.2
X2	-0.6	-1.2	3.8
X3	0.4	-2.2	-2.8
X4	2.4	0.8	-1.8
X5	-1.6	3.8	-5.8

$/(n-1)$

	M1	M2	M3
M1	2.3	-0.9	-0.6
M2	-0.9	5.7	-7.45
M3	-0.6	-7.45	65.2

C = Variance-Covariance matrix

C^{-1}

	M1	M2	M3
M1	0.48	0.1	0.02
M2	0.1	0.23	0.03
M3	0.02	0.03	0.02

Inverse of Var-Cov matrix

$(X-\overline{X})\ C^{-1}.\ (X-\overline{X})^t$

X1	X2	X3	X4	X5
3.2	-0.79	-0.81	-0.8	-0.8
-0.79	1.21	1.07	-1.24	-0.25
-0.81	1.07	1.44	-0.39	-1.32
-0.8	-1.24	-0.39	3.1	-0.68
-0.8	-0.25	-1.32	-0.68	3.05

Squared Mahalanobis distances (in diagonal)

$\sqrt{}$

X1	1.79
X2	1.10
X3	1.20
X4	1.76
X5	1.75

Mahalanobis distances

Figure 55. Numerical example illustrating the calculus of multivariate Mahalanobis distance.

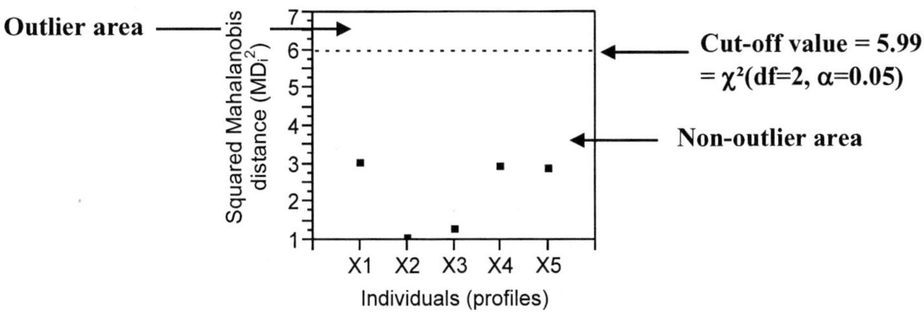

Figure 56. Graphical representation of the Mahalanobis distance by reference to a Chi-2 cut-off value with (p-1) degree of freedom.

The MD_i^2 values follow a chi-squared distribution with (p-1) degrees of freedom (Hawkins, 1980). The multivariate outliers can be identified as points having Mahalanobis distances higher than the cut-off value with a given alpha-risk (e.g. $\alpha \leq 0.05$) (Figure 56). Moreover, the most identical profiles to the centroid are those which have the least Mahalanobis distances; therefore they can be considered as the most representative of the population (Figure 56, X2, X3 points). In our simple example, the number p of variables is equal to 3, and the freedom df is equal to p-1=2. For a α risk fixed to 5% (α=0.05), the cut-off χ^2 value corresponding to df=2 is given by $\chi^2(2, 0.05)=5.99$. From the numerical example, no squared Mahalanobis distance is higher than this cut-off value; consequently, we conclude that there are not outliers at the threshold α=5%.

This first part illustrated how Mahalanobis distance is calculated and interpreted in order to detect outliers. However, the standard Mahalanobis distance suffers from the fact that it is very sensitive to the presence of outliers in the sense that extreme observations (or groups of observations) departing from the main data structure can have a great influence on this distance measure (Rousseeuw and Van Zomeren, 1990). This is somewhat unclear because the Mahalanobis distance should be able to detect outliers, but the same outliers can heavily affect the Mahalanobis distance; the reason is the sensitivity of arithmetic mean and covariance matrix to outliers (Hampel et al., 1986): the individual X_i contributes to the calculation of the mean, and this mean will be then subtracted from X_i to calculate its Mahalanobis distance. Consequently, the standard Mahalanobis distance MD_i can be biased, the outlier X_i can be masked and other points can appear more outlying than they really are. This can be illustrated by the individual X1 which has an atypically high value for the variable M3 (M3=20) (Figure 57b), but which was not detected as outlier in spite of its higher MD value (Figure 57a). Moreover, scatter plots of variables M3 vs M1 and M2 showed that individual X1 corresponds to a relationship outlier analogous to that of point c in Figure 54.

A solution consists in inserting more robust mean and covariance estimators in equation (1): the Mahalanobis distance can be alternatively calculated by using the Jackknife technique.

V.6.2. Jackknifed Mahalanobis Distance Computation

Jackknife technique consists in computing, for each multivariate observation x_i, the distance MD_{Ji} from a mean vector and a covariance matrix which were estimated without the observation x_i. This avoids the mean and covariance to be influenced by the values of the subject i. In fact, a subject i with a high value can be more easily detected as far from the centroid if it did not contribute to the calculation of mean. Consequently, any multivariate observation x_i characterized by an atypical value x_{ij} can be more easily detected as far from the centroid and/or as discordant by reference to the multivariate distribution of the whole dataset X (Figure 58).

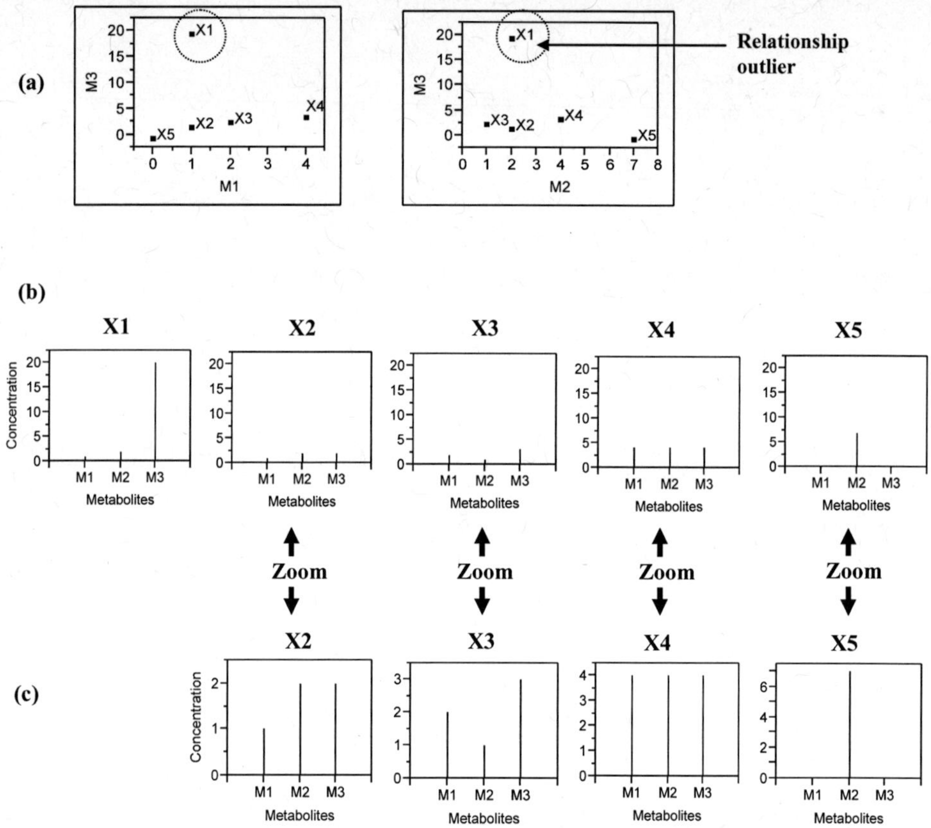

Figure 57. (a) Scatter plots between different variables showing a relationship-outlier because of atypically high coordinate for one variable M3 and ordinary coordinates for the other variables M1, M2. (b, c) Concentration profiles of the five analysed individuals X1-X5 characterized by three metabolites M1-M3.

The powerful of Jackknife technique can be illustrated by its ability to detect individual X1 as outlier because of its extreme value for the variable M3 resulting in a distorted profile compared to the four other profiles (Figure 57b). Moreover, individuals X4 and X5 were detected as outliers although their values had comparable levels to those of most of the profiles (Figure 57b). The fact that X4 and X5 are detected as outliers is not due to the levels of their values but to atypical combinations of the three values (M1, M2, M3) resulting in atypical profiles (Figure 57c): X4 had uniform profile because of equal values for the three variables, whereas X5 showed a single needle profile because of the null values of the variable M1 and M3.

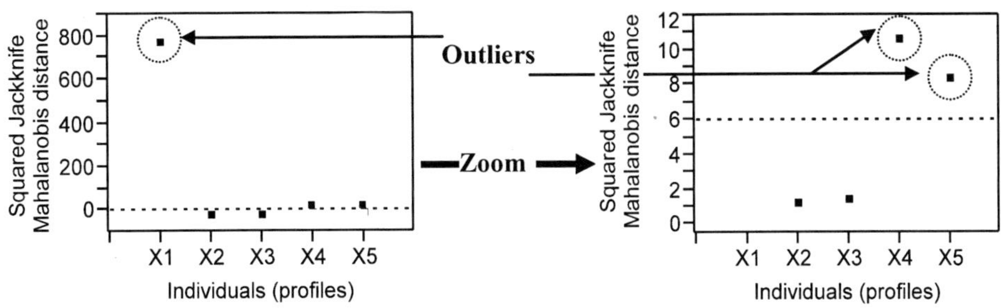

Figure 58. Outlier detection based on Mahalanobis distance calculated by the Jackknife technique. MD: Mahalanobis distance.

V.6.3. Outlier Screening from Correspondence Analysis

V.6.3.1. General Concepts of Correspondence Analysis

Correspondence analysis (CA) is a multivariate method that can be applied on a data matrix having both additive rows and columns, in order to analyze the strongest associations between individuals (rows) (e.g. patients) and variables (columns) (e.g. metabolites). On this basis, individuals strongly associated with some variables can be characterized by original or atypical profiles compared to the whole population. A strong association between an individual and a variable is highlighted by CA on the basis of a high value of the variable in the individual compared to all the values (Figure 59):

- of the other variables in the same individual on the hand, and
- for the same variable in all the other individuals on the other hand.

In other word, CA considers each value not by its absolute but by its relative level both along its row and column (Figure 59): for example, in individuals X3 and X4, the absolute values (e.g. concentration) of variable M3 (e.g. metabolite M3) are equal to 3 and 4, respectively, leading to consider the second as more important than the first. However, in terms of relative values, the 3 of X3 and the 4 of X4 represent 50% and 33%, respectively, of the total in their profiles; consequently, the value 3 of profile X3 is relatively more important than the value 4 in profile X4, leading to consider individual X3 as more associated than X4 to variable M3. However, by considering all the individuals X1 to X5, the relative level 50% of M3=3 in its profile appears to be lower than that M3=20 in X1 (87%). Individual X1 appears finally as the most associated to variable M3 by considering all the rows (profiles) and columns (variables) of the dataset. To conclude on the outlier or non-outlier state of *X1*, all the individuals X_i of the dataset must be considered according to all the variables; this allows to check if *X1* is alone to be original (a), or if the other individuals are also original under other characteristics (b). In the first case (a), the rarity of X1 makes to consider it as atypical; in the second case (b), one talks about different trends in the dataset rather than atypical cases (or outliers) (Figure 60).

V.6.3.2 Basic Computations in Correspondence Analysis

Correspondence analysis (CA) is an exploratory multivariate method which analyses the relative variations within a simple two-way table X (n rows × p columns) containing measures of correspondence between rows and columns. The matrix X consists of additive data both along the rows and columns (e.g. contingency table, concentration dataset, or any homogeneous unit matrix). Thus, CA analyses simultaneously row and column profiles.

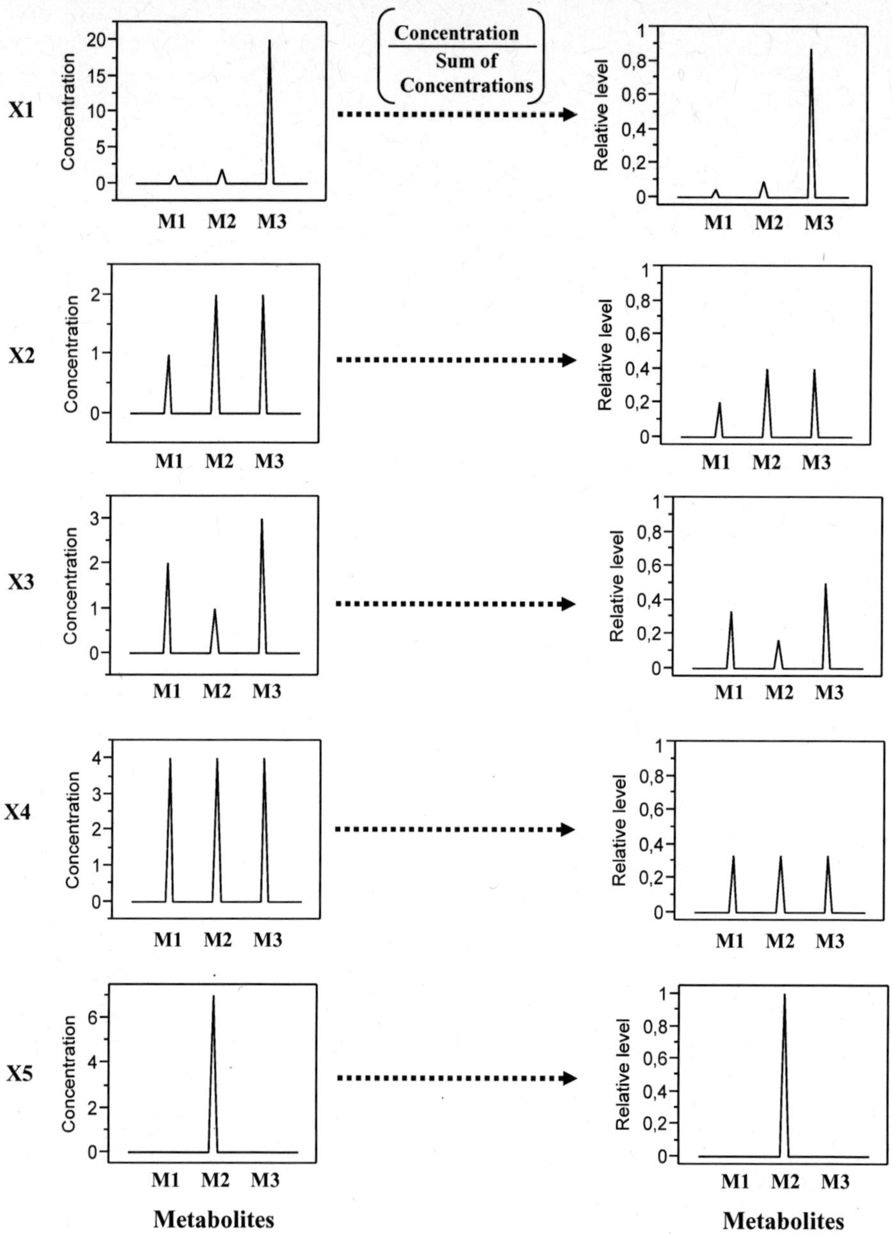

Figure 59. Standardization of concentration (absolute values) profiles into relative levels leading to data homogeneization at a scale varying between 0 and 1.

Row and column profiles are obtained by dividing each value x_{ij} (e.g. concentration of metabolite j in subject i) by its row and column sums, x_{i+} and x_{+j} respectively:

$$f_i = \frac{x_{ij}}{\sum_{j=1}^{p} x_{ij}} = \frac{x_{ij}}{x_{i+}} \ (j=1 \text{ to } p) \quad f_j = \frac{x_{ij}}{\sum_{i=1}^{n} x_{ij}} = \frac{x_{ij}}{x_{+j}} \ (i=1 \text{ to } n) \qquad \text{(eq. 4)}$$

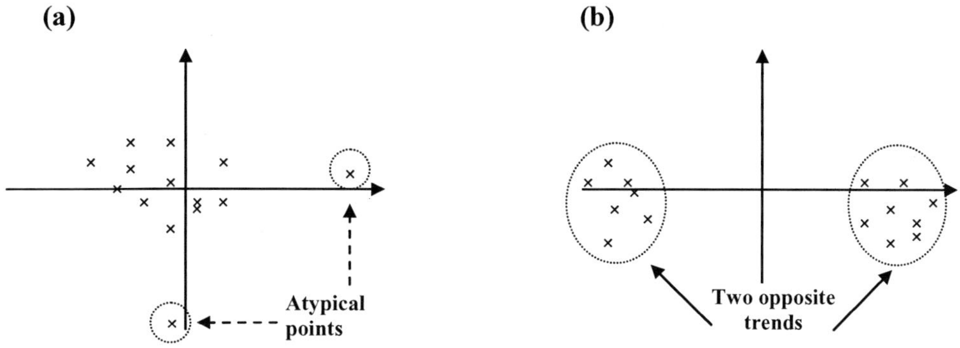

Figure 60. Illustration of two dataset structures corresponding to the presence of isolated atypical individual cases (a) and to grouped individuals into well distinct trends (b).

This transformation is appropriate to highlight the strongest associations between rows and columns: two row profiles are more similar if they show comparable relative values for the same column-variables. Reciprocally, two variables will have similar variation trends if their relative values vary in the same way in all the rows. Finally, a row i is strongly associated with a column j if it has a high value x_{ij} for this column compared with all the values both of the same row i and of the same column j. This duality along row and column leads to standardize each value x_{ij} by the square root of the product of x_{i+} and x_{+j}: $\sqrt{\frac{x_{ij}}{x_{i+}.x_{+j}}}$ (Figure 61).

From the matrix T of such standardized values, two analyses are performed to calculate new coordinates (called factorial coordinates) for rows (individuals) and columns (variables), respectively (Figure 61). Row analysis is performed on the matrix T'T, whereas column analysis is performed on the matrix TT'. One obtains two squared matrices TT' and T'T which have (p-1) eigenvalues λ_j comprised between 0 and 1; p being the smallest dimension of the dataset (generally, in a dataset ($n \times p$), there are less variables than individuals, i.e. $p<n$). Extreme eigenvalues equal to 0 or 1 are not considered because they correspond to trivial values.

The (p-1) decreasing eigenvalues λ_j are combined with the matrices T'T on the hand and TT' on the other hand, to calculate (p-1) eigenvectors V_j for the rows and for the columns, respectively. Finally, the factorial coordinates of the rows and columns are calculated from the scalar products of eigenvectors by:

- the row profiles (x_{ij}/x_{i+}) weighted by the root square of the ratio x_{++}/x_{+j},
- the column profiles (x_{ij}/x_{+j}) weighted by the the root square of the ratio x_{++}/x_{i+}.

The new coordinates resulting from row and column analyses have the characteristic to condense the variability of the initial dataset within a small dimension space (<p) consisting of independent directions (called factors). The factors have also the property to be successively shorter because they correspond to decreasing eigenvalues; this makes possible to describe the variability of the initial dataset by a minimal dimension space represented by the first factors (Escofier and Pagès, 1991): the first factor (F1) describes the maximal part of total variability followed be the second (F2) which describes a maximal part of the remaining variability not described by F1, etc. . This leads the variability of the dataset to be rapidly condensed into a small dimension space. This is particularly interesting in the case of large datasets, what is generally the case in metabolomics.

The computations of factorial coordinates are illustrated by a numerical example based on the previous dataset (Figure 55) (Figures 62, 63). After the calculus of factorial coordinates of the rows along each factor, their sign must correspond to those of the coordinates of the eigenvectors for the columns: for instance, along F1, the eigenvector of column is V1 with five coordinates (0.58, -0.12, 0.07, -0.17, -0.78) (Figure 63); the calculus of factorial coordinates of the five rows along F1 gives (-0.59, 0.27, -0.14, 0.24, 1.44); as the two sets have opposite signs, it is needed to multiply one of them by -1 to obtain appropriate superimposition between rows and columns: Thus F1 becomes F1(0.59, -0.27, 0.14, -0.24, -1.44) (Figure 62). According to the dataset, such sign correction can or can't occur.

To measure the distance between two row-profiles or two column-profiles, CA uses the chi-square distance. The distance between two row profiles (e.g. two patients) i and i' is given by (Escofier and Pagès, 1991; Greenacre, 1984; 1993):

$$d^2(i,i') = \sum_{j=1}^{p} \frac{x_{++}}{x_{+j}} \left(\frac{x_{ij}}{x_{i+}} - \frac{x_{i'j}}{x_{i'+}} \right)^2 \qquad \text{(eq. 5)},$$

where x_{++} is the total sum of the whole database, x_{i+}, $x_{i'+}$ are the sums of rows i and i', respectively, and x_{+j} is the sum of column j.

This distance is low when the profiles show similar relative values of several variables, independently of their absolute values (Figure 45). Similarly, the distance between two column profiles (e.g. two metabolite variables) j and j' is given by:

$$d^2(j,j') = \sum_{i=1}^{n} \frac{x_{++}}{x_{i+}} \left(\frac{x_{ij}}{x_{+j}} - \frac{x_{ij'}}{x_{+j'}} \right)^2 \qquad \text{(eq. 6)}$$

V.6.3.3. Graphical Interpretation of CA Results and Outlier Diagnostic

Graphical visualization of the factorial coordinates of rows helps to see how much each individual tends to be original or ordinary within the population. Moreover, the scatter plot of the factorial coordinates of columns helps to identify how the different variables are associated to original individuals: an individual which projects close to a variable means a high value in such individual for such a variable compared with all the individuals and variables of the dataset. Graphically, outliers can be highlighted by extreme points along the factors (computed axes) of CA (Greenacre, 1984, 1993). Moreover, the duality in CA allows identification of the variables responsible of the outlying states of such individuals.

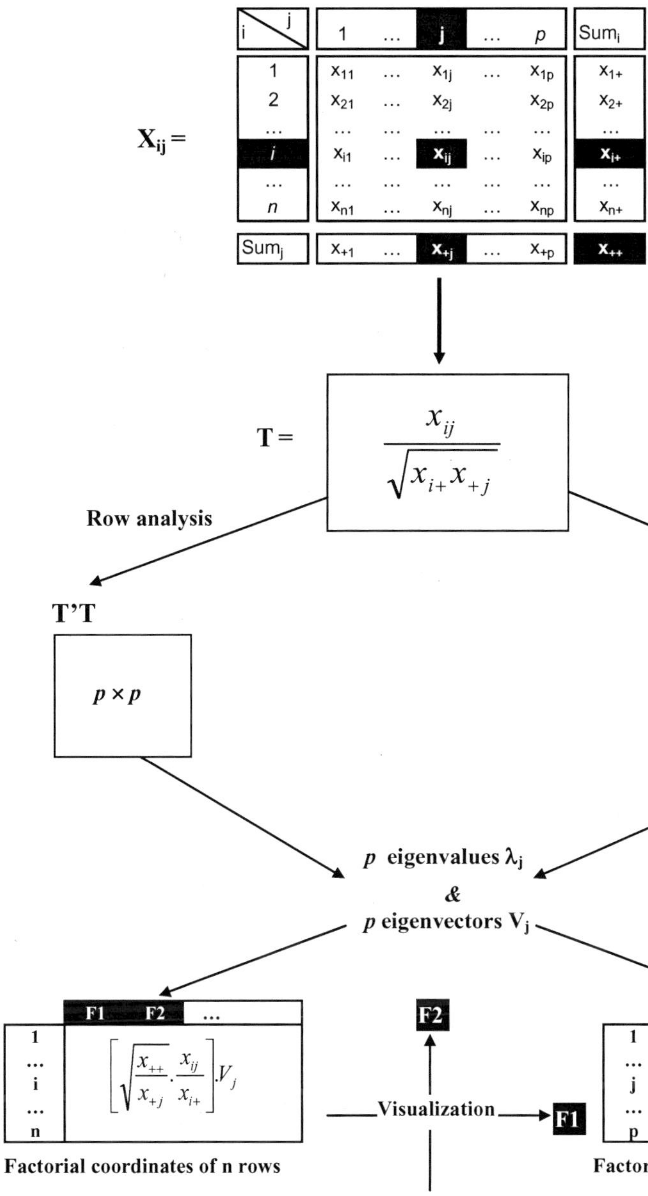

Figure 61. Principle of computation of factorial coordinates in correspondence analysis.

From the numerical example, the individuals X1 and X5 showed opposite and extreme projections along F1 (first factor) (Figure 62). Morever, long F1, the variables M3 and M2 projected in the same spaces than X1 and X5, respectively (Figure 63); this indicates that individuals X1 and X5 have relatively high values of M3 and M2, respectively, by comparison with all the values of the corresponding row and column profiles: in fact, the values: M3=20 in X1 and M2=7 in X5 represent high maxima both along their rows and columns. The opposition between X1 and X5 can be explained by an inverse variability of M2 and M3 between X1 and X5: X1 has a high M3 and a low M2, whereas X5 shows inverse characteristics. Moreover, the pair (X5, M2) appears more extreme along F1 than (X1, M3). This is due to the fact that the value 7 of M2 in X5 is relatively more important than the value 20 of M3 in X1: 100% versus 87%.

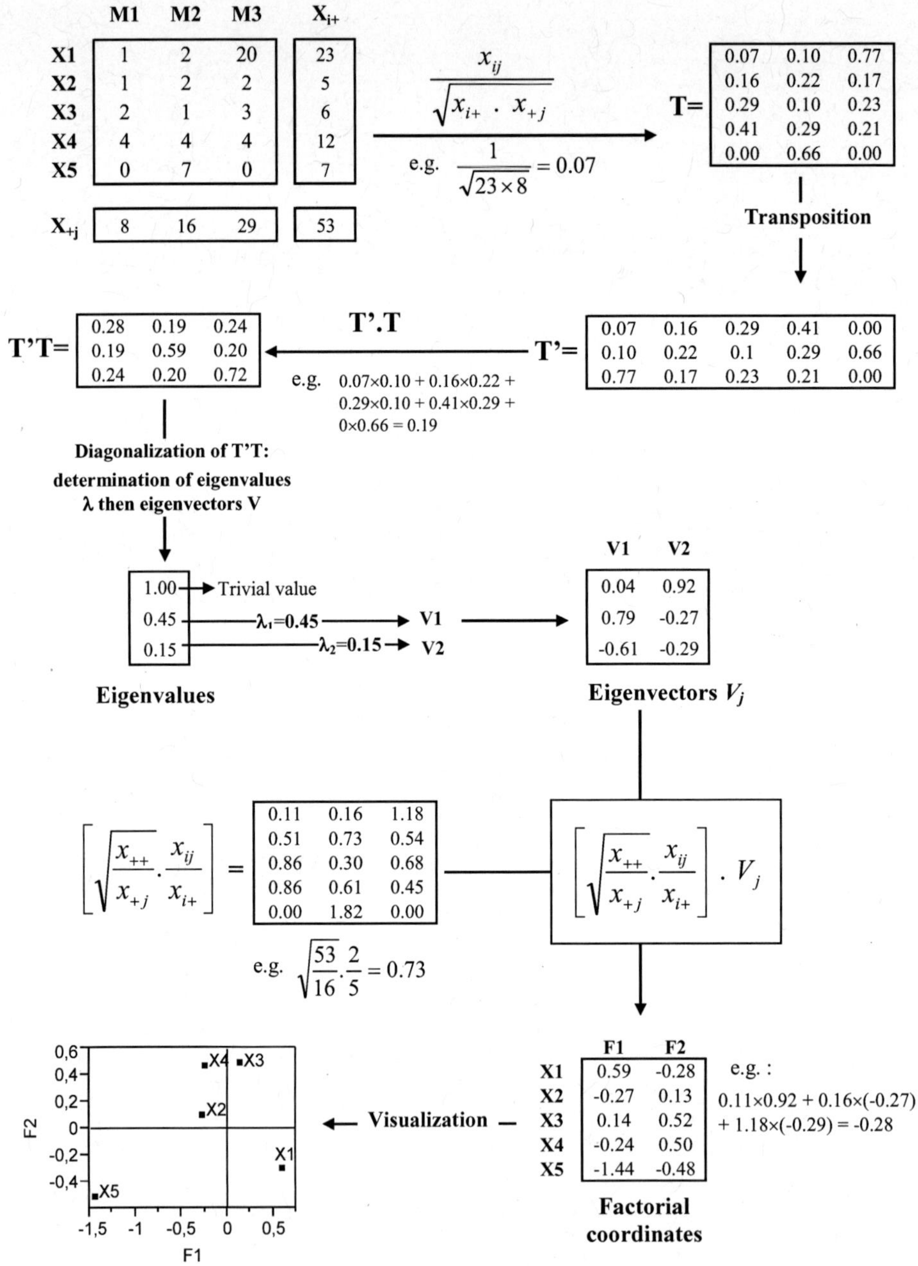

Figure 62. Numerical example illustrating the computation of factorial coordinates of rows in correspondence analysis (row analysis).

Along F2, the individuals X3 and X4 tend to form a group (Figure 62) characterized by the variable M1 (Figure 63). Taking into account the facts that F2 represent less variability than F1 on the hand, and that X3 and X4 don't represent isolate cases, this situation can't be interpreted as atypical; rather it corresponds to an original trend within the dataset: the values of M1=2 and 4 in X3 and X4 respectively are relatively more important than the other values (0≤ ≤4) of the same rows (X3 or X4) and column (M1).

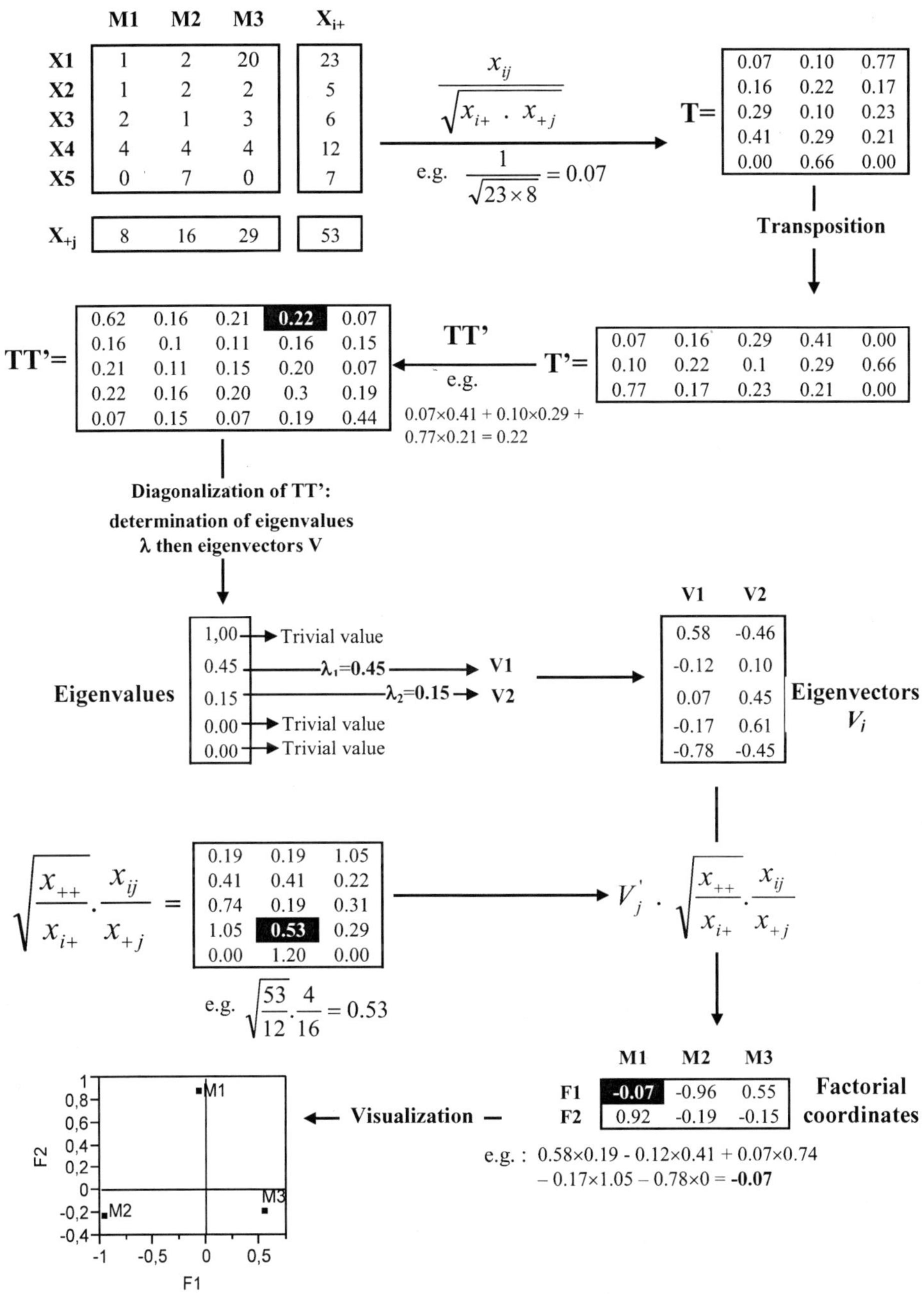

Figure 63. Numerical example illustrating the computation of factorial coordinates of columns in correspondence analysis (column analysis).

Moreover, X3 and X4 appear to be opposite to X1 and X5 along F2 which is defined by the variable M1. This can be explained by the fact that M1 has relatively high values in X3, X4 against relatively low (minimal) values in X1 and X5.

V.6.4. Outlier Diagnostic Based on Andrews Curves

V.6.4.1. General Concepts

Andrews curves represent a strong graphical tool to analyze the homogeneity and diversity of a multivariate dataset under the Euclidean distance criterion. They provide a plane representation of the multivariate distribution of the individuals based on a Fourier transformation: each individual (profile) is represented by a sine-cosine curve calculated from its initial coordinates at different rotation angle α. The resulting curve highlights the behavior of corresponding individual in the multivariate space defined by all the measured variables (e.g. metabolites). Outlier individuals can be identified by their Andrews curves isolated from the rest of the curves at a given rotation angle.

V.6.4.2. Computation of Andrews Curves

The p measured values of the p variables describing a given individual are used into a sine-cosine function to calculate a serial of values corresponding to several rotation angles α ($-\pi \leq \alpha \leq \pi$) (Figure 64a). The sine-cosine function $f_i(\alpha)$ calculated for an individual i at a rotation angle α has the form:

$$f_i(\alpha) = \frac{x_{i1}}{\sqrt{2}} + x_{i2}\sin(\alpha) + x_{i3}\cos(\alpha) + x_{i4}\sin(2\alpha) + x_{i5}\cos(2\alpha) + \ldots$$

By using q different α values, one obtains a set of q coordinates $f_i(\alpha)$ from which the Andrews curve of individual i can be plotted as $f_i(\alpha)$ versus α (Seber, 1984; Everitt and Dunn, 1992).

V.6.4.3. Graphical Outlier Diagnostic Based on Andrews Curves

By plotting the Andrews curves of all the individuals, ones can expect to see isolated bands of curves (outlying individuals) which separate from the compact mass of curves representing the homogeneous population (Figure 64b). The distances between Andrews curves are proportional to the Euclidean distances between the corresponding individuals.

A drawback of this method is that an interchange of variables leads to a different picture. However, this is not a constraint in the case of kinetic (concentration-time) database because the concentration variables are ordered in time and cannot be interchanged.

Application of Andrews curves to the previous dataset (Figure 64) shows a central zone containing condensed curves from which other curves separate gradually leading to some extreme cases: the most isolated curve concerns individual X1; it was followed by the curve of X5 then X4 which show only a slight separation from the compact centre containing the ordinary individuals X2 and X3. Individuals X1, X5 and X4 were particularly characterized by the highest values in the dataset leading to their more or less outlying states. On the basis of this Euclidean concept, individual X1 appears as the most atypical case because it has the highest value (M3=20) compared to the generally low variation range of the dataset.

(a)

Dataset X_{ij}

	M1	M2	M3
X1	1	2	20
X2	1	2	2
X3	2	1	3
X4	4	4	4
X5	0	7	0

α
-3.14
-2.51
-1.88
-1.26
-0.63
0.00
0.63
1.26
1.89
2.51
3.14

Andrews function

$$f_i(\alpha) = \frac{x_{i1}}{\sqrt{2}} + x_{i2} \times \sin(\alpha) + x_{i3} \times \cos(\alpha)$$

$$f_1(\alpha) = \frac{1}{\sqrt{2}} + 2\sin(\alpha) + 20\cos(\alpha)$$

$$f_2(\alpha) = \frac{1}{\sqrt{2}} + 2\sin(\alpha) + 2\cos(\alpha)$$

$$f_3(\alpha) = \frac{2}{\sqrt{2}} + \sin(\alpha) + 3\cos(\alpha)$$

$$f_4(\alpha) = \frac{4}{\sqrt{2}} + 4\sin(\alpha) + 4\cos(\alpha)$$

$$f_5(\alpha) = \frac{0}{\sqrt{2}} + 7\sin(\alpha) + 0\cos(\alpha)$$

f(α) \ α	-3.14	-2.51	-1.88	-1.26	-0.63	0.00	0.63	1.26	1.89	2.51	3.14
$f_1(\alpha)$	-19.30	-16.62	-7,28	4,92	15,69	20,71	18,05	8,73	-3,67	-14.25	-19.29
$f_2(\alpha)$	-1,30	-2,09	-1,81	-0,59	1,14	2,71	3,50	3,22	1,98	0,27	-1,29
$f_3(\alpha)$	-1,59	-1,60	-0,45	1,38	3,25	4,41	4,43	3,28	1,42	-0,42	-1,58
$f_4(\alpha)$	-1,18	-2,76	-2,20	0,24	3,70	6,83	8,42	7,86	5,37	1,96	-1,17
$f_5(\alpha)$	-0,01	-4,13	-6,67	-6,66	-4,12	0,00	4,12	6,66	6,65	4,13	0,01

$f_i(\alpha)$ vs α

(b)

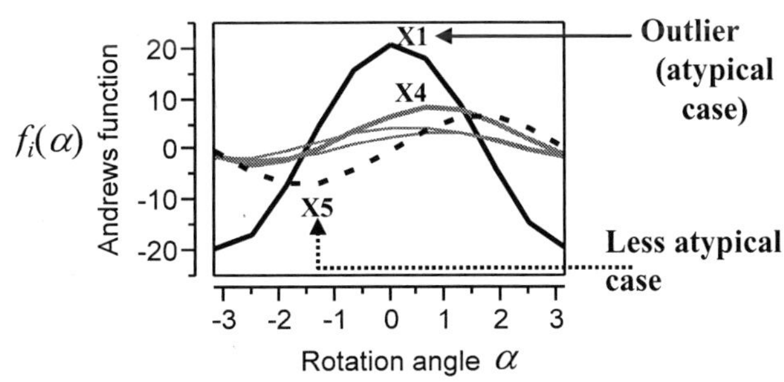

Figure 64. Numerical example illustrating computation of Andrews curves and their graphical representation and interpretation.

References

Andrews, D. F. (1972). Plots of high-dimensional data. *Biometrics,* **28**, 125-136.

Arabie, P., De Soete, G., Arabie, P., Hubert, L. J., Hubert, L. J. & De Soete, G. (Eds.) (1996). *Clustering and Classification.* World Scientific Pub. Co. Inc., River Edge, New Jersey.

Atkinson, D. E. (1977). *Cellular Energy Metabolism and its Regulation.* Academic Press, New York.

Barnett V. (1976). The ordering of multivariate data (with discussion). *J. R. Stat. Soc. A,* **139**, 318-354.

Barnett, V. (1976). The ordering of multivariate data (with discussion). *J R Stat Soc A,* **139**, 318-354.

Barnett, V. & Lewis, T. (1994). *Outliers in statistical data.* Wiley, New York.

Blackwood, C. B., Marsh, T., Kim, S. H. & Paul, E. A. (2003). Terminal restriction fragment length polymorphism data analysis for quantitative comparison of microbial communities. *Appl. Environ. Microbiol,* **69**, 926-932.

Box, G. E. P. & Cox, D. R. (1964). An analysis of transformations. *J. R. Stat. Soc. B*, **26**, 211-252.

Box, G. E. P, Hunter, W. G. & Hunter, J. S. (1978). Statistics for Experimenters: an Introduction to Design, Data Analysis and Model Building. Willey, New York.

Calik, P. & Ozdamar, T. H. (2002). Metabolic flux analysis for human therapeutic protein productions and hypothesis for new therapeutical strategies in medicine. *Biotechnol. Eng. J.*, **11**, 49-68.

Camacho, D., de la Fuente, A. & Mendes, P. (2005). The origin of correlations in metabolomics data. *Metabolomics,* **1**, 53-63.

Cerioli, A. & Riani, M. (1999). The ordering of spatial data and the detection of multiple outliers. *J Comput Graph Stat*, **8**, 239-258.

Crampin, E. J., Schnell, S. & McSharry, P. E. (2004). Mathematical and computational techniques to deduce complex biochemical reaction mechanisms. *Progress in Biophysics & Molecular Biology,* **86**, 77-112.

Cruz-Monteagudo, M., Munteanu, C. R., Borges, F., Cordeiro, M. N., Uriarte, E., Gonzalez-Diaz, H. (2008b). Quantitative Proteome-Property Relationships (QPPRs). Part 1: finding biomarkers of organic drugs with mean Markov connectivity indices of spiral networks of blood mass spectra. *Bioorg Med Chem.,* **16**, 9684-9693.

Cruz-Monteagudo, M., Munteanu, C. R., Borges, F., Cordeiro, M. N. D. S., Uriarte, E., Chou, K. C. & González-Díaz, H., (2008a). Stochastic molecular descriptors for polymers. 4. Study of complex mixtures with topological indices of mass spectra spiral and star networks: The blood proteome case. *Polymer,* **49**, 5575-5587.

Daniel, W. W. (1978). *Applied Nonparametric Statistics.* Houghton Mifflin Co. Boston, Massachussetts, **510**.

Denkert, C., Budczies, J., Weichert, W., Wohlgemuth, G., Scholz, M., Kind, T., Niesporek, S., Noske, A., Buckendahl, A., Dietel, M. & Fiehn, O. (2008). Metabolite profiling of human colon carcinoma – deregulation of TCA cycle and amino acid turnover. *Molecular Cancer,* **7**(72), 1-15.

Dimitriadou, E., Barth, M., Windischberger, C., Hornik, K. & Moser, E. (2004). A quantitative comparison of functional MRI cluster analysis. *Artif. Intell. Med.,* **31**, 57-71.

Droesbeke, J. J., Fine, J. & Saporta, G. (1997). *Plans d'expériences: applications à l'entreprise*. Technip: Paris.

Duatre, J. M., Santos, J. B. & Melo, L. C. (1999). Comparison of similarity coefficient based on RAPD markers in the common bean. *Genet. Mol. Biol.*, **22**, 427-432.

Duineveld, C. A. A., Smilde, A. K. & Doorhbos, D. A. (1993). *Chemom. Intell. Lab. Syst.*, **19**, 295.

Eide I. (1996). Strategies for Toxicological Evaluation of Mixtures. *Food Chem. Toxicol.*, **34**, 1147-1149.

Escofier B. & Pagès, J. (1991). Presentation of correspondence analysis and multiple correspondence analysis with the help of examples. In: J. Devillers, & W. Karcher (Eds.), *Applied multivariate analysis in SAR and environmental studies*. Kluwer Academic Publishers, Dordrecht, 1-32.

Estrada E. & Bodin, O. (2008). Using network centrality measures to manage landscape connectivity. *Ecol Appl.*, **18**, 1810-1825.

Estrada, E. (2006). Protein bipartivity and essentiality in the yeast protein-protein interaction network. *Journal of proteome research,* **5**, 2177-2184.

Estrada, E. (2007). Point scattering: a new geometric invariant with applications from (nano)clusters to biomolecules. *J Comput Chem.*, **28**, 767-777.

Ettenhuber, C., Radykewicz, T., Kofer, W., Koop, H. U., Bacher, A. & Eisenreich, W. (2005). Metabolic flux analysis in complex isotopolog space. Recycling of glucose in tobacco plants. *Phytochemistry,* **66**, 323-335.

Everitt, B. S. & Dunn, G. (1992). *Applied multivariate data analysis*. Wiley, New York

Everitt, B. S., Landau, S. & Leese, M. (2001). *Cluster Analysis*. Arnold Publishers, London.

Fall, C. P., Marland, E. S., Wagner, J. M. & Tyson, J. J. (2005). *Computation Cell Biology*. Springer-Verlag, NY, **488**.

Fell, D. A. (1996). *Understanding the Control of Metabolism*. Portland Press, London.

Fernie, A. R., Trethewey, R. N., Krotzky, A. & Willmitzer, L. (2004). Metabolite profiling: from diagnostics to systems biology. *Nat. Rev. Mol. Cell Biol.*, **5**, 763-769.

Filzmoser, P., Garrett, R. G. & Reimann, C. (2005). Multivariate outlier detection in exploration geochemistry. *Comput Geosci,* **31**, 579-587.

Gibbons, F. D. & Roth, P. (2002). Judging the quality of gene expression-based clustering methods using gene annotation. *Genome Res.,* **12**, 1574-1581.

Glajch, J. L., Kirkland, J. J. & Snyder, L. R. (1982). Practical optimisation of solvent selectivity in liquid-solid chromatography using a mixture-design statistical technique. *J. Chromatogr.,* **238**, 269-280.

Gnanadesikan, R. & Kettenring, J. R. (1972). Robust estimates, residuals, and outlier detection with multiresponse data. *Biometrics,* **28**, 81-124.

Gonzalez-Diaz, H. (2008). Quantitative Proteome-Property Relationships (QPPRs). Part 1: finding biomarkers of organic drugs with mean Markov connectivity indices of spiral networks of blood mass spectra. *Bioorg Med Chem.,* **16**, 9684-9693.

González-Díaz, H., González-Díaz, Y., Santana, L., Ubeira, F. M. & Uriarte, E. (2008). Proteomics, networks and connectivity indices. *Proteomics,* **8**, 750-778.

González-Díaz, H., Tenoriob, E., Castañedob, N., Santanaa, L. & Uriarte, E. (2005). 3D QSAR Markov model for drug-induced eosinophilia—theoretical prediction and preliminary experimental assay of the antimicrobial drug G1. *Bioorganic & Medicinal Chemistry,* **13**, 1523-1530.

González-Díaz, H., Vilar, S., Santana, L. & Uriarte, E. (2007). Medicinal Chemistry and Bioinformatics – Current Trends in Drugs Discovery with Networks Topological Indices. *Curr Top Med Chem.*, **7**, 1025-1039.

Gonzalez-Diaz, H., Prado-Prado, F. & Ubeira, F. M. (2008). Predicting antimicrobial drugs and targets with the MARCH-INSIDE approach. *Curr Top Med Chem.*, **8**, 1676-1690.

Goodacre, R., Vaidynathan, S., Dunn, W. B., et al. (2004). Metabolomics by numbers: acquiring and understanding global metabolite data. *Trends Biotechnol.,* **22**, 245-252.

Gordon, A. D. (1999). *Classification*. CRC Pr I Llc, Boca Raton.

Greenacre, M. J. (1984). *Theory and applications of correspondence analysis*. Academic Press, London

Greenacre, M. J. (1993). *Correspondence analysis in practice*. Academic Press, London

Guttorp, P. (1995). *Stochastic Modeling of Scientific Data*, Chapman and Hall, London, Great Britain.

Hampel, F. R., Ronchetti, E. M., Rousseeuw, P. J. & Stahel, W. (1986). *Robust statistics. The approach based on influence functions*. Wiley, New York.

Hawkins, D. M. (1980). Identification of outliers. Chapman and Hall, London.

Hayashi, K. & Sakamoto, N. (1986). *Dynamic Analysis of Enzyme Systems. An Introduction*. Springer-Verlag, Berlin.

Heinrich, R. & Schuster, S. (1996). *The Regulation of Cellular Systems*. Chapman & Hall, New York.

Hotelling, H. & Pabst, M. R. (1936). Rank correlation and tests of significance involving no assumption of normality. *Ann. Math. Statist.,* **7**, 29-43.

Ivanciuc, O., Balaban, T. S. & Balaban, A. T. (1993). Chemical graphs with degenerate topological indices based on information on distances. *Journal of Mathematical Chemistry*, **14***,* 21-33.

Jaccard, P. (1912). The distribution of the flora in the alpine zone. *New Phytol.,* **11**, 37-50.

Jain, A. K., Murty, M. N. & Flynn, P. J. (1999). Data clustering: a review. *ACM Comput.*

Janga, S. C. & Babu, M. M. (2008). Network-based approaches for linking metabolism with environment. *Genome Biology,* **9**, 239.1-239.5.

Kacser, H. & Burns, J. A. (1973). The control of flux. Symp. *Soc. Exp. Biol.,* **27**, 65-104.

Kell, D. B. (2004). Metabolomics and systems biology: making sense of the soup. *Curr. Opin. Microbiol.,* **7**, 1-12.

Kell, D. B. (2002). Metabolomics and machine learning: explanatory analysis of complex metabolome data using genetic programming to produce simple, robust rules. *Mol. Biol. Rep.,* **29**, 237-241.

Kose, F., Weckwerth, W., Linke, T. & Fiehn, O. (2001). Visualizing plant metabolomic correlation networks using clique-metabolite matrices. *Bioinformatics,* **17**, 1198-1208.

Kruger, N. J., Ratcliffe, R. G. & Roscher, A. (2003). Quantitative approaches for analysing fluxes through plant metabolic networks using NMR and stable isotope labelling. *Phytochemistry Reviews,* **2**, 17-30.

Lance, G. N. & Williams, W. T. (1967). A general theory of classificatory sorting strategies 1. Hierarchical systems. *Comput. J.,* **9**, 373-380.

Legendre, P. & Legendre, L. (2000). *Numerical Ecology*. Elsevier, Amsterdam, **853**.

Lindon, J. C., Nicholson, J. K. & Holmes, E. (Eds), (2007). *The Handbook of Metabonomics and Metabolomics*. Elsevier, Amsterdam, **561**.

Llaneras, F. & Picó, J. (2008). Stoichiometric Modelling of Cell Metabolism. *Journal of Bioscience and Bioengineering,* **105**, 1-11.

Maharjan, R. P. & Ferenci, T. (2005). Metabolomic diversity in the species *Escherichia coli* and its relationship to genetic population structure. *Metabolomics,* **3**, 235-242.

Milligan, G. W. (1980). An examination of the effect of six types of error perturbation on

Milligan, W. G. & Cooper, M. C. (1987). Methodology review: clustering methods. *Appl.*

Morgan, J. A. & Rhodes, D. (2002). Mathematical Modeling of Plant Metabolic Pathways. *Metabolic Engineering,* **4**, 80-89.

Morgenthal, K. Weckwerth, W. & Steuer, R. (2006). Metabolomic networks in plants: transitions from pattern recognition to biological interpretation. *Biosystems,* **83**, 108-117.

Morgenthal, K.,Wienkoop, S., Scholz, M., Selbig, J. & Weckwerth, W. (2005). Correlative GC–TOF–MS based metabolite profiling and LC–MS based protein profiling reveal time-related systemic regulation of metabolite–protein networks and improve pattern recognition for multiple biomarker selection. *Metabolomics,* **1**, 109-121.

Mortier, F. & Bar-Hen, A. (2004). Influence and sensitivity measures in correspondence analysis. *Statistics,* **38**, 207-215.

Nicholson, J. K., Lindon, J. C. & Holmes, E. (1999). 'Metabonomics': understanding the metabolic responses of living systems to pathophysiological stimuli via multivariate statistical analysis of biological NMR spectroscopic data. *Xenobiotica,* **29**, 1181-1189.

Nyieredy, S. z., Meier, B., Erdelmeier, C. A. J. & Sticher, O. (1985). "PRISMA": A geometrical design for solvent optimization in HPLC. *J. High Resolut. Chromatogr., Chromatogr. Communi.,* **8**, 186-188.

Oliver, S. G., Winson, M. K., Kell, D. B. & Baganz, F. (1998). Systematic functional analysis of the yeast genome. *Trends Biotechnol.,* **16**, 373-378.

Ott, K. H., Aranibar, N., Singh, B. & Stockton, G. W. (2003). Metabonomics classifies pathways affected by bioactive compounds. Artificial neural network classification of NMR spectra of plant extracts. *Phytochemistry,* **62**, 971-985.

Papin, J. A., Stelling, J., Price, N. D., Klamt, S., Schuster, S. & Palson, B. O. (2004). Comparison of network-based pathway analysis methods. *Trends Biotechnol.,* **22**, 400-405.

Papin, J. A., Price, N. D., Wiback, S. J, Fell, D. A. & Palsson, B. O. (2003). Metabolic pathways in the post-genome era. *Trends Biochem. Sci.,* **28**, 250-258.

Pattarino, F., Marengo, E., Gasco, M. R. & Carpignano, R. (1993). Experimental design and partial least squares in the study of complex mixtures: microemulsions as drug carriers. *Int. J. Pharm.* 91, pp. 157-165.

Ponce, Y. M. (2004). Total and local (atom and atom type) molecular quadratic indices: significance interpretation, comparison to other molecular descriptors, and QSPR/QSAR applications. *Bioorganic & Medicinal Chemistry*, **12**, 6351-6369. *Psych Meas.,* **11**, 329-354.

Robinson, R. B. (2005). Identifying outliers in correlated water quality data. *J Environ Eng,* **134**, 651-657.

Roessner, U., Luedemann, A., Brust, D., et al., (2001). Metabolic profiling allows comprhensive phenotyping of genetically or environmentally modified plant systems. *Plant Cell,* **13**, 11-29.

Rousseeuw, P. J. & Leroy, A. M. (1987). Robust regression and outlier detection. Wiley, New York.

Rousseeuw, P. J. & Van Zomeren, B. C. (1990). Unmasking multivariate outliers and leverage points. *J Am Stat Assoc,* **85**, 633-651.

Rouvray, D. H. (1992). The definition and role of similarity concepts in the chemical and physical sciences. *J. Chem. Inf. Comput. Sci.,* **32**, 580-586.

Sado, G. & Sado, M. Chr. (1991). *Les plans d'expériences, de l'expérimentation à l'assurance qualité* ; Afnor technique, Paris.

Savageau, M. A. (1976). *Biochemical Systems Analysis*. Addison-Wesley, Reading, MA.

Scheffe, H. (1958). *J. R. Stat. Soc.* B, **20**, 344.

Scheffe, H. (1963). *J. R. Stat. Soc.* B, **25**, 235.

Schilling, C. H., Edwards, J. S., Letscher, D. & Palsson, B. (2001). Combining pathway analysis with flux balance analysis for the comprehensive study of metabolic systems. *Biotechnol Bioeng,* **71**, 286-306.

Seber, G. A. F. (1984). Multivariate observations. Wiley, New York.

Semmar et al., (2001). Chemical diversification trends in *Astragalus caprinus* (Leguminosae) based on the flavonoid pathway. *Biochemical Systematics and Ecology*, **29**, 727-738.

Semmar, N., Bruguerolle, B., Boullu-Ciocca, S. & Simon, N. (2005b). Cluster analysis: an alternative method for covariate selection in population pharmacokinetic modeling. *Journal of Pharmacokinetics and Pharmacodynamics*, **32, 333-358**.

Semmar, N., Jay, M., Farman, M. & Chemli, R. (2005a). Chemotaxonomic analysis of *Astragalus caprinus* (Fabaceae) based on the flavonic patterns. *Biochemical Systematics and Ecology*, **33**, 187-200.

Semmar, N., Jay, M. & Nouira, S. (2007). A new approach to graphical and numerical analysis of links between plant chemotaxonomy and secondary metabolism from HPLC data smoothed by a simplex mixture design. *Chemoecology*, **17**, 139-156.

Semmar, N., Urien, S., Bruguerolle, B. & Simon, N. (2008). Independent-model diagnostics for a priori identification and interpretation of outliers from a full pharmacokinetic database: correspondence analysis, Mahalanobis distance and Andrews curves. *J Pharmacokinet Pharmacodyn,* **35**, 159-183.

Semmar, N. (2010). A New Mixture Design-Based Approach to Graphical Screening of Potential Interconnections and Variability Processes in Metabolic Systems. *Chem. Biol & Drug Design* 75, 91-105.

Shannon, C. E. (1948). A mathematical theory of communication. *Bell System Tech. J.,* **27**, 379.

Spearman, C. (1904). The proof and measurement of association between two thing. *Amer. J. Psychol.,* **15**, 72-101.

Stelling, J. (2004). Mathematical models in microbial systems biology. *Current Opinion in Microbiology,* **7**, 513-518.

Steuer, R. (2006). On the analysis and interpretation of correlations in metabolomic data. *Briefings in Bioinformatics,* **7**, 151-158.

Steuer, R. (2007). Computational approaches to the topology, stability and dynamics of metabolite networks. *Phytochemistry,* **68**, 2139-2151.

Steuer, R., Kurths, J., Fiehn, O., Weckwerth, W. (2003a). Interpreting correlations in metabolic networks. *Biochem. Soc. Trans.,* **31**(6), 1476-1478.

Steuer, R., Kurths, J., Fiehn, O. & Weckwerth, W. (2003b). Observing and interpreting correlations in metabolomic networks. *Bioinformatics,* **19**(8), 1019-1026.

Sumner, L. W., Mendes, P., Dixon, R. A. (2003). Plant metabolomics: large-scale phytochemistry in the functional genomics era. *Phytochemistry,* **62**, 817-836.

Swaroop, R. & Winter, W. R. (1971). A statistical technique for computer identification of outliers in multivariate data. *NASA Tech Notes* D-6472.

Sweetlove, L. J. & Fernie, A. R. (2005). Regulation of metabolic networks: understanding metabolic complexity in the systems biology era. *New Phytol.,* **168**, 9-24.

Tamir, A. (Ed.), 1998. *Applications of Markov Chains in Chemical Engineering*. Elsevier, Amsterdam, **604**.

Todeschini, R. & Consonni, V. (2000). *Handbook of Molecular Descriptors*: Wiley-VCH.

Vilar, S., Estrada, E., Uriarte, E., Santana, L. & Gutierrez, Y. (2005). In silico studies toward the discovery of new anti-HIV nucleoside compounds through the use of TOPS-MODE and 2D/3D connectivity indices. 2. Purine derivatives. *Journal of chemical information and modeling,* **45**, 502-514.

Waite, S. (2000). *Statistical Ecology in Practice*. Prentice Hall, Harlow, **414**.

Ward, J. H. (1963). Hierarchical grouping to optimize an objective function. *J. Am. Stat. Assoc.,* **58**, 236-244.

Weckwerth, W. (2003). Metabolomics in Systems Biology. *Annu. Rev. Plant Biol.,* **54**, 669-689.

Weckwerth, W., Loureiro, M., Wenzel, K., Fiehn, O. (2004a). Differential metabolic networks unravel the effects of silent plant phenotypes. *Proc. Natl. Acad. Sci.,* U.S.A., **101**, 7809-7814.

Weckwerth, W., Wenzel, K. & Fiehn, O. (2004b). Process for the integrated extraction, identification and quantification of metabolites, proteins and RNA to reveal their co-regulation in biochemical networks. *Proteomics,* **4**(1), 78-83.

Williams, T. C. R., Miguet, L., Masakapalli, S. K., Kruger, N. J., Sweetlove, L. J. & Ratcliffe, R. G. (2008). Metabolic Network Fluxes in Heterotrophic Arabidopsis Cells: Stability of the Flux Distribution under Different Oxygenation Conditions. *Plant Physiology*, **148**, 704-718.

Yanai, I., Baugh, L. R., Smith, J. J., Roehrig, C., Shen-Orr, S. S., Claggett, J. M., Hill, A. A., Slonim, D. K. & Hunter, C. P. (2008). Pairing of competitive and topologically distinct regulatory modules enhances patterned gene expression. *Molecular Systems Biology,* **4**(163), 1-12.

Yang, T. H., Wittmann, C. & Heinzle, E. (2004). Metabolic network simulation using logical loop algorithm and Jacobian matrix. *Metabolic Engineering,* **6**, 256-267.

Zar, J. H. (1999). *Biostatistical Analysis*. Prentice Hall, New Jersey, **663**.

In: Metabolomics: Metabolites, Metabonomics... ISBN: 978-1-61668-006-0
Editors: J.S. Knapp and W.L. Cabrera, pp. 87-119

Chapter 2

METABOLOMIC PROFILE AND FRACTAL DIMENSIONS IN BREAST CANCER CELLS

Mariano Bizzarri[1,*], Fabrizio D'Anselmi[2], Mariacristina Valerio[3], Alessandra Cucina[2], Sara Proietti[1], Simona Dinicola[1], Alessia Pasqualato[1], Cesare Manetti[3], Luca Galli[4] and Alessandro Giuliani[5]

[1]Dept. of Experimental Medicine - University La Sapienza, Rome, Italy
[2]Dept. of Surgery "Pietro Valdoni" - University La Sapienza, Rome, Italy
[3]Dept. of Chemistry - University La Sapienza, Rome, Italy
[4]Space Applications Department - Advanced Computer Systems (ACS) Rome, Italy
[5] Environment and Health Department, Istituto Superiore di Sanita', Rome, Italy

Abstract

During the last decades compelling evidence has accumulated indicating that abnormalities in metabolism of cancer cells could play a strategic role in tumour initiation and behaviour. Abnormalities in metabolism are likely a consequence of several alterations in the complex network of signal transduction pathways, which may be caused by both genetic and epigenetic factors. An aberrant energy metabolism was recognized as one of the prominent features of the malignant phenotype, since the pioneering work of Warburg. It is now well established that the majority of tumours is characterized by a high glucose consumption, even under aerobic conditions, in absence of the Pasteur Effect, i.e. the lack of inhibition of glycolysis when cancer cells are exposed to normal oxygen consumption. Several investigators provided experimental data in support of a specific structure of the metabolic network in cancer cells. The 'tumour metabolome' has been defined as the metabolic tumour profile characterized by high glycolytic and glutaminolytic capacity and a high channelling of glucose carbons toward synthetic processes.

Despite no archetypal cancer cell genotype exists, facing the wide genotypic heterogeneity of each tumour cell population, some malignant features (i.e. invasion, uncontrolled growth, apoptosis inhibition, metastasis spreading) are virtually shared by all

[*] E-mail address: mariano.bizzarri@uniroma1.it., Dept. of Experimental Medicine, University La Sapienza, Roma, Italy. (Corresponding author)

cancers. This paradox of a common clinical behaviour despite marked both genotypic and epigenetic diversity needs to be investigated by a Systems Biology approach and suggests that cancer phenotype should be considered as a sort of "attractor" in a specific space phase defined by thermodynamic and kinetic constraints. This is not the only phase space cancer cells are embedded into: in principle cancer cells, like any living entity travel along an integrated set of genetic, epigenetic or metabolomic parameters. A fractal dimension formalism can be used in a prospective reconstruction of cancer attractors. Studies conducted on MCF-7 and MDA-MB-231 breast cancer cells, exposed to different morphogenetic fields, show that metabolomic profile correlates to cell shape: modification of cell shape and/or architectural characteristics of the cancer- tissue relationships, induced through manipulation of environmental cues, are followed by significant modification of the cancer metabolome as well as of the fractal dimensions at both single cell and cell population level. These results suggest how metabolomic shifts in cancer cells need to be considered as an adaptive modification adopted by a complex system under environmental constraints defined by the non-linear thermodynamic of the specific attractor occupied by the system. Indeed, characterization of cancer cells behaviour by means of both metabolomic and fractal parameters could be used to build an operational and meaningful space phase, that could help in evidencing the transitions boundaries as well as the singularities of cancer behaviour. Hence, by revealing tumour-specific metabolic shifts in tumour cells, metabolic profiling enables drug developers to identify the metabolic steps that control cell proliferation, thus aiding the identification of new anti-cancer targets and screening of lead compounds for anti-proliferative metabolic effects.

Introduction

In the first decades of the XIX century the biochemist Otto Warburg suggested [1,2] that cancer causation might be related to an altered metabolism, i.e. a shift in energy production from oxidative phosphorylation to glycolysis, even if in presence of normal oxygen levels – the so-called "Warburg-effect". The discovery of double-helix of DNA by Watson and Crick and progress in molecular biology achieved thereafter, stated that *overall* biological information was embedded *only* within the genome sequences and – with some remarkable exceptions - the "metabolic theory" was thought as a not-specific (and not significant) "epiphenomenon" and rapidly discarded. Nevertheless, and unexpectedly, as recently pointed out by K. Garber, the "Warburg's theory is now enjoying a resurrection" [3]. So far, the specific metabolic phenotype acquired by transformed cancer cells could no longer be considered a "simple" bioproduct of cancer development and is now widely thought as a "fundamental property of cancer cells" [3].

Indeed, the high glycolytic phenotype virtually shared by all tumours, is thought to be exploited for widespread clinical applications [4]. Given anaerobic conversion of glucose to lactic acid is substantially less efficient in terms of energy yield than complete oxidation to CO_2 and H_2O, tumour cells need to sustain elevated ATP production by increasing glucose flux and further conversion to glucose-6-phosphate. This characteristic provides the biochemical rationale for tumour imaging with 2-fluoro-2-deoxy-D-glucose-positron emission tomography (FDG-PET), a technique now widely used in radiological tumour studies [5]. PET investigations revealed a significant increase uptake of glucose in both primary and metastatic cancers, showing a direct correlation between tumour aggressiveness and the rate of glucose utilization [6] These results outlined the clinical importance of metabolic studies in cancer and have moved the "glycolytic phenotype" from a laboratory oddity to the mainstream of oncology.

Alterations in cancer metabolism are not only relevant for diagnostic purposes, but also in drug discovery. Macromolecule synthesis from glucose and glucogenic precursors are critical pathways and it is now well recognized that the identification of key metabolic enzymes (relevant 'hubs' in network analysis jargon) in both glucose anabolic and catabolic processes could be of utmost importance: by revealing tumour-specific metabolic shifts, metabolomic studies could identify the key-metabolic steps controlling growth and/or apoptosis and thus acting as potential new targets for therapeutic intervention [7].

Genistein, a natural isoflavonoid with several anti-tumour properties, induces both apoptosis and inhibition of tumour proliferation [8] interfering with several signalling pathways, but mainly by altering the rate of glucose oxidation and the synthesis of nucleic acid ribose through the non-oxidative steps of the pentose cycle [9]. It is note worthy that modulating transaldolase expression and the nucleic acid ribose synthesis through the non-oxidative pentose-cycle, not only the intracellular metabolic balance but also the sensitivity to cell death signals can be significantly influenced [10]. It is intriguing that, Imatinib – a selective inhibitor of different tyrosine kinases encoded by several proto-oncogenes (KIT, PDGFR, BCR-ABL) – induces inhibition of tumour growth by altering the rate of glucose utilization and, more specifically, reducing the synthesis of nucleic acid ribose through the oxidative reactions of the pentose cycle, thus 'reverting' the 'Warburg effect' by switching from glycolysis to mitochondrial glucose metabolism [11,12]. It is a matter of debate if the interference on glucose metabolism could be considered as the major cause of cell apoptosis and if the inhibition of proto-oncogene kinases are critical steps in determining such effect, in that imatinib induces relevant modification in glucose metabolism in a akt-independent manner in imatinib-resistant cancer cells [13]. As a matter of fact, a similar growth control on cancer proliferation could be achieved by a wide variety of glucose metabolic enzyme-inhibitory compounds- like Genistein - exerting their effects directly, without the need of Bcr-Abl signal transducer pathway [14].

Moreover, the development of high-throughput techniques during the last 10-20 years, has enabled a more 'systemic' and dynamical comprehension of cell and tissue metabolism, giving further insights into anabolic and catabolic cancer pathways, thus fostering a rekindling of interest in tumour metabolism.

Metabolomics and Cancer

Since its introduction some years ago, the term 'metabolomics' states for "the complete set of metabolites/low-molecular-weight intermediates (the 'metabolome'), which are context dependent, varying according to the physiology, developmental or pathological state of the cell, tissue, organ or organism" [15].

Undoubtedly, measuring metabolite concentrations is a more sensitive approach than following the rates of chemical reactions directly. Metabolic control analysis (MCA) demonstrated that, although changes in enzyme concentrations and activities ('the proteome') could have a small impact on metabolic fluxes, changes in flux have a significant impact on metabolite concentrations [16,17]. This implies metabolomics is located at the level of the actual cell physiology as a living entity, while both proteomics and transcriptomics are, in this sense, located on a more 'remote control' layer: the metabolome of a cell can be intended as the functional end-product in terms of amplification and integration of signals coming from other functional -omic levels. However, because the concentrations of metabolites are determined by the activities of many enzymes, metabolome cannot be easily decomposed in mechanistic terms as is the case with either mRNAs or proteins both pointing to specific 'actors' of the play like a given protein or gene (even if the existence of moonlighting proteins and RNA editing start to cast doubts about the possibility of factorize into single functional entities mRNAs and protein products).

Because of the coupling of many different reactions in the metabolic network, even small perturbations in the proteome (i.e. an alteration in the concentration of a few enzymes) can cause significant changes in the concentration of many metabolites. This aspect was highlighted from MCA showing that sensitivity coefficients for metabolites are generally higher than the sensitivity coefficients for fluxes [18]. It is likely that such a special characteristic offers a biological advantage, in that it provides stability to the metabolic network with respect to mutations. Thus, the response to a decrease in the activity of an enzyme might be to increase the concentration of substrates of that enzyme, enabling the flux to be only slightly altered [19]. This 'homeostatic' modulation of metabolic fluxes is likely to be attained through a diffuse control network; indeed, the control of the metabolic flux of a pathway is spread across all the enzymes present in the pathway, rather than being controlled by a rate determining step. From these statement it follows that there is not necessarily a linear quantitative relation between mRNA concentrations and enzyme function, meanwhile, as metabolites are downstream of both genomic transcription and translation, they are potentially a better indicator of enzyme activity and thereby could provide a more reliable system's description [20]. So, as clearly stated by Griffin and Shockor, "metabolomics offers a particularly sensitive method to monitor changes in a biological system, through observed changes in the metabolic network" [21]. Moreover, examining metabolomics, or changes in metabolic profiles, can be an important part of an integrative approach for assessing gene function and relationships to phenotypes [22]. Enzymatic biochemical reactions 'encoded' by genes can be deciphered using a genomic strategy, such as that of Martzen *et al.* [23]who identified yeast genes of unknown function based on the activity of their products, or such as that of Raamsdonk, L. M. *et al.* [24] who uses metabolome data to reveal the phenotype of silent mutations.

Because of the high degree of connectivity in the metabolic network, metabolome data represent integrative information, or, in other words, a "systems property". Often, this is claimed to be the strength of metabolome analysis.

Understanding disease processes through metabolic profiling is not an entirely new concept — ^{31}P, ^{1}H and ^{13}C NMR spectroscopy, along with gas chromatography–mass spectrometry (GC–MS), have been widely used as metabolic profiling tools since the early 1970s [25,26]. Metabolomics differs, however, in that rather than analysing a single class of compounds, it involves an attempt to measure *all* the metabolites that are present within a cell simultaneously. A range of analytical techniques, including 1H NMR spectroscopy, gas chromatography-mass spectrometry (GC-MS), liquid chromatography-mass spectrometry (LC-MS), Fourier Transform mass spectrometry (FT-MS), high performance liquid chromatography (HPLC) and electrochemical array (EC-array), are required in order to maximize the number of metabolites that can be identified in a matrix. This is, however, a difficult task, and our technical possibilities are far from reaching the goal [27].

'Metabolic profiling' has been proposed as a means of measuring the total complement of individual metabolites in a given biological sample, whereas 'metabolic fingerprinting' refers to measuring a subclass of metabolites to create a 'bar code' of metabolism [28].

This idea of the 'bar code' is intrinsically holistic and redounds on the idea of attractor dynamics: if cell metabolism is a strongly interconnected network in which each metabolite is correlated to any other we do not need to assign any observed dimension (e.g. NMR profile peak) to a known metabolite, given the global characterization of the attractor is an intrinsically multidimensional feature correspondent to the 'profile as such'. This is the reason why, although NMR spectroscopy detects only a fairly small number of metabolites, it can still be used to monitor the activity of many cellular activities. NMR has been used to analyse several tumour types in humans and in animal models of cancer [29,30], and despite limitations in sensitivity and the ability to measure a broad range of metabolites, metabolomic profiles have been successfully used to distinguish between tumours types and between cell lines, both in vitro an in vivo, in animals [31] and in humans [32,33]. Although there are many different approaches to collecting metabolic profiles of cells and tumours, pattern-recognition software is needed to associate specific profiles with different cell types, tumour types or a stage of treatment [34]. Furthermore, these approaches have also been used to identify 'metabolic fingerprints' associated with breast and brain tumours. In this regard, metabolic profiles could be used to predict which tumours are most likely to respond or become resistant to a specific type of therapy [35].

Furthermore, metabolic foot-printing or exometabolome analysis [36], based on the monitoring of metabolites consumed from and secreted into the growth medium, is a valuable tool to analyse the effect of cell perturbations, such as manipulation of environmental conditions as well as genetic modifications. In fact, a living cell takes up metabolites from the medium, secretes enzymes, and excretes metabolites to the extracellular medium and hence, it leaves a highly specific metabolic footprint in the medium represented by a specific metabolite profile, that vary according to environmental conditions, species, and/or genetic backgrounds [37]. Thus, the different physiological state of wild-type cells and single-gene deletion mutants even from closely related areas of metabolism can be distinguished by differences in the profile of extracellular metabolites [36].

The measurement of extracellular metabolites present several advantages over the analysis of intracellular compounds, often referred as metabolic fingerprinting [38]. For instance, the intracellular metabolism is more dynamic and therefore, the turnover of most metabolites is extremely fast requiring an efficient quenching of cell metabolism, followed by an effective separation of intra- and extracellular metabolites and subsequent extraction of intracellular compounds [39]. Furthermore, the concentration of intracellular metabolites in cell extracts are fairly low compared with concentrations in extracellular samples. For these reasons, measurements of intracellular metabolites are time-consuming, economically demanding and subject to technical difficulties, which very often result in relatively poor reproducibility. In addition, there are several biochemical processes that are specifically related to the extracellular media, such as the degradation of complex substrates, and these can only be assessed by measuring the degradation products (secretome) in the extracellular medium [38]. Information from the secretome can be valuable in understanding the behaviour and responses of cultured cells and has the potential clinical application.

Cancer Metabolism

Proliferating and tumour-derived cells are characterized by an elevated aerobic glycolysis with an up-regulated expression of glycolytic enzymes and typically they maintain this metabolic phenotype in culture under normoxic conditions. This implies that the interplay existing in normal cells between mitochondrial respiration and glycolytic flux, by which high O_2 values inhibit the latter process (the so called Pasteur-Crabtree effect [40,41]), is lost in cancer cells. Moreover, the glycolytic rate in cultured cell lines seems to be linked to tumour aggressiveness, leading to the hypothesis that the glycolytic phenotype confers a significant proliferative advantage during the somatic evolution of cancer and is thought to be a crucial component of the malignant phenotype [42]. Nevertheless, high rate of aerobic glycolysis is not unique to tumours, as all energy-demanding cells, namely embryonic cells, utilize glycolysis, so that high glycolytic rates seems to be an hallmark of *all unspecified growing tissues* [43,44,45].

However, the phenotype that is unique to cancer is the high glycolytic fluxes *coupled* to the high lactate levels produced (mainly) via the glycolytic pathway. Indeed, lactate produced in tumour cells is partly produced also by the degradation of glutamine and serine (glutaminolysis and serinolysis) [46]. The conversion of pyruvate to lactate appears important for the maintenance of tumour cell viability. The transformation is carried out by lactate dehydrogenase (LDH), of which the A isoform is strongly upregulated in cancer tissues. Lactate production is essential for the recycling of NAD^+ in the absence of functional mithocondrial-cytoplasmic NADH shuttles due to reduced oxidative phosphorylation. Therefore, as evidenced by Fantin et al. [47], LDH-A suppression not only drives cancer cells towards a mitochondrial oxidative phenotype, but also impaired cancer cell proliferation both in vitro and in vivo.

It is still unclear why tumour cells and normal proliferating cells meet their enhanced energy requirement from glycolysis even though this pathway is far less effective in ATP production than glucose oxidation. Nevertheless, it must be emphasized that, although the yield of ATP per glucose consumed is low, if the glycolytic flux is high enough, the percentage of cellular ATP produced from glycolysis can exceed that produced from oxidative phosphorylation [48]. Secondly, glycolytic glucose degradation to lactate is the only means for the cell to produce ATP without utilization of oxygen. Wherever oxygen reacts with iron containing proteins, e.g., complexes of the mitochondrial respiratory chain, reactive oxygen species (ROS) such as superoxide anions ($^{\cdot}O2^{-}$), peroxide anions, and hydroxyl radicals can be generated. Interaction of ROS with cellular macromolecules (DNA, proteins) and lipids under steady-state conditions can lead to oxidative damage if the antioxidant defence is not fully efficient. Hence, one can hypothesize that transition to aerobic glycolysis serves as a means to minimize the production of ROS in cells during the critical phases of enhanced biosynthesis and cell division. Finally, a critical consequence of an high glycolytic phenotype is increased tumour cell acid production. Acidification of the microenvironment allow cancer cell to become more invasive and more competitive for space and substrate utilization [49].

The tumour metabolome (the complete set of metabolites/low-molecular-weight intermediates), as defined by Mazurek et al. [50], is characterized by high glycolytic and glutaminolytic capacities and a high channelling of glucose carbons toward synthetic processes. Glycolytic regulation in tumour and proliferating cells and the channelling of glucose carbons toward synthetic processes or energy production are related to the association of several enzymes in the glycolytic enzyme complex; in particular, when key enzymes like pyruvate kinase type M2 enzyme (M2-PK) or phosphoglyceromutase (PGAM) migrate out of the complex, glucose carbons are channelled towards nucleic acid synthesis through oxidative and non-oxidative pentose pathways [51,52]. In such a condition, glutamine metabolism to lactate should be increased to ensure energy production [53]. A high M2-PK activity appears to be related to the association of the enzyme within the glycolytic complex that changes in relation to the metabolic demand depending on cell cycle phases and on serine/threonine kinase activity related to oncoprotein expression [54,55].

The glycolytic activity seems to be correlate with the degree of tumour malignancy, so that glycolysis is faster and oxidative phosphorylation is slower in highly de-differentiated and fast-growing tumours than in slow-growing tumours or normal cells [56,57,58]. Furthermore, the fully transformed cell line is most dependent on glycolysis (and less dependent to oxidative metabolism) for ATP synthesis [59]. A similar pattern has been evidenced namely on breast cancer cells: non-invasive MCF7 cells have much lower aerobic glucose consumption rates compared with the highly invasive MDA-mb-231 mammary cancer cell lines [60,61]. High rate of glucose consumption correlate with both malignancy growth and response to therapy [62], meanwhile a high level of lactate (and choline phospholipids metabolites) has been proposed as a predictor of malignant evolution [63]. Moreover, there is a direct correlation between tumour progression and the HK [64,65] and PFK-1 [66,67] activities, which are increased several-fold in fast-growth tumor cells. Accordingly, it has been postulated that tumour cells which exhibit deficiencies in their oxidative capacity are more malignant than those that have an active oxidative phosphorylation [68].

Parlo and Coleman [69] proposed that the high glycolytic activity in some tumor cells is caused by mitochondrial dysfunction at the level of the Krebs cycle, which leads to a lower availability of reducing equivalents for the respiratory chain and hence a lower oxidative phosphorylation. The same authors detected that in Morris 3924A hepatoma, Pyr-derived citrate was preferentially expelled from tumor mitochondria (four times faster than in liver mitochondria) owing to a defect in the transformation of citrate into 2-oxoglutarate (i.e. failure in both aconitase and isocitrate dehydrogenase activities), which induces citrate accumulation in the mitochondrial matrix and hence citrate efflux. This aspect is of relevant importance, keeping in mind that a large availability in citrate synthesis is an absolute need for cancer cells. Indeed, tumour cells exhibit an increase of citrate from mitochondria [70], and this enhanced cytosolic release is a prerequisite for de novo tumour-lipogenesis. In the cytosol, citrate is cleaved by ATP-citrate lyase to acetyl-CoA (AcCoA) + Oxolacetate (OAA) and AcCoA is further carboxylated for incorporation into fatty acids and cholesterol, essential for de novo membranogenesis [71]. It is noteworthy, that in tumours exhibiting no increased glycolytic fluxes, lipogenesis is supported by alternative pathways. As outlined by several authors [45,72,73], glutaminolysis could provide both pyruvate and AcCoA for citrate production and lipogenesis, in the absence of glucose contribution.

Indeed, a relevant body of experimental data obtained by metabolomics studies using mass isotope distribution analysis for the simultaneous characterization of the different pathways of glucose metabolism demonstrated that the fate of glucose carbon is an increased use mainly for intracellular synthetic reactions, i.e. fatty acids and nucleic acid ribose synthesis through glutaminolysis and the non-oxidative pentose-cycle [74,75], whereas energetic purposes are secondary objectives becoming prominent only in specific phases of the cell-cycle or environmental conditions. This is an unexpected feature of cancer metabolism, in that high levels of 'aerobic glycolysis' were initially thought to explain "only" the increasing energy demand of the tumour cells. Nevertheless the re-emergence of interest in intermediary metabolism provide a timely reason to revisit this issue and address the question 'why do tumour cells glycolyse?

Metabolism and Cancer: Cause or Epiphenomenon?

Abnormalities in metabolism have been associated to several alterations in the complex network of signal transduction pathways, which may be caused by both genetic and epigenetic factors.

A great body of evidence suggests that the main mechanism by which glycolysis is substantially higher in tumour than in normal cells is the enhanced transcription of genes correspondent to enzymes pertaining to several or all metabolic and transport pathways which is accompanied by an enhanced protein synthesis [76]. Moreover, tumour cells typically maintain their metabolic phenotypes under normoxic conditions, indicating that aerobic glycolysis is constitutively up regulated through genetic and/or epigenetic changes, involving mainly the hypoxia-inducible factor 1 (HIF-1), the Akt-kinase pathway and probably many other metabolic regulatory networks [77,78]. Moreover, alterations in glycolytic enzymes have been associated with the over-expression of *c-Myc* [79] and *c-raf*, a proto-oncogene that occupies a central node in the complex network of signal transduction pathways, including the insulin-stimulated mitogen activated protein (MAP) kinase signalling cascade. So, it is likely that between profound metabolic alterations (insulin stimulation, high glycolytic rate) and oncogenes involvement it will be a tight association holds. Nevertheless, this association is far from a simple one and seems to involve the *overall* gene-network, more than only few genes. A link between altered metabolism and genome aneuploidy is formally envisaged by Metabolic Control Analysis. From this analysis, it become clear that, in order to transform "the robust normal phenotype into gain-of-flux phenotypes requires massive increases in the metabolic activity of a cell. Aneuploidy provides the necessary boost in genome dose responsible for the increased metabolic activity required for phenotypic transformation *independent of gene mutation* [...] aneuploidy readily explains the tremendous increases or decreases in metabolic activity of cancer cells compared to their normal counterparts" [80].

However, the causative link between gene mutation, genome activity and metabolism is likely to be even more complex and less obvious than previously supposed, and several data have questioned the linearity of such an association. As stated by Griffiths et al. [81], "the relationship between gene expression and metabolism is not straightforward". Moreover, Metabolic Control Analysis studies have shown that there is no general quantitative relationship between mRNA levels and cellular function, and it is widely accepted that glycolytic flux was rarely regulated by gene expression alone [20].

Indeed, it is quite surprisingly that despite no archetypal cancer cell genotype exists, facing the wide genotypic heterogeneity of each tumour cell population [82], some metabolic malignant features are virtually shared by all cancers. Namely, it is noteworthy that of all the physiological hallmarks of cancers, an altered glucose metabolism is perhaps the most common. This paradox of a *common behaviour despite marked both genotypic and epigenetic diversity*, suggests that the energy phenotype of a cancer tissue, should be considered a complex "systems property" and not a merely linearly-gene-driven phenotype, resulting from the dynamic interactions between a tissue and its microenvironment (nutrients availability, cell-to-stroma relationships, hormone flux, etc..). .

Tumour cells show exceptional dependence on glucose carbons and their level of transformation and malignancy correlates with increased metabolism of glucose. However, this metabolic phenotype is, expressed in the context of the microenvironment as being related to substrate or growth factor availability, which profoundly determines the adaptive rearrangement within and among metabolic pathways [83,84] So far, metabolic data can successfully be used in discriminating different metabolic phenotypes of the same cancer cells, evidencing that metabolic profiles, anabolic as well as energy requirements of the tumour can vary in presence of different substrate availability [85] or confluence phases [86]. Namely, as documented by our lab in a non synchronized culture of Jurkat cells, the analysis of the metabolic profile obtained using ^{13}C-NMR spectroscopy and glucose [1,2-^{13}C2] is indicative of the presence of at least two metabolic phenotypes representative of cell subpopulations in different phases of the cell cycle [87]. Furthermore, it is likely that tumour metabolism is organized in concert with the metabolic structure of the overall system composed by tumour cells, stroma, and tumour-associated fibroblasts. As stated by Koukourakis et al., "tumours survive because they are capable of organizing the regional fibroblasts and endothelial cells into a harmoniously collaborating metabolic domain" [88], and it is probable that future studies should be aimed to study tumour metabolism within the context of its microenvironment in order to acquire a more reliable knowledge of the metabolic pathway.

Moreover, even if no doubt exist about the meaningful relevance of the 'glycolytic switch', the significance, i.e. the "teleological" meaning, of this phenotypic trait is still a matter of debate.

The initial hypothesis advanced by Warburg – and generally accepted until the sixties [89] - that aerobic glycolysis results from a primary defect in mitochondrial respiration and eventually causes cancer, has been discarded by a number of investigators who interpreted the aberrations in energy metabolism as secondary events appearing only in late stage of neoplastic development.

However, some recent studies have questioned the classical interpretation of these results and have produced compelling evidence for a regular association of early carcinogenetic events with changes in energy metabolism which seem to elicit a gradual metabolic shift, eventually resulting in the malignant phenotype, prior to any identifiable modification in gene expression or genome structure.

Indeed, a meaningful change in energy as well lipid metabolism in focal preneoplastic lesions long before actual neoplasms (whether benign or malignant) become manifest, have been recorded in both kidney and liver tissues [90,91,92]. It is probable that the interplay between these metabolic changes, in conjunction with altered pH homeostasis and chronic tissue-hypoxia, could trigger some biochemical pathways, involving gene-regulatory signalling networks and finally leading to cancer initiation, with genetic abnormalities emerging only late in the course of carcinogenesis. According to this hypothesis, it has been observed that cells in a preneoplastic lesion may respond to transient episodes of hypoxia or glucose availability by switching to glycolytic metabolism [93]. In fact, cells of preneoplastic foci in the liver show a characteristic increase in the activities of key enzymes of the pentose phosphate and glycolytic pathways, i.e. glucose-6-phosphate dehydrogenase and pyruvate kinase [94]. These findings indicate the beginning of a metabolic shift in glycogenotic preneoplastic hepatocytes towards alternative metabolic pathways. The overall pattern of enzymatic changes in preneoplastic foci closely mimics the phenotypes of liver cells exposed to high insulin levels, but it can be also found in other models of hepatocarcinogenesis, i.e. in hepatocytes exposed to radiation or virus, treated with low dose of chemical hepatocarcinogens and hormones. Moreover, myeloid metaplasia induced by chronic isofenphos exposure is accompanied by increased glucose carbon deposition into nucleic acid through non-oxidative metabolic reactions, and hence followed by the rapid onset of acute myeloid leukaemia [95]. This increase in the non-oxidative metabolism of glucose in the pentose cycle and its deposition into nucleic acid represents a common metabolic phenotype observed in invasive human tumours [96].

Furthermore, a lot of experimental data obtained by biochemical and metabolomics studies, might now be claimed in support of an old carcinogenic hypothesis, previously supported only by epidemiological and clinical observations indicating that dietary habits are statistically linked to increased tumour incidence. It is generally accepted that high-fat diets as well as high dietary glycemic load (a quantitative measure of glycemic effect) are both epidemiologically related to the risk of heart disease, diabetes and several types of cancer [97, 98]. The association is significantly evidenced only in human beings with elevated body mass index (>25 kg/m2) and/or with low physical activity, indicating and increased risk in persons who already have an underlying degree of insulin resistance [99]. On the other hand, anti-diabetic drugs known to be inducers of AMPK phosphorylation, reduced the risk of cancer in diabetic patients [100]. Even if no specific defect responsible for insulin resistance and diabetes has been identified in humans, recent studies have shown that expression of genes involved in mitochondrial oxidative phosphorylation is significantly reduced in skeletal muscle of pre-diabetic and diabetic humans [101], whereas mitochondrial functions are generally impaired in diabetic patients [102]. The efficiency of mitochondrial energy conversion might be the key factor in triggering the metabolic abnormalities observed in cancer cells [103]. Reduction in the mitochondrial oxidative phosphorylation capacity is thought to facilitate the increased occurrence of tumours with ageing [104], whereas both primary or secondary impairment of mitochondrial respiratory chain enzymes may play a significant role in carcinogenesis [105]. On the other hand, disorders of the Krebs cycle activity predispose to hepatocellular carcinoma in human [106] meanwhile rare inherited deficiencies of mitochondrial succinate dehydrogenase subunits or fumarate hydratase can cause tumours in human beings [107]. Moreover, some dietetic habits or metabolic conditions that lead to cellular ATP depletion, such as fructose consumption [108, 109], or to impaired expression of oxidative-phosphorylation-related genes, mainly associated with altered phosphorylation pattern of p38 MAP kinase [110], like type 2 diabetes mellitus, have been shown to enhance growth of chemically induced tumours in rodents, or are linked to increased incidence of numerous types of cancers in humans [111]. Oxidative phosphorylation deficiency causes accumulation of radical oxygen species with limitation of nicotinamide-adenine dinucleotide regeneration and adenosine-triphosphate production, and it is likely that accumulation of these intermediary compounds [112] could be linked to tumour development [113]. In this context, a pivotal role is sustained by frataxin, a mitochondrial protein reduced in Friedreich ataxia syndrome as well as in some cancer cell lines [114]. As a matter of fact, disruption of frataxin in murine hepatocytes causes tumours and namely impairs phosphorylation of the tumour suppressor p38 MAP kinase, meanwhile over-expression of frataxin increases phosphorylation of p38 and reduces activation of a pro-proliferative MAP kinase such as ERK. Although the primary function of frataxin is still a matter of investigation, there is no doubt that reduced expression of frataxin causes impaired oxidative phosphorylation in both rodents and human, whereas over-expression of frataxin induces increased oxidative metabolism, both in non-transformed as well as in malignant cancer cells. Enhancement of the oxidative metabolism is *per se* sufficient to impairs malignant growth and reduces "the tumorigenic capacity of previously transformed cells, providing evidence for a close link between oxidative metabolism and cancer growth [...] hence, frataxin may function as metabolically active mitochondrial suppressor protein [so that] several studies come to the conclusion that impaired mitochondrial metabolism, and specifically reduced Krebs cycle activity may promote malignant growth" [114]. Conversely, increased lipidogenesis or conditions that enhance lipids synthesis and mobilization – widely recognized by epidemiological research as risk factors [115] - may further contribute in transforming the normal metabolic phenotype into a "promoting metabolic profile", therefore enhancing cancer initiation and progression [116, 117, 118]. All together, these data seem to suggest that conditions enhancing glycolytic pathways and lipidogenesis could play a relevant role in cancer initiation.

It is note worthy that several mitochondrial features of cancer cells are in common with embryonic or fetal cells, suggesting that cancer development could be considered a 'developmental disease' characterized by impaired differentiation, as already outlined and documented by increasing experimental data [119]. During both embryonic and fetal stages of development some tissue, like liver, meet most of their energy demands mainly through glycolysis [120], because both the number of mitochondria per cell and the bioenergetic activity of the existing mitochondria are lower than that present in adult tissues, despite a paradoxical increase in the cellular representation of oxidative phosphorylation transcripts. Moreover, hepatomas express isoforms of the glycolytic enzymes different from those present in adult liver, but similar to fetal isoforms [121]. It has been proposed that the aberrant mitochondrial phenotype of fast-growing hepatomas constitutes a reversion to a fetal program of expression of oxidative phosphorylation genes by activation of an inhibitor of ß-mRNA translation [122]. In fact, there are several molecular indications that mitochondria of tumour cells are undifferentiated and behave very much like foetal mitochondria [123]. These results highlight the convergence of embryonic and tumorigenic signalling pathways involved in regulating cell fate and phenotypic characteristics.

Phenotype Metabolism, Cell Shape and Microenvironment

The tumour metabolome – namely the glycolytic phenotype - by no doubt confers to the evolving cancer cell population an advantage and contributes to tissue invasion and metastasis spreading. However, such characteristics are not specific for cancer cells: embryonic tissues, as well as highly proliferating cells (like lymphocytes) [124] share a similar pattern. Moreover, cancer cell metabolism is significantly affected by cell cycle phase and confluence or sub-confluence culture conditions, displaying high plasticity to adapt in presence of adverse microenvironmental conditions. These data evidence that tumour metabolome might be considered a dynamic *reversible* phenotypic trait, likely governed by the non-linear interplays of several both genomic and non-genomic factors (epigenome, nutrient availability, oxygen and blood supply, stiffness and diffusion gradients shaping the microenvironmental constraints). On the other hand, it is reasonable to infer that the modification of microenvironmental cues, could influence tumour metabolism so to force, at least in principle, cancer cells loose (partly or entirely) their malignant features.

Tumour metabolism has been generally investigated by means of classic biochemical tools and only in the course of the last 15-20 years the availability of high-throughput techniques has enabled a dynamical and systemic understanding of the metabolic processes. Metabolic regulatory pathways are rarely completely hierarchical, i.e. the flux through steps in a metabolic pathways did not correlate proportionally with the concentrations of the corresponding enzymes or related-mRNAs, and even strategic pathways, like glycolysis, are rarely regulated by gene expression alone. Incomplete correlation may occur even when regulation is mainly hierarchical, thus indicating that the final biochemical output of a biochemical pathways is largely influenced by the internal network structure than by classical biochemical parameters, such as enzyme kinetics, substrate or protein concentration [125]. In fact, from a classical point of view, biochemical reactions are described as being under control of a "rate-limiting step", and the flux through the related pathway is finally determined by the kinetics of the "rate-limiting step". In the 1970s metabolic control analysis challenged this reductionistic approach and focused on the complex and dynamic structure of metabolic control [126]. The concentrations of metabolites are determined by the activities of many enzymes and are influenced by a lot of many intracellular as well as external factors. As a matter of fact, the individual components of the metabolome are generally far more complex functions of other components than is the case for either mRNAs or proteins. Thus, both transcriptome and proteome may be vastly incomplete monitors of regulation of cell function. This account for disappointing results obtained with targeted-gene-therapies: only few accounts of successful metabolic flux alterations as a consequence of the manipulation of gene-expression (i.e., gene-therapies) have been until now produced [127,128], because of the complex, non-linear nature of the metabolic control architecture.

How a common (and stable) biological behaviour (tumour metabolome) could be expressed by a growing tissue, despite marked both genotypic and epigenetic cell diversity? This paradox asks for Systems Biology approach. Tumour metabolome hardly could be mechanistically linked to the linear dynamics of few gene regulatory networks; otherwise it is likely to be the complex end point of several interacting non-linear pathways, involving both cells and their microenvironment. As such, tumour metabolism might be considered a "systems property", an emergent property arising at the integrated scale of the whole system and behaving like an "attractor" in a specific space phase defined by thermodynamic constraints. Here we give to the notion of attractor the most basic definition of a preferred state toward which the system converge that in principle allow for a lot of different representations: metabolic profile, gene expression patterns, thermodynamic and shape parameters. Indeed, cancer cells are complex systems, evolving according to a non-linear dynamics of gene regulatory networks. A cancer cell, like other living organisms, travels along several states. Each state can be described by an integrated set of genetic, epigenetic or metabolomic parameters: the states that are sufficiently stable (thus working as attractors of the dynamics) can be identified in terms of their fractal dimension.

As suggested by Huang et al. [129], during the carcinogenic process, cells are though to "recover" an "embryonic-like" attractor, and this specific feature could easily explain not only why tumor metabolome displays an "embryonic-like" metabolism, but also how cancer cells exposed to a embryonal morphogenetic field could be committed to apoptosis [130] or induced to differentiate, reverting their malignant phenotype, as evidenced by an increasing body of evidence [131,132, 133].

Interestingly, this morphogenetic-induced reversion is accompanied by significant shape modifications and further followed by remarkable changes in thermodynamics parameters and energy requirements. As a consequence it is not surprising that these entropic adjustments could in turn influence cell energy metabolism and, jointly with the architectural shape reorganization, could modify glucose metabolism. However, until now, this field has been only marginally a matter of investigation [134].

Cancer Cell Shape

Pathologists have long suggested, based on cell morphology, that malignant tumours represent an aberrant form of cellular development [135]: the degree of immaturity of cancer cell phenotype indeed roughly scales with malignancy.

Recently, studies on cell phenotypes and genomic functions worked on biological specimens (cells, tissues) exposed to microgravity, have evidenced a direct link between cell shape and regulatory network [136, 137 ,138] Even if little is still known about how living cells "sense" mechanical stresses – including those due to gravity – it is clear that dramatic changes in the expression of thousands of genes and of enzymatic reactions can be quickly elicited by only modifications in cell shape. Changes in the balance of forces that are transmitted across transmembrane adhesion receptors that link the cytoskeleton to other cells and to the extracellular matrix, have been demonstrated to influence cell morphology and to subsequently induce several alterations in intracellular biochemistry [139]. In this context it is unlikely that the observed wide-changes in cell phenotype and genome functions could be ascribed to a single (or few) signalling pathways operating in isolation, meanwhile it is evident that the "dramatic" twisting of the tension-dependent form of architecture promptly leads to an overall modification in both the cell shape and on thousand of cytoskeleton-linked biochemical pathways [140]. Living cells are literally "hard-wired" so that they can filter the same set of inputs to produce different outputs, and this mechanism is largely controlled through physical distortion of adhesion receptors on the cell surface that transmit stresses to the internal cytoskeleton. Thus, the switch between different cell fate could be considered dependent on cell-distortion: "by sensing their degree of extension or compression cells therefore may be able to monitor local changes in cell crowding or ECM compliance [...] and thereby couple changes in ECM extension to expansion of cell mass within the local tissue microenvironment" [141]. Local geometric control of cell functions may hence represent a fundamental mechanism for developmental regulation within the tissue microenvironment. It is worth noting that, in this perspective, microenvironment modified by space microgravity provide us an *unique* experimental opportunity, by which cell shape distortion can be thought as an *independent variable* or even *a control parameter* in itself. As stated by D.E. Ingber, "[...] cell shape is the most critical determinant of cell function [...] cell shape *per se* appears to govern how individual cells will respond to chemical signals (soluble mitogens and insoluble ECM molecules) in their local microenvironment." [142]

Yet - with some remarkable exceptions - an understandable link between shape and metabolic or genomic function never has been proposed. This is in partly due to the limited knowledge about how biochemical reactions are associated to the cytoskeleton (i.e., the internal topology of structures-linked reactions), and, on the other hand, to a lack of a standardized and wide-accepted *measure* of cell shape complexity.

The ability to correctly characterize shapes has become particularly important in biological and biomedical sciences, where morphological information about the specimen of interest can be used in a number of different ways such as for taxonomic classification and research on morphology-function relationships. A quantitative method holding promises for characterizing complex irregular structures is fractal analysis. Although classical Euclidean geometry works well for describing properties of regular smooth-shaped objects such as circles or squares is not fully adequate for complex irregular-shaped objects that occur in nature (i.e., clouds, coastlines, and biological structures). These "non-Euclidean" objects are better described by fractal geometry, which has the ability to quantify the irregularity and complexity of objects with a measurable value called the fractal dimension. Fractal dimension differs from our intuitive notion of dimension in that it can be a noninteger value, and the more irregular and complex an object is, the higher its fractal dimension relative to its topological dimension [143] Basically the non-integer value tells us about the departure of the object under analysis from the correspondent regular shape object retaining the integer part of the fractal dimension as its topological dimension. The irregular shapes of cancerous cells defy description by traditional Euclidean geometry, which is based on smooth shapes as the line, plane or sphere. In contrast, fractal geometry reveals how an object with irregularities of many sizes may be described by examining how the number of features of one size is related to the number of similarly shaped features of other sizes. Fractal geometry is well suited to quantify those morphological characteristics that pathologists have long used (and are still using today!) in a qualitative sense to describe malignancies. Despite the amazing growth in our understanding of the molecular mechanisms of cancer, as a matter of fact, most diagnosis is still done by visual examination of images and by the morphological examination of radiological pictures, microscopy of cell and tissues, and so forth [144]. A quantitative and operationally reproducible approach, such that provided by fractal analysis, will be of utmost importance and could lead to a remarkable improvement in both cyto-histological and radiographic diagnostic accuracy [145,146]

Fractal theory offers methods for describing the inherent irregularity of natural objects. Mandelbrot [147] introduced the term 'fractal' (from the Latin *fractus*, meaning 'broken') to characterize spatial or temporal phenomena that are continuous but not differentiable. In fractal analysis, the Euclidean concept of 'length' is viewed as a *process*. This process is characterized by a constant parameter D known as the fractal (or fractional) dimension. The fractal dimension can be viewed as a relative measure of complexity, or as an index of the scale-dependency of a pattern. The fractal dimension is a summary statistic measuring "overall" (morphologic) complexity [148]. One can view D "in much the same way that thermodynamics might view intensive measures as temperature" [149]. In other words, fractal dimension can be considered a systems property and, together with one or more independent variables, could enables one's in constructing a diagram of phases, like that relying on temperature, pressure and volume for gas/liquid/solid phase-transitions. This has to do with the generalization of an intuitive property of objects: the dependence of their size from a linear measurement unit, so while a 3D object like a cube increases its volume at the increase of its side following a cubic function (dimension = 3), and a square following a quadratic relation (dimension = 2), a fractal object scales following a non integer exponent. The invariance of the scaling law for a given range of the chosen 'measurement ruler' tells us that the studied object maintains its 'characteristic shape' at different scales of length and this is the case of biological objects like bronchial ramifications in the lung or even ramifications of the trees. In the case of membranes this property of scale invariance produces a dramatic increase of the surface of the system with respect to its volume so allowing for a much more efficient regime of exchange with environment

Several reviews of the applications of fractal measures in pathology and oncology [150] have appeared during the last decade, and a growing literature shows that fractals analysis provides reliable and unsuspected information [151, 152]. Fractal analysis of both cell and tissue morphology is able to differentiate benign from malignant tissues [153], low from high grade tumours [154]; it is intriguing that some aspects of the complex interplay between cancer cells and stroma have been elucidated by means of fractal studies, evidencing that tumour vascular architecture is determined by heterogeneity in the cellular interaction with the extracellular matrix rather than by gradients in diffusible angiogenic factors [155]. Moreover, fractal analysis of the interface between cancer and normal cells might provide further insight into cancer infiltrative and metastatic behaviour. It is well recognized that tumour invasion involves a variety of processes that ultimately lead to cell detachment from the primary tumour and infiltration into adjacent tissue. This pattern formation process is thought as the result of a non-genetic mechanism [156], leading to the amplification of growth instabilities at the tumour/host tissue interface, where a global switch between 'smooth margin' and 'fingering protrusions' surface patterns could allow tumour cells to acquire a metastatic phenotype [157].

So the question arise: "how important shape is" [158]? This problem, firstly proposed by Folkman and Moscona [159], has long remained unanswered, first of all, because most methods used in the past did not account for strict measures of complexity. Secondly, because no satisfactory explanatory framework was available to correlate modifications in shape to gene-regulatory functioning. As outlined by the seminal work done by D.E. Ingber and his co-workers, "the importance of cell shape appears to be that it represents a *visual manifestation of an underlying balance of mechanical forces that in turn convey critical regulatory information to the cell*" [142]. This mechanism implies that cell distortion influence citoskeleton function and cell's adhesion to ECM. Cell shape and cytoskeletal structure are tightly coupled to cell growth, with highly distorted (stretched) cells exhibiting an enhanced sensitivity to soluble mitogens [141]. Within this framework it seems that "function follows form, and not the other way around" [160].

In fact, fractal dimension and the existence of an attractor-like behaviour of dynamical system are linked by the Bendixon-Poincaré theorem [161]. Without going in depth into physico-mathematical subtleties, here it is sufficient to remind the naïve notion of an attractor as a particular configuration the system tends to, given the maintenance of a specific shape implies an energetic cost, we can easily understand that the maintenance of a well defined shape (and consequently a given fractal dimension) in time corresponds to the reach of an attractor, i.e. of a stable regime of energy expenditure .We have already stated the system phase space can be expressed in a lot of different ways ranging from shape, metabolic profile, gene expression pattern, thermodynamic parameters but all these descriptions refer to the same system, under this heading shape can be considered as a privileged observatory for the ease of obtaining complexity descriptors and for its time honoured relation with cancer diagnosis. Shape is thus optimal from both theoretical (dynamical system theory) and clinical (diagnosis) points of view. The link between shape and the metabolic phenotype of cells can thus be considered as a sort of 'circle closure' allowing to relate the morphological observations with clinical outcome by means of biochemistry.

A basic definition of degree of complexity in terms of information dimension is now needed to understand how the changes in shape (and consequently in fractal dimension) can be crucial for system evolution. The information dimension has to do with the number of undamped dynamical variables which are active in the motion of the system; this has to do with the ratio between the number of degrees of freedom that the system exploits and the number of degrees of freedom that are in principle present

Generally, it is imperative to distinguish nominal degrees of freedom from effective (or active) degrees of freedom. Although there may be many nominal degrees of freedom available, the physics of the system may organize the motion into *only a few* effective degrees of freedom. This collective behaviour is often termed self-organization and it arises in dissipative dynamical systems whose post-transient behaviour involves fewer degrees of freedom than are nominally available. The system is attracted to a lower-dimensional phase space, and the dimension of this reduced phase space represents the number of active degrees of freedom in the self-organized system. A similar trend can be observed during the shift from a morphotype to another in the course of the differentiation of a cell lineage: a cell-type proceeds along a discrete number of morphotype along its differentiating pathway, and every morphotype could be considered as a *stable steady-state* [162]. In a similar way, morphological characterization of a cell population by means of fractal analysis could provide at least one independent variable though to be used to construct a (measurable) space phase of the evolving system, in order to evidence the characteristics of the attractors and the location of singularities.

From these statements it is likely that a specific metabolic phenotype could be associated to each of these stable steady-state. Moreover, each morphotype can be described by means of a space-phase - behaving on it like an attractor - and possess specific fractal dimensions. Well-defined distinct cell morphotypes have been experimentally associated – within the same cell population – to the activation of specific gene-regulatory networks and with a specific cell fate (apoptosis, quiescence, proliferation) [163]. Therefore, it is tempting to speculate that each phenotype, as specifically defined by a shape fractal structure, could thereby be associated with a well-defined metabolic phenotype.

Cell Shape and Metabolic Phenotype

In a previous study [164] we showed that breast cancer cells (MCF7 and MDA) growing in a experimental morphogenetic field (EMF) progressively undergoes dramatic changes recorded by both cell shape modifications and metabolome reversion, analysed by NMR spectroscopy (exometabolome analysis). After 48 h, in both MDA-MB-231 and MCF-7 breast cancer cells growing in EMF, both nuclear and membrane profiles changes, evolving into a more rounded shape, loosing spindle and invasive protrusions; these features, for MDA-MB-231 cells, become very evident after 96 hours (Fig. 1).

Fractal analysis was carried out by calculating the Bending Energy (B.E.) of both nuclear and cell membrane. Data were reported for cell profile in Fig. 2.

Bending Energy is a very effective global shape characterization that express the amount of energy needed to transform the specific shape under analysis into its lowest energy state (i.e. a circle) [165] thus immediately linking the geometrical and energetic features of the observed morphologies. The "curvegram" which can be accurately obtained by using digital signal processing techniques (more specifically through the Fourier transform), provides multiscale representation of the curvature. As such, the bending energy provides and interesting resource for translation and rotation-invariant shape classification, as well as a means of deriving quantitative information about the complexity of the shapes being investigated [166]. For biological shapes (membranes, nucleus, mitochondria) the B.E. provides a particularly meaningful physical interpretation in terms of the energy that has to be applied in order to produce or modify specific objects [167].

MDA-MB-231

ctrl EMF

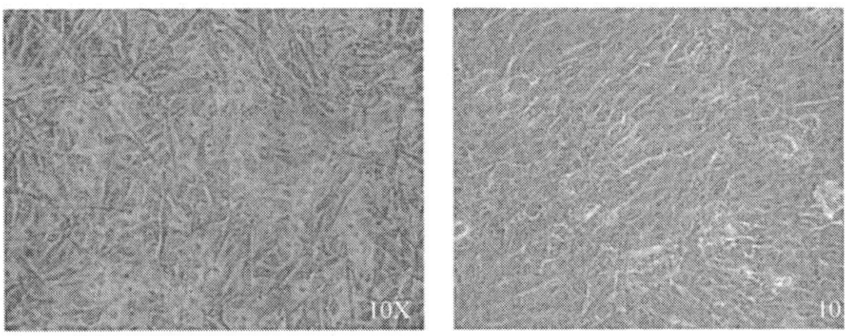

96 hours

Figure 1. MDA-MB-231 cells optical micropictures after 96 hours of treatment. The magnification is 10X.

In our study, control cancer cells exhibit high B.E. values, calculated on both membrane and nuclear profiles. EMT treatment induces a dramatic two-fold reduction on cell membrane B.E. levels, followed by a concomitantly normalization of nucleus shape, statistically significant already from the first 48 hours. Indeed, studies focusing on nuclear shape and structure have revealed strong correlations between shape change and changes in cellular phenotype. By controlling the cellular environment with microfabricated patterning, studies on mammary epithelial cell tissue morphogenesis suggest that altering nuclear organization can modulate the cellular and tissue phenotype [168]. Moreover, microenvironmental-induced shape changes in chondrocyte nuclei correlate with collagen synthesis [169] or changes in cartilage composition and density [170]. This correlative behaviour becomes even more striking when pathological states are observed. Aberrations in nuclear morphology, such as increase in nuclear size, changes in nuclear shape, and loss of nuclear domains, are often used to identify cancerous tissue [171]. It is noteworthy that a strong correlation between a cancerous phenotype and nuclear morphology has been found in breast cancer cells growing in different mechanical and structural environments [172]. Changes in nuclear stiffness could be considered a prerequisite of the increased motility observed in metastatic cancer cells [173]. In turn, these observed changes in nuclear shape may interfere with chromatin structure and could modulate gene accessibility and nuclear elasticity required for translocation, leading to a large scale reorganization of genes within the nucleus [174]. Therefore it is not surprising that EMF-induced "normalization" of nuclear shape could be followed by a subsequent change in tumour metabolome.

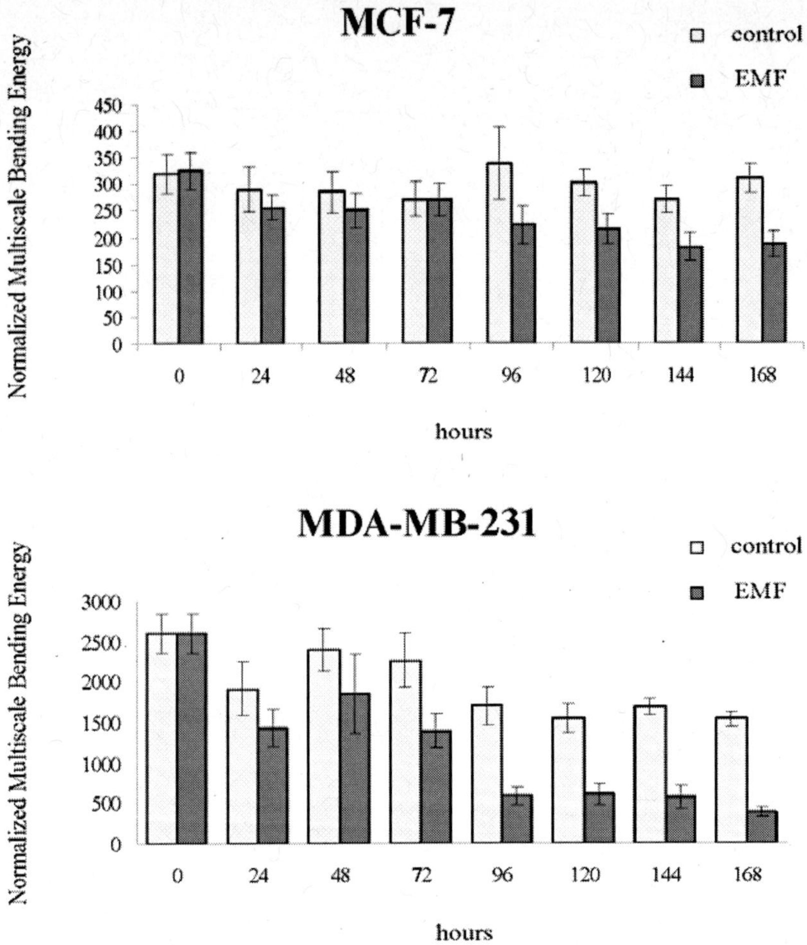

Figure 2. Bar charts showing the Bending Energy values (calculated for cell membrane) in MCF-7 and MDA-MB-231 cells, respectively in controls (yellow bars) and treated conditions (red bars).

Indeed, in EMF-treated breast cancer cells undergoing cell shape modification, glycolytic fluxes were concomitantly reduced, with a parallel decrease in lactate, glutathione, glutamine and other compounds. Namely for MDA-MB cell line, at 72 h, when cell proliferation slow-down and cell shape reaches a new stable configuration characterized by reduced values of Bending Energy, cancer cells exposed to the EMF undergo a complete metabolic reversion. Moreover, after an initial increase, EMF-treated cells showed a significant growth inhibition, without showing a significant apoptotic rate. Surprisingly, more later, between 144-168 hours, exposition to the experimental morphogenetic field leads to the emergence of complex structure – like hollow acini and ducts – reminiscent of the normal mammary gland architecture. These data are coupled with the concomitantly increase in β-casein and E-cadherin synthesis, suggesting that the in the experimental arm, treated cells were committed towards differentiating processes. It is worth noting that the most dramatic metabolic reversion was observed in the more aggressive cell line (MDA-MB-231), meanwhile the most remarkable differentiated structures were expressed by the less invasive MCF-7 breast cancer cells.

In order to get a concomitant representation in the metabolomic space, Principal Component Analysis (PCA) was carried out on a data set constituted by the differences between each spectrum obtained after 48, 72 and 96 h of culture for treated and non-treated samples and the corresponding average spectrum from the 0 h measurement. In this way, the obtained values are representative of net balances, with the positive ones being considered an estimate of net fluxes of production, and the negative an estimate of the utilization of metabolites. Five principal components (PCs) were calculated and the corresponding model explained 80% of the total variance. A *t*-test, applied to the component scores to compare control and treated cells, highlighted significant differences between the two groups on the first four PCs at each experimental time and on the PC5 at 48 and 96 h (Table I), so showing that the treatment is the main driving force of between samples variability.

Analysis of the PC1/PC2 score (Fig. 3), enabled us to evidence that PC1 is by far the major order parameter present in the data (42% of variation explained) and corresponds to the core energy metabolism as evident from its positive loading (correlation coefficient between original variable and component) with glucose utilization and its negative loadings with lactate (see Table II).

This correlation structure implies the samples having an higher PC1 scores correspond to those samples with a lower use of glucose, on the contrary those with high scores are the statistical units endowed with the higher glucose utilization and consequently the higher production of lactate. Given component scores are normalized, we can immediately appreciate the treatment entity that affected metabolic components by the single inspection of differences between treated and control groups in the component space. Looking at Figure 3 it is evident that the by far maximal difference between control and treated groups correspond to the 96h point where control samples display a much higher glucose consumption correspondent to an highly enhanced glycolytic pathway.

Even in the other time points control samples show consistently lower values of PC1 with respect to treated samples, but the differences are much lower. This is evident by the average differences in PC1 scores between control and treated groups at different times that are: 0.6 (48h), 1.0 (72h), 2.6 (96h). Moreover, after 72 h, PC2 scores obtained from EMF-treated cells, evidenced a meaningful metabolomic reversion, characterized by increased β-oxidation fluxes and reduced fatty acids synthesis. Therefore, the two principal metabolomic features of cancer metabolism – i.e. high glycolytic flux and lipogenesis – have been abolished under EMF-treatment.

Table I. *t*-test comparing control *versus* treated cells. In parentheses the percent of variance explained by each principal component is reported (threshold p<0.05).

Experimental time	PC1 (42%)	PC2 (15%)	PC3 (12%)	PC4 (7%)	PC5 (4%)
48	< 0.00001	0.007	< 0.00001	< 0.00001	0.003
72	< 0.00001	< 0.00001	< 0.00001	< 0.00001	0.326
96	< 0.00001	< 0.00001	0.006	0.001	0.044

Table II. Most correlated regions of ^{1}H NMR spectra to PC1

ppm	Factor loading	Metabolite
(3.22-3.26)	0.97	Glucose
(3.38-3.43)	0.98	Glucose
(3.69-3.73)	0.95	Glucose
(3.73-3.77)	0.96	Glucose
(3.77-3.80)	0.98	Glucose
(3.80-3.86)	0.97	Glucose
(3.92-3.97)	0.97	Glucose
(4.62-4.70)	0.98	Glucose
(5.21-5.26)	0.98	Glucose
(1.30-1.36)	-0.89	Lactate
(4.10-4.15)	-0.80	Lactate
(2.12-2.15)	-0.80	Glutamine
(2.41-2.45)	-0.93	Glutamine

It is of outmost importance that PC1 mirrors the same diverging in time behavior of the control/treated differences observed as for the shape analysis, so pointing to an empirical correlation between the shape and metabolomic descriptions. What is worth noting is that the differentiation in shape between the control and treated groups seem to happen between 48 and 72 hours, while in the case of metabolic description the two experimental groups diverge between 72 and 96 hours. This seems to indicate a causative effect of shape on metabolism more likely than *viceversa*. This is clearly an extremely preliminary result but could be profitably related to the evidence presented by Meadows et al. [134]. These authors measured glucose uptake in 48R normal human mammary epithelial cells, and MCF7 cells, and then correlate this measure to biomass, cell number and medium exposed surface demonstrating that medium exposed surface was the main driving force of glucose uptake in cells. In our experiments, having stated the increased glycolytic flux in control cells, it is worth noting that the treated cells present an increased glutamine use with respect to control ones. This increase in glutamine utilization does not correlate with a simultaneous increase in lactate (as expected if the difference between control and treated cell metabolism should confined to a mere diversification of energy sources for treated cells) nor to an increase in fatty acid synthesis (as expected when de novo cell membrane production is required to sustain cell proliferation). Indeed, EMF-treated cells showed a statistically significant growth-inhibition, confirming that glutaminolysis cannot be explained by energetic or proliferation needs: this implies the treated cells devote an higher portion of chemical energy to the other anabolic work (construction of cellular structures) than control cells. Excess of glutamine is then preferentially transformed into proteins and does not appear as lactate. This interpretation is given a proof-of-concept by the observation of the development of both differentiating pathways (as evidenced by the increased synthesis of E-cadherin and β-casein) and differentiated structures (ducts and hollow acini, mainly in MDA-MB-231 cells) in treated cells at later times (96-168 h).

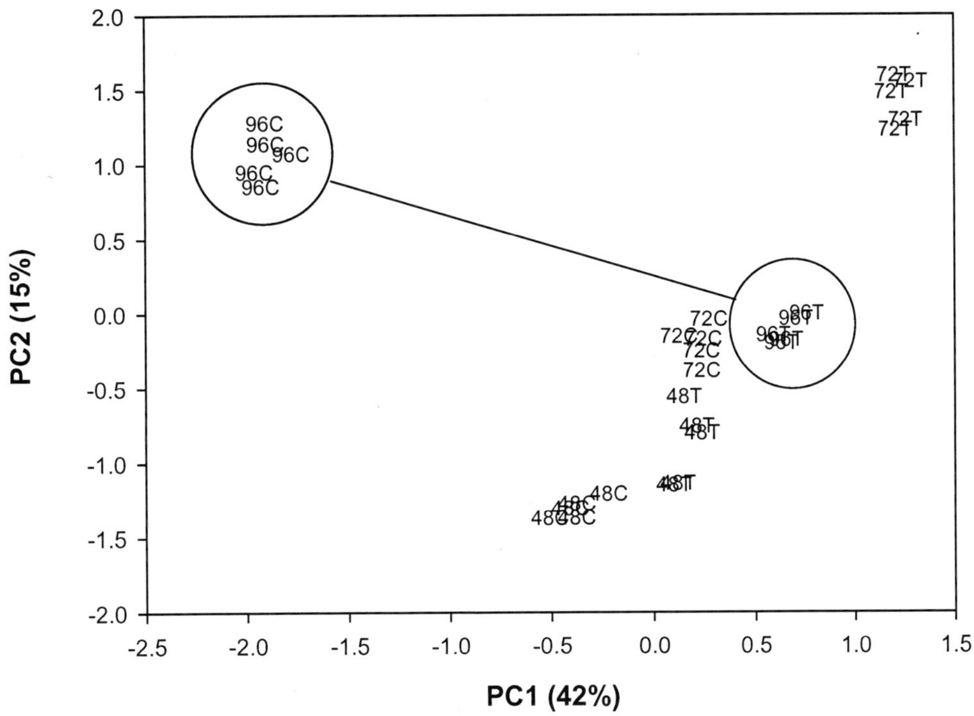

Figure 3. Overview of the PCA model built on the NMR dataset of medium samples collected from MDA-MB-231 untreated and treated cell cultures at 48, 72 and 96 hours. The score plot of the first two components (PC1 *versus* PC2) showing differentiation among groups are shown. The major metabolic difference between control and treated groups at 96 h is highlighted by the black line.

It should be emphasized that the metabolome reversion is *preceded* by significant modifications in cell shape and fractal dimensions. Namely in the more invasive cell line (MDA-MB-231), metabolome reversion attains a stable configuration without any further change, even if cancer cell population undergo several structural modifications characterized by re-establishment of cell-to-cell junction, increased expression of differentiating (such as E-cadherin) and functional molecules (casein production). These preliminary data suggest that the structural reorganization fostered by EMF through shape reorganization, induces an adaptive metabolomic reversion: EMF-treated cells loose both the glycolytic and lipogenic malignant phenotype, meanwhile differentiating processes took place.

It is worth noting that shape modification leads to a less-dissipative architecture, as it is documented by a measurable significant reduction in B.E. values. Therefore, fractal measures enable us to highlight the neglected link between cell morphology and thermodynamics. According to the Prigogine-Wiame theory of development [175], during carcinogenesis, a living system constitutively deviates from a steady state trajectory; this deviation is accompanied by an *increase* in the system dissipation function (Ψ) at the expense of coupled processes in other parts of the organism [176], where $\Psi = q_0 + q_{gl}$ (meaning, respectively, q_0 oxygen consumption and q_{gl} glycolysis intensity). Keeping in mind that B. E. represents a "dissipative" form of energy, meanwhile metabolomic data evidenced a significant reduction in glycolysis activity (in presence of unchanged values of oxygen consumption), it follows that in our experimental conditions Ψ decreased significantly, until a stable state was attained, characterized by a minimum in the rate of energy dissipation (principle of minimum energy dissipation) [177]. This behaviour is exactly the opposite to what expected in growing cancer cells and experimentally observed in our tumour control cells.

Conclusion

The re-visitation of "Warburg theory, shed light into some basically aspects of cancer cells and offers alternative hypothesis about the carcinogenic process. High glycolytic rate provides several advantages for proliferating cells. First, it allows cells to use glucose to produce abundant ATP, allowing the high energy needs of a growing tissue to be satisfied. Secondly, glucose degradation – jointly with glutaminolysis – provides cells with intermediates needed for biosynthetic pathways, including citrate for lipidogenesis and ribose sugars for nucleotides. As stressed by DeBerardinis et al., "a further advantage of the high glycolytic rate is that it allows cells to fine tune the control of biosynthetic pathways that use intermediates derived from glucose metabolism. When a high flux metabolic pathway branches into a lower-flux pathway, the ability to maintain activity of the latter is maximized when flux through the former is highest; [therefore] the very high rate of glycolysis allows cells to maintain biosynthetic fluxes during rapid proliferation but results in a high rate of lactate production" [178].

Following this perspective, the "Warburg effect" is not merely a linear consequence of gene deregulation or an adaptation to hypoxia, but a "systems property" of cancer cells, influenced by both internal and microenvironmental constraints. Even if there is hardly consensus on that viewpoint, undoubtedly knowledge acquired in recent years by means of metabolomic studies have significantly contributed to a more general and critical appraisal of the widely accepted carcinogenic theory [179].Cell energy metabolism differs in function of the cell cycle phase of activity, namely being more "dissipative" during wound healing, fast growth (specifically during embryonic development), and cancer progression. Keeping in mind that thermodynamic dissipative function is correlated with both glucose metabolism and cell shape, we suggest that the latter could interfere with metabolic pathways. Cell shape has proven to influence through architectural rearrangement several gene-regulatory pathways, thereby representing a relevant *independent factor* controlling tissue fate and cell commitment to quiescence, apoptosis or proliferation. Our preliminary data evidenced that an embryonic morphogenetic field is capable in inducing dramatic changes in breast cancer cell shape. Fractal analysis reveal that B.E. of both nuclear and cell membrane decrease significantly after 48 h of treatment. Consequently, meaningful changes in "tumour metabolome" were observed by means of NMR-spectroscopy and PCA flux analysis. Tumour cells begin to loose their glycolytic phenotype after 48 h, leading to reduced lactate accumulation, and, after 72 h, fatty acids and citrate synthesis slow-down. These data indicate that cell shape "normalization" is followed by a reversion in tumour metabolic phenotype.

Further metabolomic studies are clearly warranted in order to better correlated metabolism and shape morphology in order to handle these two set of parameters into a dynamical description of tumour cell biology.

References

[1] Warburg, O. (1926). *Über den Stoffwechsel der Tumoren*, Springer. Berlin;.

[2] Warburg, O. (1956). On the origin of cancer cells. *Science*, **123**, 309-314.

[3] Garber, K. (2004). Energy boost: the Warburg effect returns in a new theory of cancer. *J Natl Cancer Inst.*, **96**, 1805-06.

[4] Hsu, P. P. & Sabatini, D. M. (2008). Cancer cell metabolism: Warburg and beyond. *Cell*, **134**, 703-707.

[5] Hawkins, R. A. & Phelphs, M. E. (1988). PET in clinical oncology. *Cancer metastasis rev.*, **7**, 119-142.

[6] Kunkel, M. et al. (2003). Overexpression of Glut-1 and increased metabolism in tumours are associated with a poor prognosis in patients with oral squamous cell carcinoma. *Cancer*, **97**, 1015-1024.

[7] Kroemer, G. & Pouyssegur, J. (2008). Tumour cell metabolism: Cancer's Achilles' heel. *Cancer Cell*, **13**, 472-482.

[8] Alhasan, S. A., Pietrasczkiwicz, H. & Alonso, M. D. (1999). Genistein induced cell cycle arrest and apoptosis in a head and neck squamous cell carcinoma cell line. *Nutr Cancer*, **34**, 12-19.

[9] Boros, L. G., Bassilian, S., Lim, S. & Lee, W. N. P. (2001). Genistein inhibits non-oxidative ribose synthesis in MIA pancreatic adenocarcinoma cells: a new mechanisms of controlling tumor growth. *Pancreas*, **22**(1), 1-7.

[10] Banki, K., Hutter, E. & Colombo, E. (1996). Glutathione levels and sensitivity to apoptosis are regulated by changes in transaldolase expression. *J. Biol. Chem.*, **271**, 2994-3001.

[11] Gottschalk, S., Anderson, N., Hainz, C., Eckardt, S. G. & Serkova, N. J. (2004). Imatinib (STI571)-mediated changes in glucose metabolism in human leukaemia BCR-Abl-positive cells. *Clin Cancer Res.*, **10**, 6661-6668.

[12] Boren, J., Cascante, M., Marin, S., Comin-Anduix, B., Centelles, J. J., Lim, S., Bassilian, S., Ahmed S., Lee, W. N. P. & Boros, L. G. (2001). Gleevec (STI571) influences metabolic enzymes activities and glucose carbon flow toward nucleic acid and fatty acid synthesis in myeloid tumor cells. *J. Biol. Chem.*, **276**(41), 37747-37753.

[13] Tarn, C., Skorobogatko, Y. V., Tagichi, T., Eisenberg, B., Von Mehren, M., Godwin, A. K. (2006). Therapeutic effect of imatinib in gastrointestinal stromal tumors: AKT signaling dependent and independent mechanisms. *Cancer Res.*, **66**(10), 5477-5486.

[14] Peng, B., Hayes, M., Drucker, B., Talpaz, M., Sawyers, C., Resta, D., Ford, J., Man, A. (2000). *Proc. Am. Ass. Cancer Res.*, **41**, 544.

[15] Oliver, S. G. (2002). Functional genomics: lessons from yeast. *Phil. Trans. R. Soc. Lond. B.*, **357**, 17-23.

[16] Mendes, P., Kell, D. B. & Westerhoff, H. V. (1996). Why and when channeling can decrease pool size at constant net flux in a simple dynamic channel. *Biochim. Biophys. Acta*, **1289**, 175-186.

[17] Fell, D. A. (1996). *Understanding the Control of Metabolism*. Portland Press. London;.

[18] Kell, D. B. & Westerhoff, H. V. (1986). Towards a rational approach to the optimization of flux in microbial biotransformations. *Trends Biotehnol.*, **4**, 137-142.

[19] Keightley, P. D. & Kacser, H. (1987). Dominance, pleiotropy and metabolic structure. *Genetics*, **117**, 319-329.

[20] Kuile, B. H. & Westerhoff, H. V. (2001). Transcriptome meets metabolome: hierarchical and metabolic regulation of the glycolytic pathway. *FEBS Letts*, **500**, 169-171.

[21] Griffin, J. L. & Shockcor, J. P. (2004). Metabolic profiles of cancer cells. *Nature Rev. Cancer*, **4**, 551-561.

[22] Griffin, J. L. (2004). Metabolic profiles to define the genome: can we hear the phenotypes? *Trans R Soc Lond B Biol Sci.*, **359**, 857-871.

[23] Martzen, M., McCraith, S., Spinelli, S., Torres, F. & Fields, S. (1999). A biochemical genomics approach for identifying genes by the activity of their products. *Science*, **286**, 1153-1155.

[24] Raamsdonk, L. M., Teusink, B., Broadhurst, D., Zhang, N., Hayes, A., Walsh, M. C., Berden, J. A., Brindle, K. M., Kell, D. B., Rowland, J. J., Westerhoff, H. V., Van Dam, K., Oliver, S. G. (2001). A functional genomics strategy that uses metabolome data to reveal the phenotype of silent mutations. *Nature Biotechnol.*, **19**, 45-50.

[25] Devaux, P. G., Horning, M. G. & Horning, E. C. (1971). Benyzloxime derivatives of steroids; a new metabolic profile procedure for human urinary steroids. *Anal. Lett.*, **4**, 151-152.

[26] Horning, E. C. & Horning, M. G. (1971). Human metabolic profiles obtained by GC and GC/MS. *J. Chromatogr. Sci.*, **9**, 129-140.

[27] Griffin, J. L. (2006). The Cinderella story of metabolic profiling: does metabolomics get to go to the functional genomics ball? *Philos Trans R Soc Lond B Biol Sci.*, **361*(*1465*)***, 147-61.

[28] Fiehn, O. (2001). Combining genomics, metabolome analysis and biochemical modeling to understand metabolic networks. *Comp. Funct. Genomics*, **2**, 155-168.

[29] Florian, C. L., Preece, N. E., Bhakoo, K. K., Williams, S. R. & Noble, M. D. (1995). Characteristic metabolic profiles revealed by 1H NMR spectroscopy for three types of human brain and nervous system tumours. *NMR Biomed.*, **8**, 253-264.

[30] Florian, C. L., Preece, N. E., Bhakoo, K. K., Williams, S. R. & Noble, M. D. (1995). Cell type-specific fingerprinting of meningioma and meningeal cells by proton nuclear magnetic resonance spectroscopy. *Cancer Res.*, **55**, 420-427.

[31] Griffin, J. L. & Kauppinen, R. A. (2007). Tumour metabolomics in animal models of human cancer. *J Proteome Res.*, **6**(2), 498-505.

[32] Griffin, J. L. & Kauppinen, R. A. (2007). A metabolomics perspective of human brain tumours. *FEBS J.*, **274**(5), 1132-9.

[33] Valerio, M., Panebianco, V., Sciarra, A., Osimani, M., Salsiccia, S., Casciani, L., Giuliani, A., Bizzarri, M., Di Silverio, F., Passariello, R. & Conti, F. (2009). Classification of prostatic diseases by means of multivariate analysis on in vivo proton MRSI and DCE-MRI data NMR *Biomed.*, [Epub ahead of print].

[34] Usenius, J. P. et al. (1996). Automated classification of human brain tumours by neural network analysis using *in vivo* 1H magnetic resonance spectroscopic metabolite phenotypes. *Neuroreport.*, **7**, 1597-1600.

[35] Gribbestad, I. S., Sitter, B., Lundgren, S., Krane, J. & Axelson, D. (1999). Metabolite composition in breast tumors examined by proton nuclear magnetic resonance spectroscopy. *Anticancer Res.*, **19**, 1737-1746.

[36] Allen, J., Davey, H. M., Broadhurst, D., Heald, J. K., Rowland, J. J., Oliver, S. G. & Kell, D. B. (2003). High-throughput classification of yeast mutants for functional genomics using metabolic footprinting. *Nat Biotechnol*, **21**, 692-6.

[37] Kell, D. B., Brown, M., Davey, H. M., Dunn, W. B., Spasic, I. & Oliver, S. G. (2005). Metabolic footprinting and systems biology: the medium is the message. *Nat Rev Microbiol*, **03**, 557-565.

[38] Villas-Bôas, S. G., Noel, S., Lane, G. A., Attwood, G. & Cookson, A. (2006). Extracellular metabolomics: a metabolic footprinting approach to assess fiber degradation in complex media. *Anal Biochem*, **349**, 297-305.

[39] Villas-Bôas, S. G., Mas, S., Åkesson, M., Smedsgaard, J. & Nielsen, J. (2005). Mass spectrometry in metabolome analysis. *Mass Spectrom Rev*, **24**, 613-646.

[40] Pasteur, L. (1861). Experiénces et vues nouvelles sur la nature des fermentations. *C. R. Acad. Sci.*, **52**, 344-347.

[41] Crabtree, H. (1928). The carbohydrate metabolism of certain pathological growths. *Biochem J.*, **22**, 1289-1298.

[42] Gatenby, R. A. & Gillies, R. J. (2004). Why do cancers have high aerobic glycolysis? *Nat. Rev. Cancer*, **4**, 891-899.

[43] Kondoh, H. et al. (2007). A high glycolytic flux supports the proliferative potential of murine embryonic stem cells. *Antiox. Redox Signal*, **9**, 293-299.

[44] Brand, K. (1997). Aerobic Glycolysis by Proliferating Cells: Protection against Oxidative Stress at the Expense of Energy Yield. *Journal of Bioenergetics and Biomembranes*, **29**(4), 355-364.

[45] McKeehan, W. L. (1982). Glycolysis, glutaminolysis and cell proliferation. *Cell Biol Int Rep.*, **18**, 3275-3282.

[46] Lobo, C., Ruiz-Bellido, M. A., Aledo, J. C., Marquez, J., Nunez De Castro, I. & Alonso, F. J. (2000). Inhibition of glutaminase expression by antisense mRNA decreases growth and tumorigenicity of tumour cells. *Biochem J.*, **348**, 257-261.

[47] Fantin, V. R., St-Pierre, J. & Leder, P. (2006). Attenuation of LDH-A expression uncovers a link between glycolysis, mitochondrial physiology, and tumor maintenance. *Cancer cell*, **9**, 425-34.

[48] Guppy, M., Greiner, E. & Brand, K. (1993). The role of the Crabtree effect and an endogenous fuel in the energy metabolism of resting and proliferating thymocytes. *Eur. J. Biochem.*, **212**, 95-99.

[49] Gatenby, R. A. & Gawlinski, E. T. (1996). A reaction-diffusion model of acid-mediated invasion of normal tissue by neoplastic tissue. *Cancer res.*, **56**, 5745-5753.

[50] Mazurek, S. & Eigenbrodt, E. (2003). The tumor metabolome. *Anticancer Res.*, **23**, 1149-1154.

[51] Cascante, M., Centelles, J. J., Veech, R. L., Lee, W. N. & Boros, L. G. (2000). Role of thiamine (vitamin B-1) and transketolase in tumor cell proliferation. *Nutr. Cancer*, **36**, 150-154.

[52] Mazurek, S., Grimm, H., Boschek, C. B., Vaupel, P. & Eigenbrodt, E. (2002). Pyruvate kinase type M2: a crossroad in the tumor metabolome. *Br. J. Nutr.*, **87***(Suppl.***1***)*, S23-S29.

[53] Mazurek, S., Eigenbrodt, E., Failing, K. & Steinberg, P. (1999). Alterations in the glycolytic and glutaminolytic pathways after malignant transformation of rat liver oval cells. *J. Cell. Physiol.*, **181**, 136-146.

[54] Mazurek, S., Zwerschke, W., Jansen-Durr, P. & Eigenbrodt, E. (2001). Effects of the human papilloma virus HPV-16 E7 oncoprotein on glycolysis and glutaminolysis: role of pyruvate kinase type M2 and the glycolytic-enzyme complex. *Biochem. J.*, **356**, 247-256.

[55] Le Mellay, V., Houben, R., Troppmair, J., Hagemann, C., Mazurek, S., Frey, U., Beigel, J., Weber, C., Benz, R., Eigenbrodt, E. & Rapp, U. R. (2002). Regulation of glycolysis by Raf protein serine/threonine kinases. *Advan. Enzyme Regul.*, **42**, 317-332.

[56] Pedersen P. L. (1978). Tumor mitochondria and the bioenergetics of cancer cells. *Prog Exp Tumor Res*, **22**, 190-274.

[57] Zu, X. L. & Guppy, M. (2004). Cancer metabolism: facts, fantasy, and fiction. *Biochem Biophys Res Commun*, **313**, 459-465.

[58] Krieg, R. C., Knuechel, R., Schiffmann, E., Liotta, L. A., Petricoin, E. F. & Herrmann, P. C. (2004). Mitochondrial proteome: cancer-altered metabolism associated with cytochrome *c* oxidase subunit level variation. *Proteomics*, **4**, 2789-2795.

[59] Ramanathan, A., Wang, C. & Schreiber, S. L. (2005). Perturbational profiling of a cell-line model of tumorigenesis by using metabolic measurements. *Proc. Natl. Acad. Sci.*, USA, **102**(17), 5992-5997.

[60] Schomack, P. A. & Gilles, R. J. (2003). Contributions of cell metabolism and H^+ diffusion to the acidic pH of tumours. *Neoplasia* (New York), **5**, 135-145.

[61] Mazurek, S., Michel, A. & Eigenbrodt, E. (1997). Effect of extracellular AMP on cell proliferation and metabolism of breast cancer cell lines with high and low glycolityc rates. *J Biol Chem*, **272**, 4941-4952.

[62] Smith, T. A. (2001). The rate-limiting step for tumor [18F] fluoro-2-deoxy-D-glucose (FDG) incorporation. *Nucl Med Biol*, **28**, 1-4.

[63] Walenta, S., Wetterling, M., Lehrke, M., Schwickert, G., Sundfor, K., Rofstad, E. K. & Mueller-Klieser, W. (2000). High lactate levels predict likelihood of metastases, tumor recurrence, and restricted patient survival in human cervical cancers. *Cancer Res*, **60**, 916-921.

[64] Pedersen, P. L., Mathupala, S., Rempel, A., Geschwind, J. F. & Ko, Y. H. (2002). Mitochondrial bound type II hexokinase. *Biochim Biophys Acta*, **1555**, 14-20.

[65] Marin-Hernandez, A., Rodriguez-Enriquez, S., Vital-Gonzalez, P. A., Flores-Rodriguez, F. L., Macias-Silva, M., Sosa-Garrocho, M. & Moreno-Sanchez, R. (2006). Determining and understanding the control of glycolysis in fast-growth tumor cells. Flux control by an overexpressed but strongly product-inhibited hexokinase. *FEBS J*, **273**, 1975-1988.

[66] Sanchez-Martınez, C. ; Estevez, A. M. & Aragon, J. J. (2000). Phosphofructokinase C isozyme from ascites tumor cells: cloning, expression, and properties. *Biochem Biophys Res Commun*, **271**, 635-640.

[67] Meldolesi, M. F., Macchia, V. & Laccetti, P. (1976). Differences in phosphofructokinase regulation in normal and tumor rat thyroid cells. *J Biol Chem*, **251**, 6244-6251.

[68] Soderberg, K., Nissinen, E., Bakay, B. & Scheffler, I. E. (1980). The energy charge in wild-type and respiration deficient Chinese hamster cell mutants. *J Cell Physiol*, **103**, 169-172.

[69] Parlo, R. A. & Coleman, P. S. (1984). Enhanced rate of citrate export from cholesterol-rich hepatoma mitochondria. The truncated Krebs cycle and other metabolic ramifications of mitochondrial membrane cholesterol. *J Biol Chem*, **259**, 9997-10003.

[70] Parlo, R. A & Coleman, P. S. (1986). Continuous pyruvate carbon flux to newly synthesized cholesterol and the suppressed evolution of pyruvate-generated CO2 in tumours: further evidence for a persistent truncated Krebs cycle in hepatomas. *Biochim Biophys Acta*, **886**, 69-176.

[71] Memendez, J. A., Colomer, R. & Lupu, R. (2005). Why does tumour-associated fatty acid synthase (oncogenic antigen 519) ignore dietary fatty acids? *Med. Hypoth.*, **64**, 342-349.

[72] Moreadith, R. W. & Lehninger, A. L. (1984). The pathways of glutamate and glutamine oxidation by tumour cell mithocondria. Role of mithocondrial NAD(P)+-dependent malic enzyme. *J. Biol Chem.*, **259**, 6215-6221.

[73] Costello, L. C. & Franklin, R. B. (2005). 'Why do tumor glycolyse?': from glycolysis through citrate to lypogenesis. *Mol Cell Biochem.*, **280**, 1-8.

[74] Richardson, A. D., Yang, C., Osterman, A. & Smith, J. W. (2008). Central carbon metabolism in the progression of mammary carcinoma. *Breast Cancer Res Treat.*, **110**, 297-307.

[75] Boros, L. G., Torday, J. S., Lim, S., Bassilian, S., Cascante, M. & Lee, W. N. P. (2000). Transforming Growth Factor β_2 promotes glucose carbon incorporation into nucleic acid ribose through the non-oxidative pentose cycle in lung epithelial carcinoma cells. *Cancer Res.*, **60**, 1183-1185.

[76] Dang, C. V., Lewis, B. C., Dolde, C., Dang, G. & Shim, H. (1997). Oncogenes in tumor metabolism, tumorigenesis, and apoptosis. *J Bioenerg Biomembr*, **29**, 345-354.

[77] Hyun, J. Y., Chun, Y. S., Kim, T. Y., Kim, H. L., Kim, M. S. & Park, J. W. (2004). Hypoxia-Inducible Factor 1alpha- Mediated Resistance to Phenolic Anticancer. *Chemotherapy*, **50**, 119-126.

[78] Elstrom, R. L., Bauer, D. E., Buzzai, M., Karnauskas, R., Harris, M. H., Plas, D. R., Zhuang, H., Cinalli, R. M., Alavi, A., Rudin, C. M. & Thompson, C. B. (2004). Akt Stimulates Aerobic Glycolysis in Cancer Cells. *Cancer Res.*, **64**, 3892-3899.

[79] Shim, H., Dolde, C., Lewis, B. C., Wu, C. S., Dang, G., Jungmann, R. A., Dalla-Favera, R. & Dang, C. V. (1997). C-Myc transactivation of LDH-A: implications for tumor metabolism and growth. *Proc. Natl. Acad. Sci.*, USA, **94**, 6658-6663.

[80] Rasnick, D. & Duesberg P. (1999). How aneuploidy affects metabolic control and causes cancer. *Biochem J.*, **340**, 621-630.

[81] Griffiths, J. R. & Stubbs, M. (2003). Opportunities for studying cancer by metabolomics: preliminary observations on tumors deficient in hypoxia-inducible factor 1. *Advan. Enzyme Regul.*, **43**, 67-76.

[82] Kerangueven, F., Noguchi, T., Coulie, R. F., Allione, F., Wargniez, V., Simony-Lafontaine, J., Longy, M., Jacquemier, J., Sobol, H., Eisinger, F. & Birnbaum, D. (2000). Genome wide-search for loss of heterozygosity shows extensive genetic diversity of human breast carcinomas. *Cancer Res.*, **60**, 6503-6509.

[83] Griffiths, J. R., McIntyre, D. J. O., Howe, F. A. & Stubbs, M. (2002). In the tumor microenvironment: causes and consequences of hypoxia and acidity. *Novartis Foundation Symposium*, vol. **240**. Wiley. Chichester, 46-67.

[84] Yamajy, Y., Shiotani, T., Nakamura, H., Hata, Y., Hashimoto, Y., Nagai, M., Fujita, J. & Takahara, J. (1994). Reciprocal alterations of enzymic phenotype of purine and pyrimidine metabolism in induced differentiation of leukemic cells, *Adv. Exp. Med. Biol.*, **370**, 747-751.

[85] Rossignol, R., Gilkerson, R., Aggeler, R., Yamagata, K., Remington, S. J. & Capaldi, R. A. (2004). Energy Substrate Modulates Mitochondrial Structure and Oxidative Capacity in Cancer Cells. *Cancer Res.*, **64**, 985-993.

[86] Tomassini, A., Miccheli, A., Di Clemente, R., Valerio, M., Coluccia, P., Bizzarri, M. & Conti, F. (2006). NMR-based metabolic profiling of human hepatoma cells in relation to cell growth. *Biochimica Biophysica Acta*, **1760**(11), 1723-1731.

[87] Miccheli, A., Tomassini, A., Puccetti, C., Valerio, M., Peluso, G., Tuccillo, F., Calvani, M., Manetti, C. & Conti, F. (2006). Metabolic profiling by ^{13}C-NMR spectroscopy: [**1**,2-$^{13}C_2$] glucose reveals a heterogeneous metabolism in human leukemia T cells. *Biochimie*, **88**, 437-448.

[88] Koukourakis, M. I., Giatromanolaki, A., Harris, A. L. & Sivridis, E. (2006). Comaprison of metabolic pathways between cancer cells and dtromal cells in colorectal carcinomas: a metabolic survival role for tumor-associated stroma. *Cancer Res.*, **66**(2), 632-637.

[89] Warburg, O. (1966). *Molekulare Biologie des malignen Wachstums*. In: Holzer, H. & Holldorf, A. W., editors, Berlin: Springer; 1-16.

[90] Bannasch, P., Jahn, U. R., Hacker, H. J., Su, Q., Hofmann, W., Pichlmayr, R. & Otto, G. (1997). *Int. J. Oncol.*, **10**, 261-268.

[91] Bannasch, P., Hacker, H. J., Tsuda, H. & Zerban, H. (1986). *Adv. Enzyme Regul.*, **25**, 279-296.

[92] Mayer, D., Klimek, F., Rempel, A. & Bannasch, P. (1997). *Biochem. Soc. Trans.*, **25**, 122-127.

[93] Gatenby, R. A. & Gawlinski, E. T. (2003). The glycolytic phenotype in carcinogenesis and tumour invasion: insights through mathematical models. *Cancer res.*, **63**, 3847-3854.

[94] Bannasch, P., Klimek, P. & Mayer, D. (1997). Early Bioenergetic Changes in epatocarcinogenesis: Preneoplastic Phenotypes Mimic Responses to Insulin and Thyroid Hormone *Journal of Bioenergetics and Biomembranes.*, **29**(4), 3003-313.

[95] Boros, L. G. & Williams, R. D. (2001). Isofenphos induced metabolic changes in K562 myeloid blast cells *Leukemia Research*, **25**, 883-890.

[96] Boros, L. G., Torday, J. S., Lim, S., Bassilian, S., Cascante, M. & Lee, W. N. (2000). Transforming growth factor-2 promotes glucose carbon incorporation into nucleic acid ribose through the nonoxidative pentose cycle in lung epithelial carcinoma cells. *Cancer Res.*, **60**, 1183-5.

[97] Salmeron, J., Manson, J. E., Stampfer, M. J., Colditz, G. A., Wing, A. L; Willett, W. C. (1997). Dietary fiber, glycemic load and risk of non-insulin-dependent diabetes mellitus in wome. *JAMA*, **277**, 472-477.

[98] DeMeo, M. T. (2001). Pancreatic cancer and sugar diabetes. *Nutr. Rev.*, **59**, 112-115.

[99] Michaud, D. S., Liu, S., Giovannucci, E., Willett, W. C., Colditz, G. A. & Fuchs, C. S. (2002). Dietary sugar, glycemic load and pancreatic cancer risk in a prospective study. *J. Natl. Cancer Inst.*, **17**, 1293-1300.

[100] Evans, J. M. M., Donnelly, L. A., Emslie-Smith, A. M., Alessi, D. R. & Morris, A. D. (2005). Metformin and reduced risk of cancer in diabetic patients. *BMJ*, **330**, 1304-1305.

[101] Patti, M. E., Butte, A. J., Crukhorn, S., Cusi, K., Berria, R., Kashyap, S., Miyazaki, Y., Kohane, I., Costello, M., Saccone, R., Landaker, E. J., Goldfine, A. B., Mun, E., DeFronzo, R., Finlayson, J., Kahn, R. C. & Mandarino, L. J. (2003). Coordinated reduction of genes of oxidative metabolism in humans with insuilin resistence and diabetes: potential role of PGC1 and NRF1. *Proc. Nat. Acad. Sci. USA*, **100**, 8466-8471.

[102] Mootha, V. K., Handschin, C., Arlow, D., Xie, X., St. Pierre, J., Sihag, S., Yang, W., Altshuler, D., Puigserver, P., Patterson, N., Willy, P.J., Schulman, I. G., Heyman, R. A., Lander, E. S. & Spiegelman, B. M. (2004). Errα and Gabpa/b specificy PGC-1α-dependent oxidative phosphorylation gene expression that is altered in diabetic muscle. *Proc. Nat. Acad. Sci. USA.*, **101**, 6570-6575.

[103] Modica-Napolitano, J. S. & Singh, K. K. (2002). Mithocondria as targets for detection and treatment of cancer. *Expert Rev. Mol. Med.*, **4**, 1-19.

[104] Graff, A., Clayton, A. & Larsson, A. G. (1999). Mitochondrial medicine-recent advances. *J. Intern. Med.*, **246**, 11-23.

[105] Yin, P. H., Lee, H. C., Chau, G. Y., Wu, Y. T., Li, S. H. & Lui, W. Y. (2004). Alteration of the copy number and deletion of mitochondrial DNA in human hepatocellular carcinoma. *Br. J. Cancer*, **90**, 2390-2396.

[106] Scheers, I., Bachy, V., Stephenne, X. & Sokal, E. M. (2005). Risk of hepatocellular carcinoma in liver mitochondrial respiratory chain disorders. *J. Pediatr.*, **146**, 414-417.

[107] Rustin, P. (2002). Mitochondria, from cell death to proliferation. *Nat. Genet.*, **30**, 352-353.

[108] Terrier, F., Vock, P., Cotting, J., Ladebeck, R., Reichen, J. & Hentschel, D. (1989). Effect of of intravenous fructose on the P-31 MR spectrum of the liver: dose response in healthy volunteers. *Radiology*, **171**, 557-563.

[109] Enzhmann, H., Ohlhauser, D., Dettler, T. & Bannasch, P. (1989). Enhancement of hepatocarcinogenesis in rats by dietary fructose. *Carcinogenesis*, **10**, 1247-1252.

[110] Koistinen, H. A., Chibalin, A. V. & Zierath, J. R. (2003). Aberrant p38 mitogen-activated protein kinase signalling in skeletal muscle from Type 2 diabetic patients *Diabetologia*, **46**, 1324-1328.

[111] Mori, M., Saitoh, S., Takagi, S., Obara, F., Ohnishi, H., Akasaka, H., Izumi, H., Sakauchi, F., Sonoda, T., Nagata, Y. & Shimamoto, K. (2000). A Review of Cohort Studies on the Association Between History of Diabetes Mellitus and Occurrence of Cancer. *Asian Pac. J. Cancer Prev.*, **1**, 269-276.

[112] Coleman, W. B. (2003). Mechanisms of human hepatocarcinogenesis. *Curr. Mol. Med.*, **3**, 573-588.

[113] Weinberg, A. G., Mize, C. E. & Worthen, H. G. (1976). The occurrence of heaptoma in the chronic form of hereditary tyrosinemia. *J. Pediatr.*, **88**, 388-434.

[114] Schulz, T. J., Tierbach, R., Voigt, A., Drewes, G., Mietzner, B., Steinberg, P., Pfeiffer, A. F. H. & Ristow, M. (2006). Induction of oxidative metabolism by mitochondrial Frataxin inhibits cancer growth. *J. Biol. Chem.*, **281**, 977-981.

[115] Calle, E. E. & Kaaks, R.(2004). Overweight, obesity and cancer: epidemiological evidence and proposed mechanisms. *Nat Rev Cancer.*, **4**(8), 579-91.

[116] Shureiqi, I. & Lippman, S. M. (2001). Lipoxygenase modulation to riverse carcinogenesis. *Cancer Res.*, **61**, 6307-6312.

[117] Setty, B. N., Dubowy, R. L., Stuart, M. J. (1987). Endothelial cell proliferation may be mediated via the production of endogenous lipoxygenase metabolites. *Biochem. Biophys. Res. Commun.*, **144**, 345-351.

[118] Gercel-Taylor, C., Doering, D. L., Kraemer, F. B. & Taylor, D. D. (1996). Aberrations in normal systemic lipid metabolism in ovarian cancer patients. *Gynec. Oncol.*, **60**, 35-41.

[119] Soto, A. M., Maffini, M. V. & Sonnenschein, C. (2008). Neoplasia as development gone awry: the role of endocrine disruptors. *Int J Androl.*, **31**(2), 288-93.

[120] Jones, R. H. & Ozanne, S. E. (2009). Fetal programming of glucose-insulin metabolism. *Mol Cell Endocrinol.*, **297**(**1-2**), 4-9.

[121] Pedersen, P. L. (1978). Tumor mitochondria and the bioenergetics of cancer cells. *Prog. Exp. Tumor Res.*, **22**, 190-274.

[122] Cuezva, J. M., Ostronoff, L. K., Ricart, J., de Heredia, L. M., Di Liegro, C. M. & Izquierdo, J. M. (1997). Mitochondrial biogenesis in the liver during development and oncogenesis. *J. Bioener. Biomem.*, **29**(4), 365-377.

[123] Capuano, F., Varone, D. & D'Eri, N. (1996). Oxidative phosphorylation and F(O)F(1) ATP synthase activity of human hepatocellular carcinoma. *Biochem Mol Biol Int.*, **38**, 1013-1022.

[124] Wang, T., Marquardt, C. & Foker, J. (1976). Aerobic glycolysis during lymphocite proliferation. *Nature*, **261**, 702-705.

[125] Sweetlove, L. J. & Fernie, A. R. (2005). Regulation of metabolic networks: understanding metabolic complexity in the systems biology era. *New Phytol.*, **168**(1), 9-24.

[126] Kacser, H. & Burns, J. A. (1973). The control of flux. *Symp. Soc. Exp. Biol.*, **27**, 65-104.

[127] Sthephanopoulos, G. & Valin, J. J. (1991). Network rigidity and metabolic engineering in metabolite overproduction. *Science*, **252**, 1675-1681.

[128] Bailey, J. E. (1999). Lessons from metabolic engineering for functional genomics and drug discovery *Nat. Biotechnol.*, **17**, 616-618.

[129] Huang, S. & Ingber, D. E. (2007). A non-genetic basis for cancer progression and metastasis: self-organizing attractors in cell regulatory networks. *Breast Dis.*, **26**, 27-54.

[130] Cucina, A., Biava, P. M., D'Anselmi, F., Coluccia, P., Conti F., Di Clemente, R., Miccheli, A., Frati, L., Gulino, A. & Bizzarri, M. (2006). Zebrafish embryo proteins induce apoptosis in human colon cancer cells (Caco2). *Apoptosis*, **11**, 1617-1628.

[131] Kasemeier-Kulesa, J. C., Teddy, J. M., Postovit, L. M., *et al.* (2008). Reprogramming multipotent tumor cells with the embryonic neural crest microenvironment. *Dev Dynam*, **237**, 2657-2666.

[132] Lee, L. M., Seftor, E. A., Bonde, G., Cornell, R. A. & Hendrix, M. J. C. (2005). The fate of human malignant melanoma cells trasplanted into zebrafish embryos: assesment

of migation and cell division in the absence of tumor formation. *Dev Dynam*, **233**, 1560-1570.

[133] Postovit, L. M., Maragaryan, N. V., Seftore, E. A., *et al.* (2008). Human embryonic stem cell microenvironment suppress the tumorigenic phenotype of aggressive cancer cells. *Proc Natl Acad Sci USA*, **105**, 4329-4334.

[134] Meadows, A. L., Kong, B., Berdichevsky, M., Roy, S., Rosiva, R., Blanch, H. W. & Clark, D. S. (2008). Metabolic and Morphological Differences between Rapidly Proliferative Cancerous and Normal Breast Epithelial Cells. *Biotechnol. Prog.*, **24**, 334-341.

[135] Virchow, R. L. K. (1859). Cellular pathology. special ed. London, UK, John Churchill; **1978**, 204-7.

[136] Bizzarri, M. (2008). Consequences of space exploration for mankind. *G Ital Nefrol.*, **25**(6), 686-689.

[137] Boonstra, J. (1999). Growth factor-induced signal transduction in adherent mammalian cells is sensitive to gravity *FASEB*, **13**, S35-S42.

[138] Carmeliet, G. & Bouillon, R. (1999). The effect of microgravity on morphology and gene expression of osteoblasts *in vitro FASEB*, **13**, S129-134.

[139] Pourati, J., Maniotis, A., Speigel, D., Schaffer, J. L., Butler, J. P., Fredberg, J. J., Ingber, D. E., Stamenovic, D. & Wang, N. (1998). Is cytoskeletal tension a major determinant of cell deformability in adherent endothelial cells? *Am. J. Physiol.*, **274**, C1283-C1289.

[140] Wang, N., Tytell, J. D. & Ingber, D. E. (2009). Mechanotransduction at a distance: mechanically coupling the extracellular matrix with the nucleus. *Nat Rev Mol Cell Biol.*, **10**(1), 75-82.

[141] Chen, C. S., Mrksich, M., Huang, S., Withesides, G. M. & Ingber D. E. (1997). Geometric control of cell life and death. *Science*, **276**, 1425-1428.

[142] Ingber, D. E. (1999). How cells (might) sense microgravity. *FASEB*, **13**, S3-S15.

[143] Mandelbrott, B. B. (1982). *The fractal geometry of the Nature*, W.H. Freeman. New York;.

[144] Rosai, J. (2001). The continuing role of morphology in the molecular age. *Modern Path.*, **14**, 258-260.

[145] Rangayyan, R. M. & Nguyen, T. M. (2007). Fractal analysis of contours of breast masses in mammograms. *J. Dig. Imag.*, **20**(3), 223-237.

[146] Rangayyan, R. M., El-Faramawy, N. M., Desautels, J. E. L. & Alim, O. A. (1997). Measures of acutance and shape for classification of breast tumours. *IEEE Trans Med Imag.*, **16**(6), 799-810.

[147] Mandelbrot, B. B. (1975). Stochastic models for the Earth's relief, the shape and the fractal dimension of the coastlines, and the number-area rule for islands. *Proc. Nat. Acad. Sci.U.S.A.*, **72**, 3825-3828.

[148] Cutting, J. E. & Garvin, J. J. (1987). Fractal curves and complexity. *Percept. Psicophys.*, **42**, 365-370.

[149] Smith, T. G., Lange, G. D. & Marks, W. B. (1996). Fractal methods and results in cellular morphology - dimensions, lacunarity and multifractals. *J. Neurosci Methods.*, **69**, 123-136.

[150] Losa, G. A., Merlini, D., Nonnenmacher, T. F. & Weibel, E. R. (2002). (Eds.) *Fractals in Biology and Medicine*, Birkhauser Verlag. Basel;.

[151] Baish, J. W. & Jain, R. K. (2000). Fractals and Cancer. *Cancer Res.*, **60**, 3683-3688.

[152] Cross, S. S. (1997). Fractals in pathology. *J. Pathol.*, **182**, 1-8.

[153] Cross, S. S., McDonagh, A. J. G., Stephenson, T. J., *et al.* (1995). Fractal and integer-dimensional analysis of pigmented skin lesions. *Am J Dermatol.*, **17**, 374-378.

[154] Claridge, E., Hall, P. N., Keefe, M., *et al.* (1992). Shape analysis for classification of malignant melanoma. *J Biomed Eng.*, **14**, 229-324.

[155] Gazit, Y., Berk, D. A., Leunig, M., Baxter, L. T. & Jain, R. K. (1995). Scale-invariant behavior and vascular network formation in normal and tumour tissue. *Phys Rev Lett.*, **75**, 2428-2431.

[156] Michaelson, J. S., Cheongsiatmoy, J. A., Dewey, F., *et al.* (2005). Spread of human cancer cells occurs with probabilities indicative of a nongenetic mechanism. *Br J. Cancer.*, **93**, 1244-1249.

[157] Tracqui, P. (2009). Biophysical model of tumor growth. *Rep. Prog. Phys.*, **72**, 1-30.

[158] Landini, G. & Rippin, J. W. (1996). How important is tumour shape? Quantification of the epithelial connective tissue interface in oral lesions using local connected fractal dimension analysis. *J Pathol.*, **179**, 210-217.

[159] Folkman, J. & Moscona, A. (1978). Role of cell shape in growth control. *Nature*, **273**, 345-349.

[160] Ingber, D. E. (2005). Mechanical control of tissue growth: function follows form. *Proc Natl Acad Sci USA*, **102**(33), 11571-11572.

[161] Scheck, F. (1990). *Mechanics*, Springer. Verlag, Heidelberg, Germany;.

[162] Toussaint, O. & Schneider, E. D. (1998). The thermodynamics and evolution of complexity in biological systems *Comp. Biochem. Physiol.*, **120**, 3-9.

[163] Ingber, D. E. (2008). Can cancer be reversed by engineering the tumour microenvironment? *Sem Cancer Biol.*, **18**, 356-364.

[164] D'Anselmi, F., Valerio, M., Cucina, A., Galli, L., Proietti, S., Dinicola, S., Pasqualato, A., Manetti, C., Ricci, G., Giuliani, A., Bizzarri, M. Metabolism and cell shape in cancer: a fractal analysis. *Int J Biochem Cell Biol.* (in press).

[165] Bowie, J. E. & Young, I. T. (1977). An analysis technique for biological shape. *Acta Cytol.*, **21**, 739-746.

[166] Cesar, R. M. Jr. & Costa, L. & da F. (1997). The application and assessment of multiscale Bending Energy for Morphometric characterization of neural cells. *Rev. Sci. Instrum.*, **68,** 2177-2186.

[167] Castleman, K. R. (1996). *Digital Image Processing*, Prentice-Hall. NJ, Engelewood Cliffs;.

[168] Lelièvre, S. A., Weaver, V. M., Nickerson, J. A., Larabell, C. A., Bhaumik, A., Petersen, O. W. & Bissell, M. J. (1998). Tissue phenotype depends on reciprocal interactions between the extracellular matrix and the structural organization of the nucleus. *Proc Natl Acad Sci U S A.*, **95**, 14711-14716.

[169] Thomas, C. H., Collier J. H., Sfeir C. S. & Healy, K. E. (2002). Engineering gene expression and protein synthesis by modulation of nuclear shape. *Proc Natl Acad Sci* U S A, **99**, 1972-1977.

[170] Guilak, F. (1995). Compression-induced changes in the shape and volume of the chondrocyte nucleus. *J Biomech.*, **28**, 1529 -1541.

[171] Zink, D., Fischer, A. H. & Nickerson, J. A. (2004). Nuclear structure in cancer cells. *Nat Rev Cancer*, **4**, 677- 687.

[172] Paszek, M. J., Zahir, N., Johnson, K. R., Lakins, J. N., Rozenberg, G. I., Gefen, A., Reinhart-King, C. A., Margulies, S. S., Dembo, M., Boettiger, D., Hammer, D. A. & Weaver, V. M. (2005). Tensional homeostasis and the malignant phenotype. *Cancer Cell*, **8**, 241-254.

[173] Wolf, K. & Friedl, P. (2006). Molecular mechanisms of cancer cell invasion and plasticity. *Br J Dermatol.*, **154**, 11-15.

[174] Dahl, K. N., Ribeiro, A. J. S. & Lammerding, J. (2008). Nuclear Shape, Mechanics, and Mechanotransduction. *Circ Res.*, **102**, 1307-1318.

[175] Prigogine, I. & Wiame, J. M. (1946). Biologie et Thermodynamique des phenomenes irreversibles. *Experientia*, **2**, 451-453.

[176] Zotin, A. I. (1990). *Thermodynamic bases of biological processes: physiological reactions and adaptations*. Walter de Gruyter. Berlin,.

[177] Zotin, A. A. & Zotin A. I. (1997). Phenomenological theory of ontogenesis. *Int J Dev Biol.*, **41**, 917-921.

[178] DeBerardinis, R., Lum, J. J., Hatzivassiliou, G. & Thompson, C. B. (2008). The Biology of Cancer: Metabolic Reprogramming Fuels Cell Growth and Proliferation. *Cell Metabolism*, **7**, 11-20.

[179] Cascante, M., Boros, L. G., Comin-Anduix, B., de Atauri, P., Centelles, J. J. & Lee, P. W. N. (2002). Metabolic control analysis in drug discovery and disease. *Nat. Biotech.*, **20**, 243-249.

In: Metabolomics: Metabolites, Metabonomics... ISBN: 978-1-61668-006-0
Editors: J.S. Knapp and W.L. Cabrera, pp. 121-161

Chapter 3

FROM METABOLIC PROFILING TO METABOLOMICS: FIFTY YEARS OF INSTRUMENTAL AND METHODOLOGICAL IMPROVEMENTS

Chiara Cavaliere, Eleonora Corradini, Patrizia Foglia, Riccardo Gubbiotti, Roberto Samperi and Aldo Laganà *
SAPIENZA Università di Roma, Rome, Italy

Abstract

Molecular biology has recently concentrated on the determination of multiple gene-expression changes at the RNA level (transcriptomics), and into determination of multiple protein expression changes (proteomics). Similar developments have been taking place at metabolite small-molecule level, leading to the increasing expansion in studies now termed metabolomics. This approach can be used to provide comprehensive and simultaneous systematic profiling of metabolite levels in biofluids and tissues, and their systematic and temporal changes. Analysis of metabolites is not a new field; long prior to the development of the various ''omics'' approaches, the simultaneous analysis of the plethora of metabolites seen in biological fluids had been carried out largely, but historically it has been limited to relatively small numbers of target analytes. However, the realization that metabolic pathways do not act in isolation but rather as part of an extensive network has led to the need for a more holistic approach to metabolite analysis.

The main analytical techniques employed for metabolomics studies are based on NMR spectroscopy and mass spectrometry (MS), that, in turn, can be considered complementary each other. Neverthless, MS measurement following chromatographic separation offers the best combination of sensitivity and selectivity, so it is central to most metabolomics approaches. Either gas chromatography after chemical derivatization, or liquid chromatography (LC), with the newer method of ultrahigh-performance LC being used increasingly, can be adopted. Capillary electrophoresis coupled to MS has also shown some promises. Analyte detection by MS in complex mixtures is not as universal as for NMR and quantitation can be impaired by variable ionization and ion-suppression effects. A LC chromatogram is generated with MS detection, usually using electrospray ionization (ESI),

* E-mail address: aldo.lagana@uniroma1.it. Phone: +39-06-49913679 Fax: +39-06-490631. Dipartimento di Chimica, SAPIENZA Università di Roma, Box n° 34 - Roma 62, Piazzale Aldo Moro 5, 00185 Rome, Italy. (Corresponding author)

and both positive- and negative-ion chromatograms can be recorded. The utilization of nano-ESI can reduce ionization suppression effects due to the increased ionization efficiency. Mass analyzer able to produce high mass resolution, mass accuracy, and tandem MS, such as quadrupole-time-of-flight (Q-TOF) or high-resolution ion trap instruments, are employed.

Direct infusion (DI)-MS/MS using Fourier transform ion cyclotron resonance mass spectrometers provides a sensitive, high-throughput method for metabolic fingerprinting. Unfortunately, DI-MS analysis is particularly susceptible to ionization suppression arising from competitive ionization. In metabolomics, matrix assisted laser desorption-ionization (MALDI) has largely been confined to the targeted analysis of high-molecular weight metabolites due to the substantial signals generated by the matrix in the low-molecular-weight region (<1,000 *m/z*). Recent advancements in laser desorption techniques include desorption-ionization MS from porous silicon chips and matrices that have minimal background signals in the low-molecular-weight region. These offer new opportunities for the utilization of MALDI ionization in metabolite screening and fingerprinting employing MALDI-TOF/TOF. However, the technique is still subject to ion suppression and yields poor quantitative detection. Desorption ESI (DESI), a new ambient, soft-ionization technique that combines features from both ESI and desorption-ionization methods, allows the direct analysis of animal and plant tissues. However, DESI experimental conditions typically require optimization for each sample type, so time must be invested initially in optimizing the experimental parameters.

Introduction

In the post-genomic era, the attention of molecular biology has concentrated more and more on the determination of multiple gene-expression changes at the RNA level, (transcriptomics), the determination of multiple protein expression changes in a cell or tissue (proteomics), and at the small-molecule metabolite level (metabolomics). The general aim of these techniques is to gain new insights and a better understanding of the biological functioning of a cell or organism [1,2]. Their practical applications include virtually all aspects of the system biology of cellular (and also sub-cellular) compartments, and complex organisms, ranging from the identification of differences between certain sets of organisms (e.g., differences in genotypes) to the identification of differences between subjects affected by various diseases and disease-free controls, or the elucidation of factors that influence biochemical events following external stimuli such as exposure to environmental toxins or other stressors.

The main limitation associated with interpreting transcriptomics and proteomics is often the difficulty of relating observed gene-expression fold changes or protein-level (not activity) changes to conventional disease and pharmaceutically relevant end-points. In other words, changes in the transcriptome and proteome do not always result in altered biochemical phenotypes (the metabolome) [3,4]. It has been suggested that metabolomics, among the 'omics' technologies, may in fact provide the most "functional" information [3]. The metabolome can be considered as the final stage in the chain of events from genes to metabolism, and the metabolic phenotype is the most direct reflection of the actual state of a biological system. Metabolites have a well-defined function in the life of the biological system and are also contextual [5], reflecting the surrounding environment. Thus, quantitative global analysis of endogenous metabolites from cells, tissues, fluids, etc. is becoming an integral part of functional genomics effort [4,6,7] as well as a tool for discovering diagnostic biomarkers [8-11].

As reported by D. Ryan and K. Robards [12], terminology relating to metabolomics has been (and is still) controversial. The term "metabolome" was first used by Olivier *et al.* in 1998 [13] to describe the entire set of metabolites synthesized by an organism, on analogy of "genome" and "proteome". More recently, this definition has been limited to "the quantitative complement of all of the low molecular weight molecules present in cells in a particular physiological or developmental state" [14]. The term "metabolomics" was coined by O. Fiehn and defined as a comprehensive analysis in which all metabolites of a biological system were identified and quantified [15]. The confusion in the terminology arises from the similar term "metabonomics", which was coined earlier by J.K. Nicholson *et al.* [16]. Later, metabonomics has been described as a subset of metabolomics [17] which, in contrast, seeks to measure those metabolites which change in response to a stimulus of one sort or another. Although the authors who first used the term metabonomics accepted Fiehn's distinction [18], the two terms have been often used interchangeably. Whatever the accepted definition is, the two fields employ similar methodologies and have the common aim of analyzing the metabolome. Hence, most of the literature supports the use of the term metabolomics to describe a comprehensive, non-targeted analytical approach that is universally applicable to identify and quantify all metabolites of a biological system and the term metabonomics will no longer be used in this chapter.

Metabolomics is a rapidly maturing field: it is increasingly being applied to study biological systems including microorganisms [19,20], plants [21,22], mammals [23-26], and environment [27,28]. Metabolomes are complex systems composed of hundreds or thousands of metabolites (1,168 for yeast [29], 200,000 for plants in the total vegetable kingdom [17], >6,500 for mammals [30]) with a wide range of physical and chemical properties, and a large dynamic range. The study of these systems requires an integrated approach or metabolome pipeline [31] and a number of strategies are applied [32,33]. From an analytical perspective, metabolomics is a huge analytical challenge that needs the ability to perform high-throughput experiments with relatively low operating costs after the initial investment for the purchase of the necessary instruments [32].

Nuclear magnetic resonance spectroscopy (NMR) and mass spectrometry (MS) are the primary analytical techniques used for the identification and quantification of a large set of metabolites present in a given biological system [34-39]. Each technology shows some advantages and they are essentially complementary [37,38]. Although ^{1}H and ^{13}C NMR are capable of measuring most aspects of the metabolome, the extremely large dynamic range typically encountered in biological systems added to the difficulties in coupling NMR to chromatography, are a drawback for NMR since important aspects of the metabolome composition may potentially go unmeasured. Thus, because of its high sensitivity, the capability to analyze highly complex samples (especially when hyphenated with chromatographic separations) and a good dynamic range throughout its measurement, the MS-based approaches have started to take the leadership in metabolomic research. This chapter will be focused only on MS based methodologies.

Because of the huge amount of data produced by a single metabolomics experiment, computational tools are crucial for analyzing the data. A visual inspection of the results can quickly reveal errors, so it is often needed to validate the output of analysis. On the other hand, comparing such complex multi-dimensional data manually to find corresponding signals is a tedious task, as each experiment usually consists of thousands of individual scans, each containing hundreds or even thousands of distinct signals. Long before the development of the various ''omics'' approaches, the simultaneous analysis of the plethora of metabolites seen in biological fluids had been carried out largely by MS and it was shown that these complex data sets could be interpreted using multi-variate statistics [40]. Now MS instrument software could be used for most of these tasks, but this has several drawbacks. Instrument software is generally expensive, and most of the instrument softwares cannot import data from other instruments of other brands. Therefore, different software has to be used for different data sets. A viable alternative could be freely available, open source viewers (see for example [41]). In parallel, bioinformaticians have attempted to compute the size of the metabolome on the basis of genome information [42]; however, such approaches are considerably restrained by the quality of genome annotation. To gain ultimate value from large-scale experiments the data and associated metadata that they produce need to be incorporated in databases. Large databases of NMR spectra are already available, while the situation for MS is not so advanced; however, recent years have brought major developments and a large number of web-based tools that can be of great help in the interpretation of chromatography/MS data are now currently available.

Origins and Development: Looking Back

Analysis of metabolites is not a new field; but historically it has been limited to relatively small numbers of target analytes as in the study of a particular metabolic pathway. However, in the long run, the realization that metabolic pathways do not act in isolation but rather as part of an extensive network has led to the need for a more holistic approach to metabolite analysis. Historical approaches in MS-based methods include metabolite profiling, metabolite fingerprinting, and target analysis. Metabolite profiling involves the identification and quantitation of a predefined set of metabolites of known or unknown identity and belonging to a selected metabolic pathway [15,43]. The aim of metabolite fingerprinting is the rapid classification of numerous samples using multivariate statistics, typically without differentiation of individual metabolites or their quantitation. Target analysis is limited exclusively to the qualitative and quantitative analysis of a particular metabolite or metabolites. As a result, only a very small fraction of the metabolome is focused upon, signals from all other components being ignored [44]. Because of their nature, these approaches provide a restrictive non-comprehensive view of the metabolome. Nevertheless, metabolite profiling represents the oldest and most established approach and can be considered the precursor for metabolomics.

The concept that individuals might have a "metabolic pattern" that would be reflected in the constituents of their biological fluids was first developed and tested by Roger Williams and his associates during the late 1940s and early 1950s. Utilizing data from over 200,000 paper chromatograms, many run with techniques developed in his own laboratory for this purpose, Williams was able to show convincingly that the taste thresholds and the excretion patterns for a variety of substances varied greatly from individual to individual [45]. The work of Williams and his group, however, was apparently not duplicated by others, hence his ideas about the utility of metabolic pattern analysis remained essentially dormant until the late 1960s, when gas chromatography (GC) [46-49] and liquid chromatography (LC) [50,51] were sufficiently advanced to allow such studies to be carried out with considerably less effort. The concept of metabolite fingerprint as specie-specific GC profile reflecting different metabolic patterns was reported the first time by R. Kuntzman in 1966 [52]. The concept of metabolic profiles was introduced finally by E.C. and M.G. Horning [53,54], who coined the term to refer to qualitative and quantitative analyses of complex mixtures of physiological origin "metabolic profiles are multicomponent GC analyses that define or describe metabolic patterns for a group of metabolically or analytically related metabolites." At the beginning; researchers involved in the field were not aware of the difference between metabolite (or metabolic) profiling and metabolic fingerprint, owing the enormous difficulties in obtaining quantitative data, difficulties now solved with the development of computerized data handling.

From an instrumental point of view, although metabolomics is a relatively new term, its origin can be traced back to 1956-59, when Golay presented his lecture on the "Theory of chromatography in open and coated tubular column with round or rectangular cross-sections" at symposium on GC held in Amsterdam [55], Zlatkis developed Golay's idea [56], Beynon realized the high resolution MS of organic compounds [57,58], and Gohlke coupled GC with a Time of Flight (ToF) mass spectrometer [59]. Almost from the birth of GC, people involved in organic MS saw the potential advantage of separating a complex mixture into its components followed by structural analysis by MS. As investigation toke place it was evident that GC-MS was different from both GC and MS. Three major hurdles had to be overcome: i) the large amount of gas leaving the column (working with packed columns), while MS separates the ions in high vacuum condition; ii) the need for rapid mass spectral acquisition; iii) the enormous amount of data collected during a GC-MS analysis.

The first problem was solved by a device named "jet separator" [60] that eliminates selectively most of the carrier gas. Undoubtedly the open tubular columns were much more amenable for GC-MS, owing the much smaller flow-rate, but their use become popular only in the mid '70s, after Horning's group developed a method for the preparation of thermostable capillary columns which provided extremely high resolution [54]. When GC-MS was at its beginning, magnetic sector mass spectrometer did not have a rapid data acquisition capability as the ToF instruments which, on the other hand, presented other problems, and low resolution. The first magnetic sector instrument built as a GC-MS was commercially available in the mid '60s; although the problem of rapid acquisition was solved, this instrument did not tackle the issue of the amount of data acquired in a GC-MS analysis. It used a light beam oscilloscope to record the mass spectra that were manually selected for recording. Minicomputers were also developed in the mid '60s, and in few years the automated collection of GC-MS data became possible. The contribution of Hites and Biemann [61-63] was particularly important. The generation of several different types of chromatograms was made possible by processing the *m/z* signal and its intensity recorded by the data system: the reconstructed total ion current, the specific *m/z* value current and the MS spectrum recorded in a certain scan time.

GC-MS began to become a routine technology with the introduction of quadrupolar instruments. Quadrupole technology, which includes the transmission quadrupole (TQ) and the quadrupole ion trap (QIT), was explored by Paul [64]. The first GC-QTMS was developed in the late '60s and rapidly substituted the magnetic sector based instruments [65], owing to the simplicity of operation and the continued advancement of the data station. The TQ mass spectrometer was very well suited for GC-MS: compared to a magnetic sector it was smaller in size and much more suitable for rapid scanning and electronic control. The limit of TQ instruments was (and still is) the fact that they do not consent high resolution and accurate mass measurements, since, throughout the scan range, ions with a certain *m/z* are resolved from those of *m/z* +1. The coupling of high resolution (HR) GC to HR mass spectrometry (HR-MS) was realized for the first time by the Burlingam's group in 1975 with a double focalization magnetic-electrostatic analyzer [66]. Although the QIT analyzer was born from the same Paul research that produced TQ, its history is much more complicated owing the intrinsic complexity of the ion motion inside it, and the first GC-QITMS was commercially viable at the end of 1985 [67]. QITs were initially developed as low-resolution, low-mass detectors for use with GC, but their performance has been considerably improved and several versions are available with enhanced resolution. The use of QIT was promoted by their capability of operating in multiple MS (MS^n). In the meanwhile, there had been a significant hardware development with the introduction by Yost and Henke in 1978 [68] of the triple quadrupole (QqQ), that included two quadrupole mass analyzers (Q) separated by a quadrupolar not-separating collision cell (q). Precursor ions separated by the first Q were converted to fragments by collision with a gas molecule in q, and the second Q separates the ions produced. Tandem MS (MS^2), and MS^n will become the key of compound identification after separation by high performance LC (HPLC).

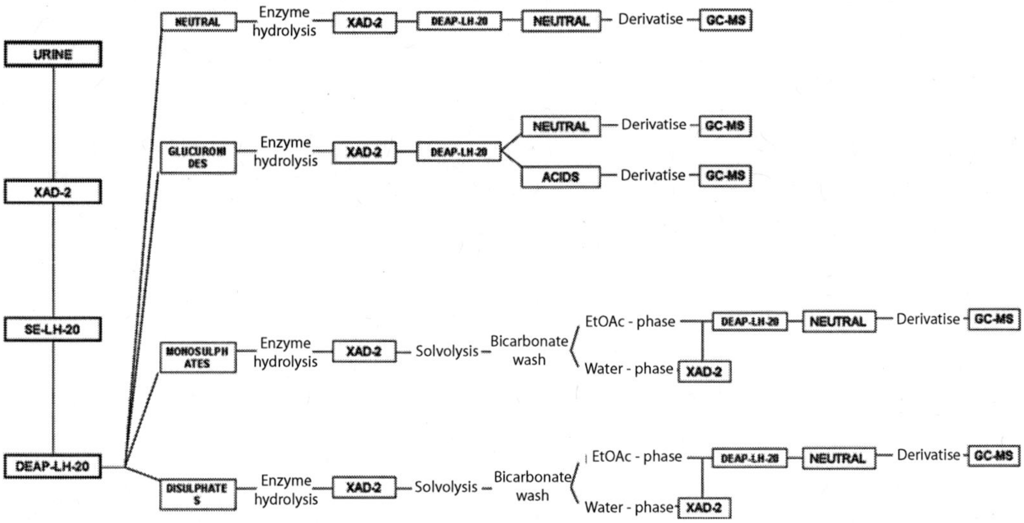

Figure 1. General scheme for the analysis of urinary steroid conjugates including group separation. XAD2: Amberlite XAD2 poly(styrene-divynilbenzene) resin. SE-LH-20: Sulphoethyl Sephadex LH-20 cation exchange resin. DEAP-LH-20: Diethylaminohydroxypropyl Sephadex-LH-20. Seventy-seven different deconjugated steroid metabolites (51 glucuronides, 44 monosulphates, and 22 disulphates) were detected by GC-MS from 25 mL male urine as metoxime-trimetylsylil derivates. (Reproduced from reference 74 by permission.)

Early applications of metabolic profiling by GC and GC-MS were in the field of volatile metabolites in serum, plasma, urine, and breath [69-72]; steroidal hormones and their metabolites in plasma and urine [53,73,74]; organic acids in urine [75-81]; amino acids in urine [82]; aliphatic alcohols [83]. However, some attempts were also made to obtain multiclass metabolic profiles, [54,66,77] by using sample preparation/fractionation and a suitable derivatization step for non-volatile compounds. Almost the whole body of literature concerned the detection of human disease, whereas studies regarding plants were rather rare

[84,85]. During the early '70s, despite the high resolution that could be achieved with capillary GC columns, the profiles thus obtained were exceedingly complex, making identification and quantitative analysis of individual peaks correspondingly more difficult because, even with the highest available resolution, capillary GC columns do not completely separate all components of physiological fluids. This problem was tackled by a spectra library search for matches [86-88] and quantitation based on mass chromatogram areas relative to that of an internal standard [89,90].

Limited available technology was often compensated by the skillfulness of researchers as illustrated in Figure 1 [74]. Seventy-seven steroids metabolites were determined in a male urine sample by GC-MS after extraction, group separation, deconjugation, clean up and derivatization. Steroid metabolites were extracted from urine (25 mL) by solid phase extraction (SPE) with a column of poly(styrene-divynilbenzene) XAD-2 resin, cationic compounds were eliminated by cation exchange, and neutral, glucuronides, sulphates and disulphates fractionated by an anionic exchange resin. Free steroids were obtained by enzymatic hydrolysis (followed by solvolysis for sulphates), extracted by SPE and cleaned up by the anion exchange column. Each fraction was derivatized by mothoxyamine and trimethylsilylimidazole and analyzed separately.

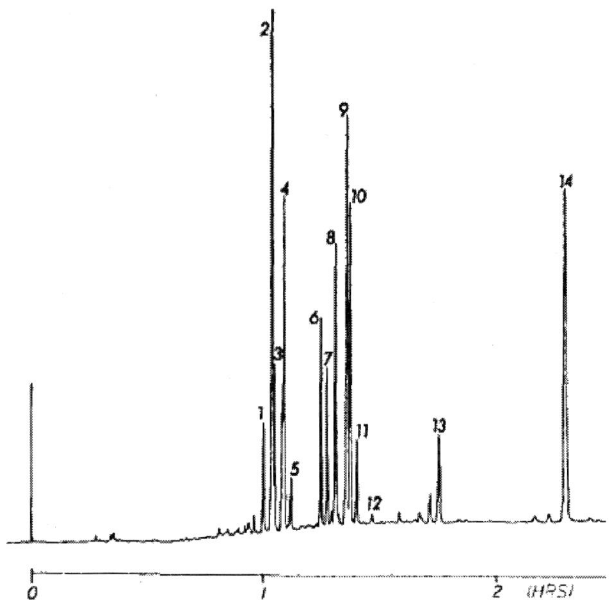

Figure 2. Separation of 14 benzoylated steroid standards. The column used was a fused silica capillary (1 m X 0.24 mm id.) packed with 3 μm bonded spherical particles. Flow rates used for the separations were approximately 1 μL/min. Stepwise gradient conditions: 80% acetonitrile (ACN)/H,O (15 min); 85% ACN/H_2O (14 min); 90% ACN/H_2O (15 min); 95% ACN/H_2O (18 min); 100% ACN (held). Key: (1) 11-hydroxyandrosterone; (2) 11-hydroxyetiocholandone; (3) allotetrahydrocortisol; (4) tetrahydrocortisol; (5) tetrahydrocortisone; (6) β-ortolone; (7) β-cortol; (8) α-cortolone; (9) α-cortol; (10) etiocholanoione; (11) androsterone; (12) dihydroepiandrosterone; (13) pregnanetriol; (14) androstandiol.

The major limitation of GC identification is the need for thermostable, volatile analytes; derivatization of the polar functional group can improve volatility, but a derivatiation step introduces bias and it is not always possible. This limitation is overcome by LC, which is virtually suitable for the separation of all kind of molecules. The modern HPLC started with the introduction of stationary phases chemically bonded on a silica surface [91,92]; however, two limitations delayed the widespread use of HPLC in metabolic profiling and metabolomics. The first was the about one order of magnitude lesser efficiency than GC; the second were the initial difficulties on coupling with MS. The solution of the first problem was tackled at the beginning of the '80s by the Novotny research group which obtained micropacked columns as efficient as the GC columns [93,94]. Drawbacks of this technology were the long analysis times (2 to 3 hours) and the reproducibility of columns. Figure 2 shows the separation of 14 benzoylated steroid standards originally reported in reference [93].

Interfacing LC with MS is not so straightforward as GC-MS. The primary problem was the elimination of solvent while preserving sufficient amounts of analyte, as the liquids increased 500-1000 times their volume in the vapor phase; the second arises from the fact that many analytes are minimally volatile and may also be thermally labile. The road which led to a successful coupling of LC to MS was long and meandering [95]. After many attempt, a technological success for an interface between LC and MS was attained with the Atmospheric Pressure Chemical Ionization (APCI) developed by the Horning group [96] and the EletcroSpray Ionization (ESI), introduced by J.A. Fenn [97]. Both interfaces coincide, in physical place, with the respective ion sources, but the mechanisms are different. To these interfaces/sources the Matrix Assisted Laser Desorption-Ionization (MALDI) source, developed by Hillenkamp and Karas [98] should be added to complete the triad of desorption-ionization sources. ESI is at present time the most used ionization method in metabolomics, mostly because of the range of analytes that can be ionized [99]. Although MALDI is better suited for analysis of compounds having molecular weight >1 kDa, and cannot interfaced with LC, recent developments in this technique offer exciting new opportunities for the utilization of MALDI ionization in metabolite screening and fingerprinting. With ESI, APCI and MALDI, ionization in the positive-ion mode is via proton addition to give $[M+H]^+$ ions, or via the attachment of some other cation C^+ to give $[M+C]^+$ ions. By reversing the polarity of the ion-source, ionization can be achieved in the negative-ion mode; this is usually accomplished by the loss of a proton to give $[M-H]^-$ ions. ESI can give multiply charged ions for molecules having more than one ionizable site, whereas APCI and MALDI give substantially monocharged ions.

MALDI development renewed the interest in the ToF-MS, because this ion analyzer does not present any upper limitation in the *m/z* range that can be analyzed. This interest resulted in new developments such as improved resolving power and very rapid acquisition of data. The need for instruments showing high resolution, large *m/z* acquisition range and MS^n capability also promoted the upgrade of old instruments, such as the Fourier Transform Ion Cyclotrone Resonance (FT-ICR) ion trap developed by Marshall and Comisarow in the middle '70s [100], and the introduction of new ones such as the linear-QIT [101], the electrostatic (orbi)trap [102], and the hybrid Q-ToF [103].

Although capillary electrophoresis (CE) was introduced in 1981 as a high performance separation technique [104], and the first successful coupling of CE with MS was reported in 1987 [105], only at the end of the '90s it was applied to metabolic profiling [106]. This was probably due to the limited loadability of CE that poses high demands on the sensitivity of the detector.

Chromatographic Separation Techniques – Mass Spectrometry

GC-MS

The introduction in 1979 of fused-silica capillary columns resulted in higher resolution, higher efficiency, better reproducibility, and smaller sample size [107] than ever before, and during the '80s and '90s open tubular column technology improved even more significantly. Also GC-MS coupling experienced remarkable improvements with the introduction of QIT instruments and more sensitive TQ instruments. Most of the literature regarding metabolic profiles published till 1999 deals with analysis of human body fluids for biomedical investigations and diagnostics [108,109], with particular attention devoted to the biochemistry of steroids [110-113]. Surprisingly, only few studies published in these twenty years deal with plants and microorganisms, such as fungi and bacteria [85,114-117]. This tendency was reversed with the passage from the 20th to the 21th century.

The idea of metabolomics, born in the early 21th century [15], was the evolution of the concept of metabolic profiling, focusing on an improved understanding of biological networks by systematic and comprehensive analysis of metabolism. Contrarily to metabolic profiles early history, metabolomic research initially focused on plants but rapidly expanded to other areas. The large increase in the number of reports since 2002 goes to show the level of maturity of GC-MS, that lends itself to be used for a large variety of biological investigations [118]. Although such a comprehensive coverage is not yet possible, significant advancements in the large-scale GC-MS profiling of metabolites have been achieved and offer unique insight into the metabolic biochemistry of organisms. Today, GC-MS-based metabolite profiling in plants is regarded as a standard tool in plant research and is routinely applied in a variety of laboratories; compared to biomedical research or microbiology, plant-science papers still form the majority of published papers on GC-MS metabolite profiling. Recent metabolomics investigations of clinical interest using GC-MS include the development of analysis strategies for the plasma metabolome [119], urinary metabolite profiling [120], and biomarker investigation in disease such as heart failure [24], pre-eclampsia [121], diabetes [122], ovarian and kidney cancer, using carcinoma tissue [123,124].

Sample Preparation

Accurate determination of metabolite levels by GC-MS requires well-validated procedures for sampling and sample treatment. However, despite half a century of experience, the procedures used for biological sample preparation remain an issue. Quantitative extraction of all the metabolites in a biological sample would require multiple extractions with different solvent systems. Sample preparation for metabolomic studies depends not only on the method of analysis, but also on the type of sample being analyzed, and whether specific metabolites are of interest, or the profiling of all metabolites. For example serum and plasma contain proteins, glycoproteins, and lipoproteins; urine contains a high concentration of salts and urea; while plants contain a large amount of polymeric insoluble compounds.

Blood plasma contains a wide variety of chemically diverse low molecular weight substances, which vary widely in concentration and stability and are non-covalently bound to proteins, thus a protein precipitation step is introduced using an organic solvent, heat, or acid. A factorial experimental design was used to test the deproteinization and extraction efficiency of five organic solvents commonly used for serum metabolomics (methanol, ethanol, acetonitrile, acetone and chloroform), and a mixture of these solvents [119]. The results of the study suggested that methanol alone was the best of the tested solvents for extracting the metabolites quantitatively, then the optimal sample/solvent ratio was also determined. From the results of the designs, an extraction method was developed in which 100 μL of blood plasma is extracted with a 900 μL mixture of methanol and water (8:1 v/v) containing all the internal standards, followed by centrifugation. Another study underlined the fact that often only a slow protein precipitation can avoid unwanted sample loss, and suggested the use of acetonitrile, added slowly at 8 °C until the acetonitrile:plasma ratio was 8:2 (v/v) [125].

Metabolic profiles of urine have been studied since the '60s. The sample preparation was very laborious, involving extraction and fractionation to obtain fractions containing different metabolite classes [74,77]. With the much more reliable and sensitive technology now available, several GC-MS-based analytical methods have been developed for the metabolic profiling of compounds belonging to different chemical classes in urine samples [120,126-129]. These methods were derived from the study by Shoemaker [130] which eliminates excess urea by urease, and urease excess by ethanol precipitation; then the sample is evaporated and derivatized for GC-MS

Extraction protocols on plant tissues focused on the integration of metabolite levels with protein and transcript data use ternary solvent compositions at low temperatures [131]. Plant samples were harvested, immediately frozen in liquid nitrogen, crushed and extracted with the solvent. Solubilized metabolites were recovered after centrifugation, while proteins remained in the pellets. A range of metabolomic applications utilize very similar protocols [132-135]; a double step extraction (methanol/water followed by the ternary mixture) could be used to separate polar from non-polar compounds.

Applications in microbial biology often focus on metabolic engineering with emphasis on primary metabolism. The analysis of microbial samples is challenging: large amounts and numbers of components derived from the growth medium and the buffer used for quenching may be present, and their concentration may vary significantly from sample to sample, for instance, when comparing microorganisms grown on different growth media or harvested at different times during growth. Due to the high concentrations, these matrix compounds can be a potential disturbance during derivatization or analysis and influence the performance of the complete analysis. Interestingly, different protocols for microbial sample preparations were suggested regarding the optimal temperature required to achieve a fast quenching of metabolism and efficient metabolite extraction [136-141]. Very recently, quantitative extraction techniques of intracellular metabolites have been compared [142], and boiling ethanol or a chloroform/methanol mixture were found to give the best performance in terms of recovery and precision.

GC-MS Analysis

A basic requirement for GC-MS analysis is analyte volatility and thermal stability. Few metabolites meet these requirements, however the majority of metabolites can be made volatile through chemical derivatization prior to GC-MS analysis. Relatively little work has been performed on improving derivatization reactions for GC-MS-based metabolite profiling [143]. The most commonly utilized derivatizing procedure for GC-MS metabolite profiling includes a two-step derivatization scheme. The first step uses alkoxyamines to convert carbonyl groups to oximes in order to stabilize the reducing sugars in the open-chain conformation and also to prevent the decarboxylation of α-ketoacids. The second step replaces the active hydrogen in polar functional groups, such as carboxylic acid, alcohols and amines, with a trimethylsilyl group using N-methyl-N-trimethylsilyltrifluoroacetamide. This scheme is essentially the same as that used by the metabolic profiling pioneers. Other derivatization reactions, such as alkylation and esterification, derivatize a narrower range of metabolites than silylation. Recently, the dialkildithioacetal acetate derivatives to overcome current limitations in flux analysis of sugars [144], and derivatization of urine samples with ethyl cloroformate [127] were suggested.

GC-MS using electron impact (EI) ionization coupled to quadrupole analyzer combines very high separation power and reproducible retention times with a versatile, sensitive, and selective mass detection. As the full scan response of the EI ionization mode for quadrupole instruments is approximately proportional to the amount of compound injected, i.e., more or less independently of the compound, all compounds suitable for GC analysis are detected non discriminatively. This makes the technique very suitable for comprehensive analysis of a wide range of metabolites. Also the assignment of the identity of peaks detected with GC-MS using EI ionization via a database of mass spectra is straightforward, due to the extensive and reproducible fragmentation patterns obtained. If the MS spectrum is not present in the database, the fragmentation pattern can be used to obtain more information about the identity or compound class of a metabolite.

Quadrupole MS provides high sensitivity and large dynamic range, but low resolution, only nominal mass accuracy and relatively slow scan speeds. The most abundant metabolites suffer least from spectral overlapping, while low-abundant or novel metabolites require efficient separation for positive detection and structural characterization. 50,000–100,000 theoretical plates are regularly achieved in GC separations however, depending on the complexity of the sample, more than 1000 metabolites may be present in detectable quantities in a given sample. Most recently, using ToF mass analyzers an acquisition rate of 10-20 Hz can be routinely used [145]. Such data provide the possibility of deconvolving the mass spectra of closely eluting chemical species if the spectra are sufficiently distinct.

Average mass spectral purity for such a number of peaks is dramatically improved if two-dimensional GC is used for separation. Comprehensive 2D-GC was first introduced by Liu and Phillips in 1991 [146]. It is an online method in which the entire effluent from the first column is sent to the second column [147]. This kind of technique is especially useful in global metabolomic studies. The use of comprehensive 2D-GC offers a multiplicative increase in peak capacity by combining two columns with orthogonal separation characteristics by means of a thermal modulator, which focuses the effluent from the first column periodically in small segments that are then transferred to the second column. Thermal modulation carries the additional benefit of creating narrow second dimension peaks and, thereby, increasing peak heights that increase detection sensitivity [148-150]. The enhanced peak capacity and sensitivity make 2D-GC-ToF-MS highly suited for metabolic fingerprinting. Some reports have been published on the use of this technique for metabolomic purposes [158]; however, a range of practical problems remain before comprehensive 2DE-GC separations can become routine applications for metabolically complex samples. Modulation-period times inevitably reduce some of the chromatographic resolution that is achieved in the first dimension, so first-dimension retention times are less well defined than in truly one-dimensional separations. In addition, existing software solutions for peak picking, integration and alignment can not yet cope with the issues over data export and analysis. Among the possible solutions proposed there are those by Shellie *et al.* [152] and Almstetter *et al.* [153].

LC-MS

Despite some limitations, LC-MS have the potential to become the packhorse of metabolomic analysis, largely because of the availability of the technology, and the ready compatibility of reversed-phase (RP) separations with biological samples. The widespread use of LC-MS for global metabolic profiling is relatively recent, but over the past few years there has been a rapid and continuing increase in the number of publications based on this approach [38,154-155]. LC is a more universal separation technique than GC, and can be tailored for the targeted analysis of specific metabolite groups or utilized in a broader non-targeted manner. LC-MS operates at lower analysis temperatures than GC-MS, which enables the analysis of heat-labile metabolites which are commonly degraded during GC analysis. LC-MS analysis does not require sample derivatization, and this simplifies the sample-preparation steps as well as identification of metabolites, which can be complicated by chemical modifications of unknowns prior to GC-MS. However, a major disadvantage of LC-MS compared to GC-MS is the lack of transferable LC-MS libraries for metabolite identification. The mass-spectral variability between LC-MS systems in terms of the relative ion abundances associated with adduct formation, in-source fragmentation, tandem mass spectra fragment ions, hinder the direct comparison of LC-MS data between laboratories [155]. Moreover, LC-MS-based techniques are less advanced for metabolomic applications than GC-MS based methods, which have been shown in numerous studies to be reliable and reproducible. LC-MS metabolic studies appeared later in the literature than the GC-MS ones and were devoted mainly to targeted metabolites or metabolite classes in plants [44,156-158]. More recently, LC-MS has become a standard approach for many metabolomic analyses due to its ability to separate, ionize and detect a wide range of chemicals [35,154,159-164].

Sample Preparation

LC-MS-based methods, especially those employing RP separation are ideal for metabolomic analysis of samples, such as urine, which can be injected directly onto the column, without any pre-treatment other than removal of particulates, as seen in many of the reported applications [161,162,165,166]. Blood plasma can also be analyzed with minimal sample pre-treatment, based typically on the removal of proteins via solvent precipitation [159,167,168], and tissue extracts are also amenable to LC-MS-based analysis [169]. Plant specimens are usually frozen and extracted by polar solvents such as methanol, acetonitrile or their mixture with water [163,164]; a valid alternative to frizzing could be lyophilization [158]. When only selected classes of metabolites have to be analyzed, sample extracts can be cleaned up by solid phase extraction (SPE) [157,158,169,170].

LC-MS Analysis

The bulk of applications use reversed RP-HPLC and gradient elution, with run times lasting from a few minutes to several hours. For HPLC-MS analysis, conventional column formats, typically 2.1 mm i.d., 15–25 cm in length and packed with 3–5 μm particles have been used for years [35,154].

Electrospray ionization (ESI), preferably in both positive and negative mode of ionization, is the most commonly used ionization technique for LC-MS, but APCI is also used to a lesser extent [171,172]. One of the disadvantages in utilizing ESI for interfacing LC to MS in metabolic profiling and metabolomics studies is the occurrence of ionization suppression. Contributing factors to this phenomenon include: 1- solvent matrix effects (i.e. where solvent components, especially buffers, ''compete'' with analytes for ionization); 2- erratic electrospray behavior as a result of increased liquid conductivity from various salts and charged species; 3- competition for the limited number of charges during co-elution of two or more compounds with dramatic differences in proton affinities or surface activities, particularly if high analyte concentrations are present [173-176]. This can produce signal intensities that are not linearly related to the analyte concentrations or lead to inability to detect some analytes. Thus, metabolite analysis is complicated by their chemical diversity, and dynamic ranges. It is estimated that the metabolome extends over 7-9 order of magnitude of concentration [43].

APCI is less prone to matrix effects, but also a less universal ion source than ESI. They could be considered more or less complementary to each other, being APCI suitable for moderately polar compounds. Although a simultaneous ESI/APCI ionization source, referred to as multimode ionization (MM), is commercially available [177], MM ionization has been rarely used for metabolomics [159]. This paper reports also the combination of in line (+)-ESI/APCI with LC fraction collection, and off line MALDI and Desorption/Ionization on Silicon (DIOS). Complementing the (+)-ESI analysis with (+)-APCI resulted in an additional 20% increase in the number of detected ions, and, by combining inline (+)-ESI with (+)-APCI and off line (+)-MALDI/DIOS analysis, the information content more than doubled compared to ESI only.

The effect of ionization suppression on analyte molecules can be greatly minimized through improved LC separation and reduced LC operating flow rates (both of which lead to more efficient ESI), as well as decreased sample loading to the LC column. Thus the separation efficiency, quantified by the separation peak capacity defined as "the theoretical number of resolved peaks that can be fitted into the separation space" [178] determines the coverage and the completeness of the analysis. This increase is generally due to improved detection of the lower abundance species, which are ultimately better resolved from species that are present either in higher abundance or that have higher proton affinities or surface activities. Decreased flow rate and sample loading (with a concomitant increased analyte concentration in the eluting phase), potentially reduce ionization suppression, resulting in an overall increase in the dynamic range of the measurement. The first improvement can be obtained by increasing specific efficiency (decreasing HETP) and the second one by decreasing the internal diameter of the column. Thus, a key area for further innovation in metabolic profiling is the use of higher resolution and miniaturized separation systems.

A HETP decreasing can be reached by decreasing the diameter of the packing particles, but the pressure drop increases exponentially. Jorgenson's group introduced ultrahigh pressure

LC (UPLC) where columns are packed with sub 2 μm particles and operating at 60,000–100,000 psi. Such a system resulted in 200,000–730,000 theoretical plates/m, extremely sharp peaks, high sensitivity and high resolution at unprecedented velocity [179,180]. This alternative to conventional LC is currently readily available with the trade names UPLC and RR (rapid resolution) LC and widely used [161,162,166-168,181]. A means of reducing the back pressure associated with the use of small particle-size stationary phases and increasing efficiency, is to perform separations at elevated temperatures, as such conditions result in reduced solvent viscosity and thereby lower back pressures as far as in reduced interdiffusion coefficients. Figure-3a shows the total ion profile in positive ESI obtained by injecting 5 μL of an urine sample in a RRLC system consisting in a two C18 columns (100 × 2.1 mm I.D.; 1.8 μm particle size) connected in series. The elution gradient time and column temperature were adjusted to obtain the largest positive features (recorded spectra) and the best retention time reproducibility. More than 15,000 different spectra were recorded by the Q-ToF mass spectrometer, with a 50% increase respect to a single 10 cm column chromatography.

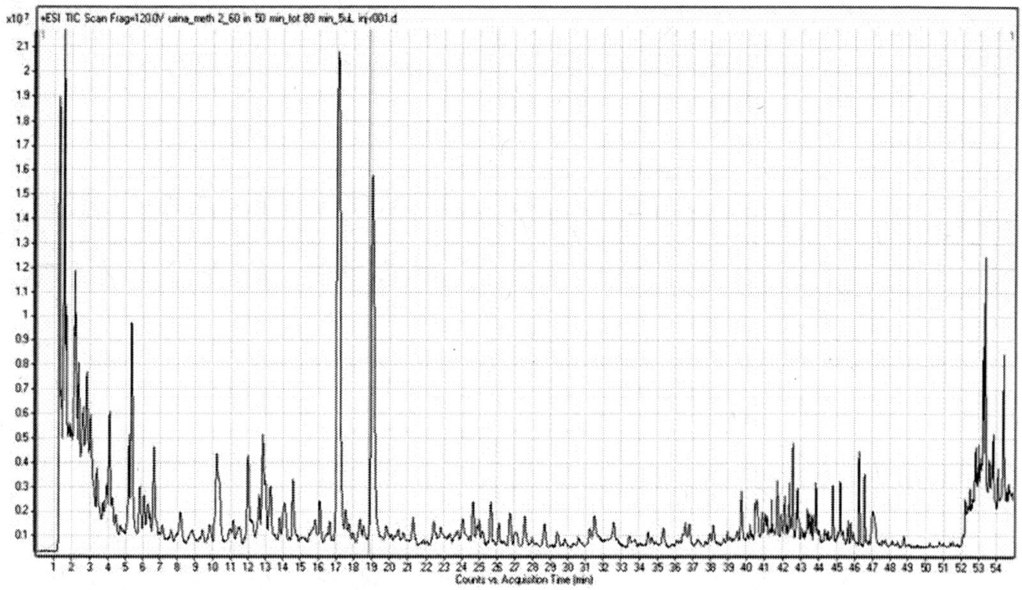

Figure 3a. Total ion profile in positive electrospray ionization. Sample: 5 μL of urine filtered through 10k centrifugal filter device. Liquid chromatography-tandem mass spectrometry was performed using a rapid resolution binary pump, two Zorbax Eclipse Plus C18 columns (100 × 2.1 mm I.D.; 1.8 μm particle size) connected in series and a Q-TOF series 6520 mass spectrometer (all from Agilent). The mobile phase was (A) H_2O, and (B) CH_3CN, both 0.1% (v) formic acid, and the solvent gradient program was 2% B at time 0, 2% B at time 5 min, 20% B at 35 min, 60% B at 65 min, 95% B at 65.1 min and 95% at 70 min. Stop time was 75 min and the re-equilibration time was 25 min. The flow rate was 0.3 mL/min and column temperature was set at 50°C.

High-temperature (HT) chromatography can be used either to deliver the mobile phase at higher flow rates, thereby reducing analysis times or to increase the length of the column to obtain higher resolution separations [181-183]; temperatures up to 90°C have been used [183]. HT chromatography poses the question of both analytes and packing stability, and its reliability was carefully checked for the studied samples. Very recently the so-called porous shell or fused core particles have been introduced [184]. These 2.7 μm particles, consisting of a 1.7 μm solid core and a 0.5 μm porous shell of high-purity silica are designed to allow very fast and efficient separation without some of the disadvantages of conventional columns with small, totally porous particles. The characteristics of these fused core particles represent a fortunate compromise between separation speed and modest operating pressures. They have been recently applied for lipid profiles in plasma, and more than 160 lipids belonging to eight different classes were detected in a single LC-MS run [185]. In the near future, metabolomics may benefit from the use of relatively long columns packed with fused core particle to increase efficiency.

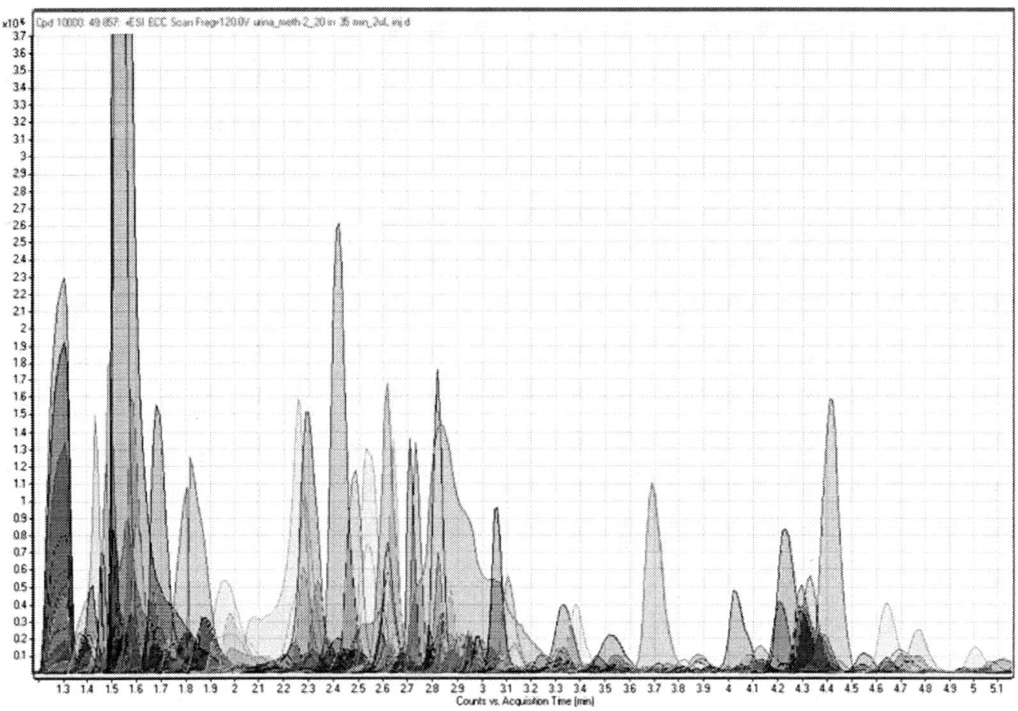

Figure 3b. Part of the chromatogram showing the peaks automatically selected for MS/MS acquisition.

Capillary LC (200-50 μm I.D.) can also be used to greatly increase the performance of LC-MS system. Whilst long capillaries can be used to increase resolution, this increased separation power comes at the cost of long analysis times and high operating pressure [186]. However, the utility of this approach is amply demonstrated as a very high number of features can be obtained. The use of normal length (10-30 cm) capillary LC can reduce the amount of sample required for analysis. This may be particularly valuable when only small volumes can be obtained and, in addition, they increase detection sensitivity and detected metabolite dynamic range [187]. Comparison with a conventional LC-MS analysis of the same samples made on a column of the same length and packed with the same stationary phase showed that the capillary system generated twice as many ions as the conventional system, presumably due to reduced ion suppression, and was up to 100-fold more sensitive for some metabolites. Alternatively, silica-based and polymer-based RP monolithic capillary columns have been utilized in metabolomics applications [188,189]. An advantage of the monolithic systems is their relatively low back pressure (compared to conventional packed capillaries), enabling either comparatively high flow rates or the use of long capillaries.

Various types of RP phases with different polarities have been used in metabolite research; such RP-stationary phase are suitable for the analysis of compounds of medium and low polarity but do not give particularly good results for polar and/or polar ionic metabolites. A multi-column approach, applied to human plasma analysis, involving the use of three different stationary phase chemistries with separations performed on C18, amino and phenyl-hexyl columns has been used to increase coverage [9]. For such polar/ionic compounds, separation using hydrophilic interaction chromatography (HILIC) is an option [124,190-192]. HILIC is performed on a pure silica column or very polar chemically bonded silica, and acetonitrile is used as weak solvent, while water is the strong one. A drawback of HILIC is the very long equilibration time needed after a gradient is performed. Two dimensional separation (2D chromatography), as for proteomics, may increase the number of metabolites detected but, up to the present time it has not been used for metabolomics.

CE is considered a highly efficient, flexible separation technique. One of its main assets for fingerprinting, where samples must undergo the minimum possible manipulation, is the capability to analyze complex matrices such as urine without previous treatment. CE-ESI-MS interface development has been an active area of investigation for over 20 years [105,193,194]. However, completing the electrical circuit required for CE in a manner that results in a stable electrospray and suitable detection limits has been a challenge: system stability is essential for sensitivity. Recently, approaches based on CE-MS [195-197] have emerged as powerful tools for the comprehensive analysis of charged metabolites and have played a critical role in understanding intricate biochemical and biological systems [197-204]. Because the scaling laws of CE make it amenable to small-volume sampling, it has been used extensively for single-cell and subcellular analyses of metabolites [205-206]. Compared to both GC and LC, CE is much less utilized in metabolomics (a recent exhaustive review reports the present state of the art [207]), and an increase of its importance in the field is to be expected when some of the problems still present in coupling with MS have been overcome.

The mass-to-charge ion analyzers used in metabolomics follow strictly the technical improvements in instrumentation. Although TQ, used as GC detector in many studies in the past, did not allow high resolution and accurate mass measurements, it represents a very robust system, and it is still used sometimes [142]. QIT, although more sensitive than TQ, is used only occasionally [208], probably because of its scarce dynamic range. QqQ analyzer consents a MS/MS acquisition and its fourth generation models are very sensitive in the Multi Reaction Monitoring (MRM) acquisition mode. This characteristic, together with a rapid scan capability (about 50 μs per scan), makes this instrument very valuable in targeted metabolites analysis by LC-MS [170,209].

Recently, GC time-of-flight MS (ToF-MS) has become more popular for metabolite profiling due to its higher mass accuracy and mass resolution relative to quadrupoles [134,145,153,183]. Further, ToF-MS offers very high scan speeds, necessary for adequate sampling of chromatographic peak widths in the range of 0.5–1 s. Thus, the use of high scan speeds facilitates the implementation of fast GC methods, which can reduce the analysis time and increase productivity. LC-ToF-MS and LC-Q-ToF-MS/MS are also increasingly used in metabolite analysis [162,183,187,210]. The mass accuracy of ToF instruments has historically been in the 5–10ppm range, technological advances in recent years have shown that ToF can achieve a mass accuracy of 1–2 ppm when internally calibrated [211].

Another hybrid instrument used in the metabolic profile is the linear ion trap-triple quadrupole mass spectrometer (Q-Trap) [157,158,168,212]. The Q-Trap mass spectrometer is a modified triple quadrupole where the Q3 region can be operated either as a conventional quadrupole mass filter or as a linear ion trap with axial ion ejection. Thus, the instrument encompasses the functionality of an ion trap mass spectrometer, with its associated high sensitivity for product ion scanning, and that of a triple-quadrupole mass spectrometer. The system also has MS^3 capabilities which are useful to determine the origin of the fragments [213]. The most powerful MS ion analyzers, in terms of resolution, are the FT-ICR and the Orbitrap; these instruments can also perform MS/MS. Nevertheless their application, coupled with LC in metabolomics is rare [214-217], and an increasing number of applications, especially for LC-Orbitrap can be possible in a near future.

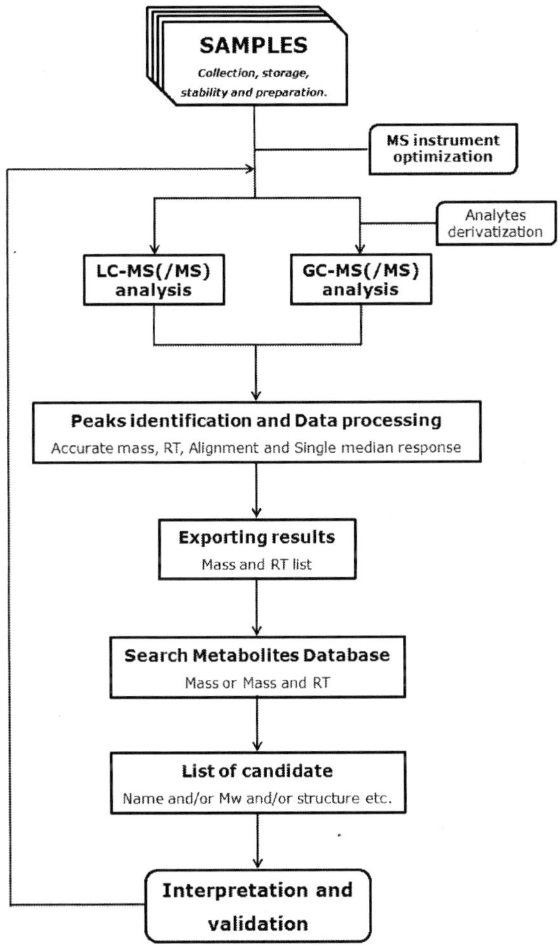

Figure 4. Workflow of a metabolomics GC-MS or LC-MS(/MS) platform.

Ion mobility spectrometry (IMS) was introduced in the '70s as a low resolution ion-separation and detection device [218]. IMS is based on the fact that ions with different shapes travel at different speeds when they are pulled by a weak electric field through a drift cell filled with a buffer gas. Coupling of IMS to MS results in a 2D, orthogonal separation technique. IMS was interfaced with ToF-MS in the late '90s [219], permitting the simultaneous acquisition of ion mobility spectra and mass spectra in a single run. The combined LC-IM-MS approach using Q-ToF with electrospray ionization, has been recently tested for urine metabolome study, and demonstrates the potential for high coverage, high throughput analysis [220].

Whatever the technological platform (hardware) was chosen, a metabolomic experiment include, in addition to the experimental design, a series of operational steps as reported, for example, in Figure4.

Direct Infusion MS and MS/MS

Direct MS (DMS) analyses of complex metabolite mixtures in conjunction with chemometric data analysis offer a viable solution when high-throughput screening is mandatory. The high-throughput capacity of DMS fingerprinting of complex mixtures is similar to that of NMR fingerprinting.

Direct infusion MS (DIMS), or flow injection of complex metabolic extracts without chromatographic separation via ESI provides a sensitive, high-throughput method for metabolic fingerprinting. The greater sensitivity compared to NMR makes it a very useful approach for large-scale screening. Obviously, DI-ESI-MS analysis is much more susceptible to ionization suppression than LC-ESI-MS, thus it is not usually advocated as a quantitative method. The utilization of nano-ESI reduces ionization suppression effects due to the increased ionization efficiency of nano-DIMS, and chip-based nanospray emitters provide a fully automated platform for high-throughput DIMS metabolite measurements [221]. DIMS is unable to differentiate among isomeric compounds, however $DIMS^n$ using ion traps produces fragment ions that often enable the differentiation of isomeric structures. Recently a rapid metabolic profiling method using both untargeted and targeted $DIMS^3$ with a relatively low resolution linear ion trap mass spectrometer has been shown to yield sufficient precision and accuracy for application in genetical metabolomics provided that suitable software tools were developed [222].

FT-ICR-MS is a powerful tool for DIMS because of its very high mass resolution (10^6) and mass accuracy (<1 ppm), and it has been successfully applied for metabolite-fingerprinting studies [223,224]. However, large ion populations influence negatively the mass-measurement accuracy and limit the dynamic range of FT-ICR when a wide *m/z* range is tapped. Narrow *m/z* ranges are therefore commonly acquired separately to increase the dynamic range and mass accuracy for metabolic profiling. An optimized strategy, namely high sensitivity selected ion monitoring (SIM)-stitching approach, is usually followed. Each wide-scan mass spectrum is recorded as a series of overlapping selected ion monitoring (SIM) windows that are stitched together using novel algorithms [225]. This, reducing space-charge effects, increases the dynamic range and maintains high mass accuracy.

Ion suppression during ESI, ion–ion interactions in the detector cell, and thermally-induced white noise remain major challenges for this approach [226]. Orbitraps may be a good alternative to the more expensive and higher maintenance FT-ICR mass spectrometers, with similar resolving powers and mass accuracies of 2–5 ppm.

Desorption and Imaging MS

MALDI-MS is a popular analytical technique for protein and peptide analysis. Although the high throughput nature of MALDI-MS makes it an ideal tool for large-scale metabolomic studies, its application in the field has been rather limited. Due to the elevated chemical background signals generated by the matrix in the low molecular-weight region that obscures the detection of metabolites in the range, MALDI-MS has been confined to the targeted analysis of high-molecular weight metabolites [227,228]. Recent advancements in laser desorption techniques include matrices that have minimal background signals in the low-molecular-weight region (''ionless matrices''), yet still assisting an efficient ionization/desorption of the analytes [229,230], and DI-MS from porous silicon chips [231]. These offer exciting new opportunities for the utilization of desorption-ionization in metabolite screening and fingerprinting; however, the technique is still subject to ion suppression and yields poor quantitative detection of metabolites [232].

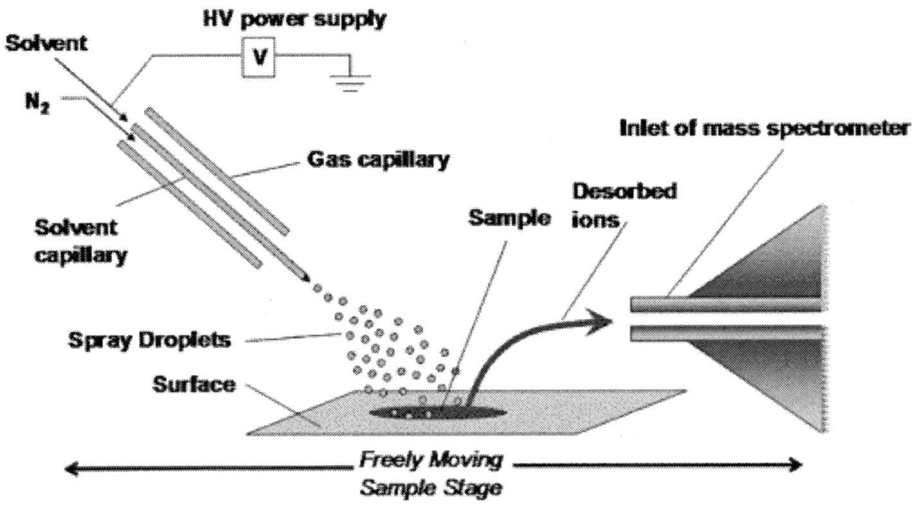

Figure 5a. DESI (desorption electrospray) scheme.

Desorption ESI (DESI) is a ionization technique that combines features of ESI and desorption ionization to permit analysis directly from a surface with virtually no sample preparation. An electrospray emitter is used to generate a spray of charged micro droplets that is directed towards an ambient sample surface. Molecules on the surface are subsequently desorbed, ionized, desolvated and directed to the MS inlet [233]. Virtually no sample preparation is required for DESI, thus allowing the direct analysis of tissues or biological fluids that can be deposited on an inert surface and then analyzed. However, DESI experimental conditions typically require optimization for each sample type, so time must be invested initially in optimizing the experimental parameters such as the chemical and physical nature of the surface and the nature of the spraying solvent [234]. Also the geometry of the system sample surface- sprayer tip- MS inlet, directly affects the ionization efficiency and sensitivity. DESI seems to have a higher tolerance to sample-matrix effects than ESI, however the quantitative precision of DESI, as of other surface ionization techniques, is less than that of ESI. The application of DESI in metabolomics is relatively new, but its ambient DI properties as well as its high throughput capacity make it an attractive tool for metabolomics. Figure 5a shows a DESI scheme, and Figure 5b the images produced from an analysis of lipids in a rat brain tissue sample.

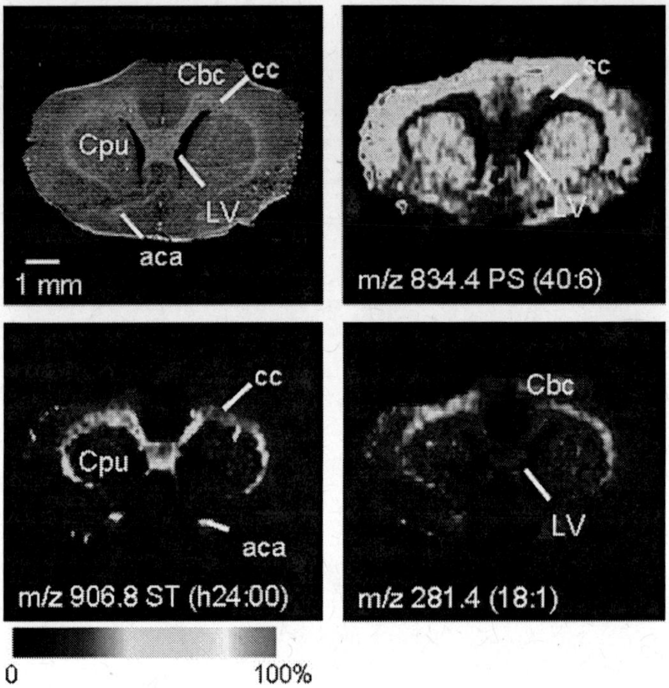

Figure 5b. Images produced from an analysis of lipids in a rat brain tissue sample. The first (up left) is an optical image, and the others are ion images created from desorption electrospray ionization analysis.

Extractive ESI (EESI) is another new ESI technique that uses two separate sprayers. One sprayer nebulizes the sample solution that intersects with a second electrospray containing charged micro-droplets of the ionizing reagent solvent, usually an acidic aqueous methanol. Analyte molecules are ionized following collision with the reagent micro-droplets and then mass analyzed. EESI is related to DESI, but was developed for the direct analysis of trace compounds in the solution phase, especially when complex mixtures are of interest [235]. EESI utilizes two spray sources, eliminating the use of a surface on which the analyte is first collected. One spray source nebulizes the sample while the other provides charged solvent droplets. The two spray sources are set at an angle to each other and to the mass spectrometer inlet so as to introduce the analyte of interest directly to the source [236]. The advantage of EESI is its ability to analyze complex biological samples, such as urine and serum, directly, with minimum or no sample preparation for an extended period of time. The direct infusion of such complex biological samples to conventional ESI ion sources causes an irrecoverable loss in signal intensity due to the formation of salt adducts, sample carry-over, or cumulative build-up of non-volatile components in the ion source. The long-term spray stability of untreated biological samples in EESI is very promising for high throughput metabolomics, as EESI significantly reduces data-collection interruptions due to frequent ion source cleaning to remove non-volatile accumulations associated with the ESI-DIMS of crude biological samples [143]. EESI may represents a valid, more sensitive alternative to NMR in metabolomics; recently, it has been demonstrate that NMR data and EESI mass spectral data can be cross-validated [237].

Imaging MS (IMS) is an other emerging technology that permits the direct analysis and determination of the distribution of molecules in tissue sections. Tissues are analyzed intact and thus spatial localization of molecules within a tissue is preserved [238]. To investigate the spatial distribution of specific biomolecules, an increasing number of work groups are aiming to combine the sensitivity and specificity of mass spectrometry with imaging capabilities. This has included both MALDI and ESI, and has provided fresh impetus to secondary ion mass spectrometry (SIMS) imaging. The potential of IMS has long been recognized. The first elemental imaging experiments using SIMS were made in the '60s [239,240]. However, widespread bimolecular IMS had to wait for the advent of MALDI and SIMS methodologies capable of generating ions from biomolecules with sufficient sensitivity. The high mass capabilities of MALDI enabled it to be readily applied for imaging protein distributions within tissue sections [241], whereas the principle SIMS application in semi-conductor research directed its development toward higher spatial resolution. The success of imaging MALDI has begun to steer SIMS developments toward high mass molecules while retaining the high spatial resolution capabilities. However, as a rule, MALDI imaging and SIMS imaging provide spatial information about different classes of biological compounds. MALDI provides proteomics information, and SIMS that of lipids and other surface active, relatively low MW species. Hence, MALDI can record the spatial distribution of high mass molecules using the chemically specific molecular ions; however, typical spatial resolutions are approximately 25 μm or more (10 μm sources are now available). SIMS is able to provide high spatial resolution images, sub-micron is routine and 50 nm is commercially available [242]; however, the molecular ion mass range is much lower than that of MALDI, most imaging experiments use ions of m/z <500.

MALDI can analyze hundreds of proteins directly from tissue sections, and combining this with spatial coordinates, allows the spatial distribution of these proteins to be determined in parallel, without a label and within practical time-scales [243]. The sample preparation, spatial resolution and sensitivity of the ionization step, are all important parameters that affect the type of information obtained. Recently, significant progress has been made in each of these steps for both SIMS and MALDI imaging of biological samples. Mass resolution is an important feature because it defines the degree of chemical specificity and, through improving the precision of mass measurement, helps improve mass accuracy. The ultra-high mass resolution and mass accuracy of the Fourier transform ion cyclotron resonance mass spectrometer, resolution >100,000 and sub part-per-million mass accuracy, is beginning to be developed for imaging mass spectrometry experiments [244,245].

The advantage of a SIMS analysis is that tissue sections did not need any further sample preparation steps: the sample plate can be mounted and the analysis performed. Impressive biological images of atomic ions and low mass fragments can be obtained directly from tissue. The sensitivity improvements provided by polyatomic primary ions, added as very thin films, show sufficient intensity for imaging of intact biomolecules [246-248].

DESI, being an atmospheric pressure (AP) surface ionization technique, can be used as a chemical IMS. Many lipid species are easily ionized by DESI, making them attractive target molecules from which to create molecular images of thin tissue sections. DESI-MS has been used to construct chemical images of tissue sections of mouse pancreas, rat brain, and metastatic human liver adenocarcinoma, as well as whole tissue analysis of adipose tissue surrounding a chicken heart giving strong signals from the major lipid components of biological membranes [249-253]. Theoretically, DESI may be coupled with all the MS and MS/MS analyzer and, although the IMS application of DESI has been limited to phospholipids, the technique seems to have the potential for a wider metabolite profiling.

AP-IMS techniques are particularly attractive because, in principle, working at AP also enables the study of live specimens. Laser ablation ESI (LAESI) [254,255] is a new AP-IMS technique. LAESI IMS is realized as a combination of lateral imaging and depth profiling by tissue ablation with single laser pulses of ca. 2.94 μm wavelength. The O-H vibrations of the native water molecules in the tissue samples readily absorbed the laser pulse energy leading to ablation, i.e., the ejection of a microscopic volume of the sample in the form of neutral particulates and/or molecules. This plume was then intercepted by an electrospray, and the ablated material was efficiently post-ionized. Tandem MS was performed on numerous ions to help with metabolite structure assign. A recent, novel development of LAESI was the combination of lateral imaging and molecular depth profiling capabilities, which enabled a truly 3D metabolite distribution imaging. Although 3D ambient imaging with LAESI has been proven feasible in plant tissues, further improvements are needed in spatial resolution [255].

Data Handling

A great drawback of all "omics" sciences where a large number of qualitative and quantitative data can be generated per single run is the problem of data analysis. Hence, there is a demand for computational tools to handle and interpret the large amounts of data. For metabolomics to be successful, raw analytical data must be converted into metabolites (named chemicals) and their concentrations, or variation of concentration, data that can be usefully interpreted for biological research.

The data-processing techniques for the deconvolution of chromatographic peaks, library search, and GC retention index developed originally by Kovats [256] have been in use since the '70s to help identification of peaks recorded by GC-MS. [257] Metabolite profiling by chromatography-MS and statistical analysis still relies on efficient data-processing procedures, and minimum reporting requirements have recently been suggested [258,259]. In all published metabolomic studies based on GC-MS, the starting point of the data analysis has been the deconvolution of chromatograms, followed by peak matching, and then identification of the differences between samples. Multivariate methods have been developed and used to clarify chromatographic and spectral profiles from overlapping chromatographic peaks obtained using various types of hyphenated chromatography systems. The multivariate curve resolution methods can be divided into iterative, non-iterative, and hybrid approaches, and all present both advantages and disadvantages [134]. Software programs have been developed capable of spectral filtering (noise elimination), peak detection finding the peaks corresponding to the same compounds, *m/z* alignment (aiming at matching the corresponding peaks across multiple sample runs) and normalization (adjusting the intensities within each sample run by reducing the systematic error) [260-264]. Software packages have augmented their capability during the years, and many can be downloaded from the Internet.

After data alignment and correction of retention times, a data matrix can be generated from the peak lists as output for subsequent multivariate statistical analysis, including supervised classification for fingerprinting and principal component analysis for data visualization [153]. Visualization of complex mass spectrometric data sets is becoming increasingly important in metabolomics. In a recent paper a versatile tool suitable for many frequently occurring tasks handling LC-MS data is described [265]. Depending on the task at hand, different views may be used: single spectra visualized as a plot of intensity against mass-to-charge ratio (1D view); LC-MS maps displayed in a 2D view from a bird's-eye perspective with color-coded intensities; selected regions of LC-MS maps displayed in a 3D view.

The chemical identification of metabolites is fundamental for the extraction of biological context from the data. It is not easy to identify a metabolite in a metabolome, especially low level metabolites which are at or slightly above the noise level or are masked by other metabolites. An estimated total number of possible metabolites ranges from 200,000 to 1,000,000 [4,266]. A number of strategies are being tried out to assist in the chemical identification of the unknowns, including the development of metabolite-specific mass spectral libraries and databases [267-270]. The basis on which metabolites are identified varies among the metabolomics community. In an effort to standardize the reporting and interpretation of metadata, the Metabolomics Standard Initiative (MSI) [271] has formed five working groups that follow the general workflow model in metabolomics: biological context; chemical analysis; data processing; ontology; and data exchange.

Two types of identification are possible: putative or preliminary identification and definite identification. A global metabolomic study involves the search for all metabolites in a biological specimen. If the search is for known metabolites, the identification involves comparing the experimental data with that of pure standards. Experimentally determined accurate mass or electron-impact mass spectrum are typically applied for putative identification. In LC-MS and DIMS the accurate mass is used to define molecular formulae from which suitable metabolites can be derived by searching electronic resources. However, isomers have the same accurate mass and therefore require a separate, orthogonal property for definite identification of all potential isomers. Most metabolite identifications reported are typically non-novel as they have been previously characterized, identified, and reported at a rigorous level in literature. Thus, non-novel metabolites not being identified for the first time are typically identified through the co-characterization with authentic chemical standards. The Chemical Analysis Working Group has recently proposed a guideline for the identification of non-novel metabolites [258], in which a minimum of two independent and orthogonal data relative to an authentic standard compound, typically retention time or index and fragmentation mass spectrum analyzed under identical experimental conditions, are considered necessary for metabolite identification.

However, if the metabolites are not known, metabolite identification is much more difficult. This effort would involve multiple chromatographic separations using different column chemistries and mobile phases. It also would require high resolution MS and MS/MS for accurate mass analysis, statistical analysis software for data mining, metabolite databases, and access to an index of hundreds of pure compounds for confirmation. Databases are still at an early level of development, and techniques, such as chemical-formula determination via accurate mass, often serve only to narrow the potential search field rather than provide an unambiguous formula. Biomarker identification therefore remains a potentially time-consuming process and a limitation for this technique. Some freely and commercially available web-resources for MS-based metabolomics are mainly reported in references [261,272] and also in the literature quoted in this section.

Future Perspectives

LC-MS is certainly going to become a key technology for the provision of global metabolomics. Major advances in available technologies and a totally new approach to analysis are essential before a true metabolomics, not achievable with the current state of development in analytical science, can be performed. This will include developments in all steps of metabolite analysis, from sampling, sample storage and preparation, to data acquisition, storage and processing, coupled with a greater understanding and application of bioinformatics.

Future developments in exhaustive mining of the metabolome will no doubt evolve through improvements in chromatographic technologies and sensitive MS instruments with a high capability for mass accuracy. Continued improvements in miniaturized MS, multi-dimensional chromatography, and multiplexed MS approaches will offer new opportunities in metabolomics. An increase in the number of chemically identifiable metabolites is a fundamental key to the interpretation and understanding of the biological context of metabolomics experiments. High throughput techniques, such as DESI and EESI, are very attractive and will certainly undergo major improvements in sensitivity, dynamic range, and high resolution MS/MS coupling. Finally, chemical imaging techniques based on MS are very exciting and, although not a novelty, still in their infancy. In vivo analysis of intact samples is an attractive proposition, and in this context, NMR-based technologies have a distinct advantage over instruments based on MS. But every gap may be overcome.

References

[1] van der Werf, MJ; Jellema, RH; Hankemeier, T. Towards replacing closed with open target selection strategies. *J. Ind. Microbiol. Biotechnol.*, 2005 32, 234-252.

[2] van der Werf, MJ. Microbial metabolomics: replacing trial-and-error by the unbiased selection and ranking of targets. *Trends Biotechnol.*, 2005 23, 11-16.

[3] Sumner, LW; Mendes, P; Dixon, RA. Plant metabolomics: large-scale phytochemistry in the functional genomics era. *Phytochemistry*, 2003 62, 817-36.

[4] Fiehn, O; Kopka, J; Dormann, P; Altmann, T; Trethewey, RN; Willmitzer, L. Metabolite profiling for plant functional genomics. *Nat. Biotechnol.*, 2000 18, 1157-1161.

[5] Raamsdonk, LM; Teusink, B; Broadhurst, D; Zhang, N; Hayes, A; Walsh, MC; Berden, JA; Brindle, KM; Kell, DB; Rowland, JJ; Westerhoff, HV; van Dam, K; Oliver, SG. A functional genomics strategy that uses metabolome data to reveal the phenotype of silent mutations. *Nat. Biotechnol.*, 2001 19, 45-50.

[6] Clish, CB; Davidov, E; Oresic, M; Plasterer, TN; Lavine, G; Londo, T; Meys, M; Snell, P; Stochaj, W; Adourian, A; Zhang, X; Morel, N; Neumann, E; Verheij, E; Vogels, JT;

Havekes, LM; Afeyan, N; Regnier,F; van der Greef, J; Naylor, S. Integrative biological analysis of the APOE3-leiden transgenic mouse. *Omics*, 2004 8, 3-13.

[7] Goodacre, R; Vaidyanathan, S; Dunn, WB; Harrigan, GG; Kell, DB. Metabolomics by numbers: acquiring and understanding global metabolite data. *Trends Biotechnol.*, 2004 22, 245-252.

[8] Brindle, JT; Antti, H; Holmes, E; Tranter, G; Nicholson, JK; Bethell, HW; Clarke, S; Schofield, PM; McKilligin, E; Mosedale, DE; Grainger, DJ. Rapid and noninvasive diagnosis of the presence and severity of coronary heart disease using 1H-NMR-based metabonomics. *Nat. Med.*, 2002 8, 1439-1444.

[9] Sabatine, MS; Liu, E; Morrow, DA; Heller, E; McCarroll, R; Wiegand, R; Berriz, GF; Roth, FP; Gerszten, RE. Metabolomic identification of novel biomarkers of myocardial ischemia. *Circulation*, 2005 112, 3868-3875.

[10] Kawashima, H; Oguchi, M; Ioi, H; Amaha, M; Yamanaka, G; Kashiwagi, Y; Takekuma, K; Yamazaki, Y; Hoshika, A; Watanabe, Y. Primary biomarkers in cerebral spinal fluid obtained from patients with influenza-associated encephalopathy analyzed by metabolomics. *Int J. Neurosci.*, 2006 116, 927-936.

[11] [11] Ippolito, JE; Xu, J; Jain, S; Moulder, K; Mennerick, S; Crowley, JR; Townsend, RR; Gordon, JI. An integrated functional genomics and metabolomics approach for defining poor prognosis in human neuroendocrine cancers. *Proc. Natl. Acad. Sci.* 2005 102, 9901-9906.

[12] Ryan, D; Robards, K. Metabolomics: The greatest omics of them all? *Anal. Chem.*, 2006 78, 7954-7958.

[13] Oliver, SG; Winson, MK; Kell, DB; Baganz, F. Systematic functional analysis of the yeast genome. *Trends Biotechnol.*, 1998 16, 373-378.

[14] Goodacre, R. Making sense of the metabolome using evolutionary computation: seeing the wood with the trees. *J. Exp. Bot.*, 2005 56, 245-254.

[15] Fiehn, O. Combining genomics, metabolome analysis, and biochemical modelling to understand metabolic networks. *Comp. Funct. Genom.*, 2001 2, 155-168.

[16] Nicholson, JK; Lindon, JC; Holmes, E. “Metabomics": understanding the metabolic responses of living systems to pathophysiological stimuli via multivariate statistical analysis of biological NMR spectroscopic data. *Xenobiotica*, 1999 29, 1181-1189.

[17] Fiehn, O. Metabolomics - the link between genotypes and phenotypes. *Plant Mol. Biol.*, 2002 48, 155-71.

[18] Lindon, JC; Holmes, E; Nicholson, JK. So what's the deal with metabonomics? *Anal. Chem.*, 2003 75, 384A-391A.

[19] MacKenzie, DA; Defernez, M; Dunn, WB; Brown, M; Fuller, LJ; de Herrera, S; Guenther, A; James, SA; Eagles, J; Philo, M; Goodacre, R; Roberts, IN. Relatedness of medically important strains of Saccharomyces cerevisiae as revealed by phylogenetics and metabolomics. *Yeast*, 2008 25, 501-512.

[20] Smedsgaard, J; Nielsen, J. Metabolite profiling of fungi and yeast: from phenotype to metabolome by MS and informatics. *J. Exp. Bot.*, 2005 56, 273-286.

[21] Allwood, JW; Ellis, DI; Heald, JK; Goodacre, R; Mur, LAJ. Metabolomic approaches reveal that phosphatidic and phosphatidyl glycerol phospholipids are major discriminatory non-polar metabolites in responses by Brachypodium distachyon to challenge by Magnaporthe grisea. *Plant J.*, 2006 46, 351-368.

[22] Hall, RD. Plant metabolomics: from holistic hope, to hype, to hot topic. *New Phytol.*, 2006 169, 453-468.

[23] Kell, DB. Systems biology, metabolic modelling and metabolomics in drug discovery and development. *Drug Discov. Today.*, 2006 11, 1085-1092.

[24] Dunn, WB; Broadhurst, DI; Deepak, SM; Buch, MH; McDowell, G; Spasic, I; Ellis, D; Brooks, N; Kell, DB; Neyses, L. Serum metabolomics reveals many novel metabolic markers of heart failure, including pseudouridine and 2-oxoglutarate. *Metabolomics*, 2007 3, 413-426.

[25] Kenny, LC; Broadhurst, D; Brown, M; Dunn, WB; Redman, CWG; Kell DB; Baker, PN. Detection and identification of novel metabolomic biomarkers in preeclampsia. *Reprod. Sci.*, 2008 15, 591-597.

[26] Lindon, JC; Holmes, E; Nicholson, JK. Metabonomics in pharmaceutical R & D. *FEBS J.*, 2007 274, 1140-1151.

[27] Tanaka, Y; Higashi, T; Rakwal, R; Wakida, S-Ii; Iwahashi, H. Quantitative analysis of sulfur-related metabolites during cadmium stress response in yeast by capillary electrophoresis-mass spectrometry. *J. Pharm. Biomed. Anal.*, 2007 44, 608-613.

[28] Viant, MR. Metabolomics of aquatic organisms: the new 'omics' on the block. *Mar. Ecol.-Prog. Ser.*, 2007 332, 301-306.

[29] Herrgard, MJ; Swainston, N; Dobson, P; Dunn, WB; Arga, KY; Arvas, M; Bluthgen, N; Borger, S; Costenoble, R; Heinemann, M; Hucka, M; Le Novere, N; Li, P; Liebermeister, W; Mo, ML; Oliveira, AP; Petranovic, D; Pettifer, S; Simeonidis, E; Smallbone, K; Spasic, I; Weichart, D; Brent, R; Broomhead, DS; Westerhoff, HV; Kirdar, B; Penttila, M; Klipp, E; Palsson, BO; Sauer, U; Oliver, SG; Mendes, P; Nielsen, J; Kell, DB. A consensus yeast metabolic network reconstruction obtained from a community approach to systems biology. *Nat. Biotechnol.*, 2008 26, 1155-1160.

[30] Wishart, DS; Knox, C; Guo, AC; Eisner, R; Young, N; Gautam, B; Hau, DD; Psychogios, N; Dong, E; Bouatra, S; Mandal, R; Sinelnikov, I; Xia, J; Jia, L; Cruz, JA; Lim, E; Sobsey, CA; Shrivastava, S; Huang, P; Liu, P; Fang, L; Peng, J; Fradette, R; Cheng, D; Tzur, D; Clements, M; Lewis, A; De Souza, A; Zuniga, A; Dawe, M; Xiong, Y; Clive, D; Greiner, R; Nazyrova, A; Shaykhutdinov, R; Li, L; Vogel, HJ; Forsythe, I. HMDB: a knowledge-base for the human metabolome. *Nucleic Acids Research*, 2009 37, D603-D610.

[31] Brown, M; Dunn, WB; Ellis, DI; Goodacre, R; Handl, J; Knowles, JD; O'Hagan, S; Spasic, I; Kell, DB. A metabolome pipeline: from concept to data to knowledge. *Metabolomics*, 2005 1, 39-51.

[32] Dunn, WB. Current trends and future requirements for the mass spectrometric investigation of microbial, mammalian and plant metabolomes. *Phys. Biol.,* 2008 5, 1-24.

[33] Lu, W; Bennett BD; Rabinowitz, JD. Analytical strategies for LC-MS-based targeted metabolomics. *J. Chromatogr. B*, 2008 871, 236-242.

[34] Griffin, JL. Metabonomics: NMR spectroscopy and pattern recognition analysis of body fluids and tissues for characterisation of xenobiotic toxicity and disease diagnosis. *Curr. Opin. Chem. Biol.*, 2003 7, 648-654.

[35] Want, EJ; Nordstrom, A; Morita, H; Siuzdak, G. From Exogenous to Endogenous: The Inevitable Imprint of Mass Spectrometry in Metabolomics. *J. Proteome Res.*, 2007 6, 459-468.

[36] Ward, JL; Baker, JM; Beale, MH. Recent applications of NMR spectroscopy in plant metabolomics. *FEBS J.* 2007 274, 1126-1131.

[37] Pan, Z; Raftery, D. Comparing and combining NMR spectroscopy and mass spectrometry in metabolomics. *Anal. Bioanal. Chem.*, 2007 387, 525-527.

[38] Dettmer, K.; Aronov, PA; Hammock, BD. Mass spectrometry-based metabolomics. *Mass Spectrom. Rev.*, 2007 26, 51-78.

[39] Hollywood, K; Brison, DR; Goodacre, R. Metabolomics: current technologies and future trends. *Proteomics*, 2006 6, 4716-4723.

[40] Gartland, KPR; Beddell, CR; Lindon, JC; Nicholson, JK. Application of pattern recognition methods to the analysis and classification of toxicological data derived from proton nuclear magnetic resonance spectroscopy of urine. *Mol. Pharmacol.*, 1991 39, 629-642.

[41] Insilicos_Viewer. http://www.insilicos.com/Insilicos Viewer.html.

[42] Nobeli, I; Ponstingl, H; Krissinel, EB; Thornton, JM. A Structure-based Anatomy of the E. coli Metabolome. *J. Mol. Biol.*, 2003 334, 697-719.

[43] Dunn, WB; Ellis, DI. Metabolomics: Current analytical platforms and methodologies. *Trends Anal. Chem.*, 2005 24, 285-93.

[44] Halket, JM; Waaterman, D; Przyborowska, AM; Patel, RKP; Fraser, PD; Bramley, PM. Chemical derivatization and mass spectral libraries in metabolic profiling by GC/MS and LC/MS/MS. *J Exp. Bot.*, 2005 56, 219-43.

[45] Cited in "Gates, SC; Sweeley, CC. Quantitative metabolic profiling based on gas chromatography. *Clin. Chem.*, 1978 24, 1663-1673".

[46] Sjövall, J. Separation and determination of bile acids. *Methods Biochem. Anal.*, 1964 12, 97-141.

[47] Horning, MG; Knox, KL; Dalgliesh, CE; Horning, EC. *Anal. Biochem.*, 1966 17, 244-257.

[48] Horning, EC. (1968) Gas phase analytical methods for the study of steroid hormones and their metabolites. In K. B. Eik-Nes & E. C. Horning Ed. *Gas Phase Chromatography of steroids*. pp 1-71 Springer Verlang, Berlin, Germany.

[49] Sandberg, DH; Sjövall, J; Sjövall, K; Turner, DA. Measurement of human serum bile acids by gas-liquid chromatography. *J. Lipid Res.*, 1965 6, 182-192.

[50] Jolley, RL; Freeman, ML. Automated carbohydrate analysis of physiologic fluids. *Clin. Chem.*, 1968 14, 538-547.

[51] Burtis, CA; Goldstein, G; Scott, CD. Fractionation of human urine by gel chromatography *Clin. Chem.*, 1970 16, 201-206.

[52] Kuntzman, R; Welch, RM; Conney, AH. Factors influencing steroid hydroxylases in liver microsomes. *Advances in Enzyme Regulation*, 1966 4, 149-160.

[53] Horning EC; Horning MG. Metabolic profiles: gas phase methods for analysis of metabolites. *Clin. Chem.*, 1971 17, 802-809.

[54] Horning, EC; Horning, MG; Szafranek, J; Van Hout, P; German, AL; Thenot, JP; Pfaffenberger, CD. Gas-phase analytical methods for the study of human metabolites. Metabolic profiles obtained by open tubular capillary chromatography. *J. Chromatogr.*, 1974 91, 367-378.

[55] Golay, MJE. (1958) Theory of chromatography in open and coated tubular column with round or rectangular cross-sections. In A. Zlatkis, ed. *Gas Chromatography* 1958, Proc. Symp. 36-53; Amsterdam, The Netherland.

[56] Zlatkis, A; Kaufman, HR. Use of coated tubing as columns for gas chromatography. *Nature*, 1959 184, Suppl. No.26, 2010.

[57] Beynon, JH. The use of the mass spectrometer for the identification of organic compounds. *Microchimica Acta*, 1956 44, 437-453.

[58] Beynon, JH; Clough, S. A mass spectrometer mass marker. *J. Scientific instruments*, 1958 35, 289-291.

[59] Golhke, RS. Time-of-flight mass spectrometry and gas-liquid partition chromatography. *Anal. Chem.*, 1959 31, 535-541.

[60] Ryhage, R. Use of a mass spectrometer as a detector and analyzer for effluents emerging from high temperature gas liquid chromatography columns. *Anal. Chem.*, 1964 36, 759-764.

[61] Hites, RA; Biemann, K. A computer-compatible digital data acquisition system for fast-scanning, single-focusing mass spectrometers. *Anal. Chem.*, 1967 39, 965-970.

[62] Hites, RA; Biemann, K. Mass spectrometer-computer system particularly suited for gas chromatography of complex mixtures. *Anal. Chem.*, 1968 40, 1217-1231.

[63] Biemann, K; Cone, C; Webster, BR. Computer-aided interpretation of high-resolution mass spectra. II. Amino acid sequence of peptides. *J. Am. Chem. Soc.*, 1966 88, 2597-2598.

[64] Paul, W; Reinhard, HP; Zahn, O. The electric mass filter as mass spectrometer and isotope separator. *Zeitschrift fuer Physik*, 1958 152, 143-182.

[65] Finnigan, RE. Quadrupole mass spectrometers. *Anal. Chem.*, 1994 66, 969A-975A.

[66] Ingame, AL. Real-time gas chromatography/high resolution mass spectrometry and its application to the analysis of physiological fluids. *J. Chromatogr. Sci.*, 1974 12, 647-55.

[67] Syca, JEP. (1995) Commercialization of the quadrupole ion trap; in R.E. March, & J.F.J. Todd eds., *Particle aspects of ion trap Mass Spectrometry*. Vol.1 CRC Press, Boca Raton, FL,USA.

[68] Yost, RA; Enke, CG. Selected ion fragmentation with a tandem quadrupole mass spectrometer. *J. Am. Chem. Soc.*, 1978 100, 2274-2275.

[69] Zlatkis, A; Poole, CF; Brazell, R; Lee, KY; Hsu, F; Singhawangcha, S. Profiles of organic volatiles in biological fluids as an aid to the diagnosis of disease. *Analyst*, 1981 106, 352-360.

[70] Zlatkis, A; Bertsch, W; Bafus, DA; Liebich, HM. Analysis of trace volatile metabolites in serum and plasma. *J. Chromatogr.*, 1974 91, 379-383.

[71] Zlatkis, A; Bertsch, W; Lichtenstein, HA; Tishbee, A; Shunbo, F; Liebich, HM; Coscia, AM; Fleischer, N. Profile of volatile metabolites in urine by gas chromatography-mass spectrometry. *Anal. Chem.*, 1973 45, 763-767.

[72] Pauling, L; Robinson, AB; Teranishi, R; Cary, P. Quantitative analysis of urine vapor and breath by gas-liquid partition chromatography. *Proc. NatI. Acad. Sci.*, 1971 68, 2374-2376.

[73] Novotny, M; Maskarinec, MP; Steverink, ATG; Farlow, R. High-resolution gas chromatography of plasma steroidal hormones and their metabolites. *Anal. Chem.*, 1976 48, 468-472.

[74] Setchel, KDR; Almé, B; Axelson, M; Sjövall, J. The multicomponent analysis of conjugates of neutral steroids in urine by lipophilic ion exchange chromatography and

computerized gas chromatography-mass spectrometry. *J. Steroid Biochem.*, 1976 7, 615-629.

[75] Witten, TA; Levine, SP; Killian, MT; Boyle PJ; Markey SP. Gas-chromatographic-mass-spectrometric determination of urinary acid profiles of normal young adults. II. The effect of ethanol. *Clin. Chem.*, 1973 19, 963-966.

[76] Thompson, JA; Markey, SP. Quantitative metabolic profiling of urinary organic acids by gas chromatography-mass spectrometry: comparison of isolation methods. *Anal. Chem.*, 1975 47, 1313-1321.

[77] Jellum, E; Stokke, O; Eldjarn, L. Combined use of gas chromatography, mass spectrometry, and computer in diagnosis and studies of metabolic disorders. *Clin. Chem.*, 1972 18, 800-809.

[78] Molnar, I; Horvath, C. Rapid separation of urinary acids by high-performance liquid chromatography. *J. Chromatogr.*, 1977 143, 391-400.

[79] Gan, I; Korth, J; Halpern, B. Use of gas chromatography-mass spectrometry for the diagnosis and study of metabolic disorders. Screening and identification of urinary aromatic acids. *J. Chromatogr.*, 1974 92, 435-441.

[80] Jakobs, C; Solem, E; Ek, J; Halvorsen, K; Jellum, E. Investigation of the metabolic pattern in maple sirup urine disease by means of glass capillary gas chromatography and mass spectrometry. *J. Chromatogr.*, 1977 143, 31-38.

[81] Gates, SC; Sweeley, CC; Krivit, W; DeWitt, D. Automated metabolic profiling of organic acids in human urine. II. Analysis of urine samples from "healthy" adults, sick children, and children with neuroblastoma. *Clin. Chem.*, 1978 24, 1680-1689.

[82] Dirren, H; Robinson, AB; Pauling, L. Sex-related patterns in the profiles of human urinary amino acids. *Clin. Chem.*, 1975 21, 1970-1975.

[83] Liebich, HM; Al-Babbili, O; Zlatkis, A; Kim, K. Gas-chromatographic and mass-spectrometric detection of low-molecular-weight aliphatic alcohols in urine of normal individuals and patients with diabetes mellitus. *Clin. Chem.*, 1975 21, 1294-1296.

[84] Phillips, RD; Jennings, DH. Succulence, cations and organic acids in leaves of Kalanchoe daigremontiana grown in long and short days in soil and water culture. *New Phytologist*, 1976 77, 333-339.

[85] Groneman, AF; Posthumus, MA; Tuinstra, LGM; Traag, WA. Identification and determination of metabolites in plant cell biotechnology by gas chromatography and gas chromatography/mass spectrometry. Application to nonpolar products of Chrysanthemum cinerariaefolium and Tagetes species. *Anal. Chim. Acta*, 1984 163, 43-54.

[86] Biller, JE; Biemann, K. Reconstructed mass spectra, a novel approach for the utilization of gas chromatography-mass spectrometer data. *Anal. Lett.*, 1974 7, 515-528.

[87] McLafferty, FW; Hertel, RH; Villivock, BD. Computer identification of mass spectra. VI. Probability based matching of mass spectra. Rapid identification of specific compounds in mixtures. *Org. Mass Spectrom.*, 1974 9, 690-702.

[88] Sweeley, CC; Young, ND; Holland, JF; Gates SC. Rapid computerized identification of compounds in complex biological mixtures by gas chromatography-mass spectrometry. *J. Chromatogr.*, 1974 99, 507-517.

[89] Reimendal, R; Sjövall, J. Computer evaluation of gas chromatographic-mass spectrometric analyses of steroids from biological materials. *Anal. Chem.*, 1973 45, 1083-1089.

[90] Gates, SC; Smisko, MJ; Ashendel, CL; Young, ND; Holland, JF; Sweeley, CC. Automated simultaneous qualitative and quantitative analysis of complex organic mixtures with a gas chromatography-mass spectrometry-computer system. *Anal. Chem.*, 1978 50, 433-441.

[91] Kirkland, JJ. High speed liquid partition chromatography with chemically bonded organic stationary phases. *J. Chromatogr. Sci.*, 1971 9, 206-214.

[92] Sebestian, I; Halasz, I. Chemically bonded monomeric stationary phases with silicon-carbon bonds for gas and liquid chromatography. *Chromatographia*, 1974 7, 371-375.

[93] Novotny, M; Alasandro, M; Konishi, M. Microcolumn liquid chromatography of benzoyl derivatives of steroid metabolites. *Anal. Chem.*, 1983 55, 2375-2377.

[94] Tsuda, T; Novotny, M. Packed microcapillary columns in high performance liquid chromatography. *Anal. Chem.*, 1978 50, 271-275.

[95] Klink, FE. [2001] Mass spectrometry. Liquid chromatography/Mass Spectrometry. In R. E. Meyers Ed. *Enciclopedia of Analytical Chemistry*, vol 13 pp 11805-11809 Ramtech Ltd, Tarzana, CA, USA.

[96] Carroll, DI; Dzidic,I; Haegele, KD; Stilllwell, RN; Horning, EC. Packed microcapillary columns in high performance liquid chromatography. *Anal. Chem.*, 1975 47, 2369-2373.

[97] Yamashita, M; Fenn, JB. Electrospray ion source. Another variation on the free-jet theme. *J. Phys. Chem.*, 1984 88, 4451-4459.

[98] Karas, M; Bachmann, D; Bahr, U; Hillenkamp, K. Matrix-assisted ultraviolet laser desorption of non-volatile compounds. *Int. J. Mass Spectrom. Ion processes*, 1987 78, 53-68.

[99] Cole, RB. (1997) *Electrospray ionization mass spectrometry: fundamentals, instrumentation, and applications.* Ed. Wiley-Interscience, New York,USA

[100] Comisarow, MB; Marshall, AG. Fourier transform ion cyclotron resonance spectroscopy. *Chem. Fis. Lett.*, 1974 25, 282-283.

[101] Hager, JW. A new linear ion trap mass spectrometer. *Rapid commun. Mass Spectrom.*, 2002 16, 512-526.

[102] Hu, Q; Noll, RJ; Li, H; Makarov, A; Hardmann, M; Cooks, RG. The Orbitrap: A new mass spectrometer. *J. Mass Spectrom.*, 2005 40, 430-443.

[103] Morris, HR; Paxton, T; Dell, A; Langhorne, J; Berg, M; Bordoli, RS; Hoyes, J; Bateman, RH. High sensitivity collisionally-activated decomposition tandem mass spectrometry on a novel quadrupole/orthogonal-acceleration time-of-flight mass spectrometer. *Rapid Commun. Mass Spectrom.*, 1996 10, 889-896.

[104] Jorgenson, JW; Lukacs, KDA. Zone electrophoresis in open-tubular glass capillaries. *Anal. Chem.*, 1981 53, 1298-1302.

[105] Olivares, JA; Nguyen, NT; Yanker, CR; Smith, RD. On-line mass spectrometric detection for capillary zone electrophoresis. *Anal. Chem.*, 1987 59, 1230-1232.

[106] Presto Elgstoen, KB; Zhao, JY; Anacleto, JF; Jellum, E. Potential of capillary electrophoresis, tandem mass spectrometry and coupled capillary electrophoresis-tandem mass spectrometry as diagnostic tools. *J. Chromatog. A*, 2001 914, 265-275.

[107] Dandeneau, RD; Zerenner, EH. An investigation of glasses for capillary chromatography. *High Resolut. Chromatogr. Chromatogr. Commun.*, 1979 1, 351–356.

[108] Mamer, OA. Metabolic profiling: a di-lemma for mass spectrometry. *Biol. Mass Spectrom.*, 1994 23, 535-539.

[109] Gelpi, E. Trends in biochemical and biomedical applications of mass spectrometry. *Intern. J. Mass Spectrom. and Ion Processes*, 1992 118-119, 683-721.

[110] Vrbanac, JJ; Sweeley, CC; Pinkston, JD. Automated metabolic profiling analysis of urinary steroids by a gas chromatography mass spectrometry data system. *Biomed. Mass Spectrom.*, 1983 10, 155-161.

[111] Vrbanac, JJ; Braselton, WE Jr; Holland, JF; Sweeley, CC. Automated qualitative and quantitative metabolic profiling analysis of urinary steroids by a gas chromatography-mass spectrometry-data system. *J. Chromatogr.*, 1982 239, 265-276.

[112] Wolthers, BG; Kraan, GPB. Clinical applications of gas chromatography and gas chromatography-mass spectrometry of steroids. *J. Chromatogr. A*, 1999 843, 247-274.

[113] Honour, JW. Steroid profiling. *Annals of Clin. Biochem.*, 1997 34, 32-44.

[114] Sauter, H; Lauer, M; Fritsch, H. Metabolic profiling of plants. A new diagnostic technique. ACS Symposium Series 1991 443 (Synth. Chem. Agrochem. 2), 288-299.

[115] Graham, TL. A rapid, high-resolution high performance liquid chromatography profiling procedure for plant and microbial aromatic secondary metabolites. *Plant Physiology*, 1991 95, 584-593.

[116] Grant, BR; Greenaway, W; Whatley, FR. Metabolic changes during development of Phytophthora palmivora examined by gas chromatography/mass spectrometry. *J. General Microbiol.*, 1988 134, 1901-1911.

[117] Zechman, JM; Aldinger, S; Labows, JN Jr. Characterization of pathogenic bacteria by automated headspace concentration-gas chromatography. *J. Chromatog. B*, 1986 377, 49-57.

[118] Fiehn, O. Extending the breadth of metabolite profiling by gas chromatography coupled to mass spectrometry. *Trends in Analytical Chemistry,* 2008 27, 261-269.

[119] Jiye, A; Trygg, J; Gullberg, J; Johansson, AI; Jonsson, P; Antti, H; Marklund, SL; Moritz, T. Extraction and GC/MS Analysis of the Human Blood Plasma Metabolome. *Anal. Chem.*, 2005 77, 8086-8094.

[120] Pasikanti, KK; Ho, PC; Chan, ECY. Development and validation of a gas chromatography/mass spectrometry metabonomic platform for the global profiling of urinary metabolites. *Rapid Commun. Mass Spectrom.*, 2008 22, 2984-2992.

[121] Renny, LC; Dunn, WB; Ellis, DI; Myers, J; Baker, PN; Kell, DB. Novel biomarkers for pre-eclampsia detected using metabolomics and machine learning. *Metabolomics*, 2005 1, 227-234.

[122] Major, HJ; Williams, R; Wilson, AJ; Wilson, IDA. Metabonomic analysis of plasma from Zucker rat strains using gas chromatography/mass spectrometry and pattern recognition. *Rapid Commun. Mass Spectrom.*, 2006 20, 3295-3302.

[123] Denkert, C; Budczies, J; Kind, T; Weichert, W; Tablack, P; Sehouli, J; Niesporek, S; Konsgen, D; Dietel, M; Fiehn, O. Mass spectrometry-based metabolic profiling reveals different metabolite patterns in invasive ovarian carcinomas and ovarian borderline tumors. *Cancer Res.*, 2006 66, 10795-10804.

[124] Kind, T; Tolstikov, V; Fiehn, O; Weiss, RH. A comprehensive urinary metabolomic approach for identifying kidney cancer. *Anal. Biochem.*, 2007 363, 185-195.

[125] Boernsen, KO; Gatzek, S; Imbert, G. Controlled Protein Precipitation in Combination with Chip-Based Nanospray Infusion Mass Spectrometry. An Approach for Metabolomics Profiling of Plasma. *Anal. Chem.*, 2005 77, 7255-7264.

[126] Fancy, SA; Beckonert, O; Darbon, G; Yabsley, W; Walley, R; Baker, D; Perkins, GL; Pullen, FS; Rumpel, K. Gas chromatography/flame ionisation detection mass spectrometry for the detection of endogenous urine metabolites for metabonomic studies and its use as a complementary tool to nuclear magnetic resonance spectroscopy. *Rapid Commun. Mass Spectrom.*, 2006 20, 2271-2280.

[127] Qiu, Y; Su, M; Liu, Y; Chen, M; Gu, J; Zhang, J; Jia, W. Application of ethyl chloroformate derivatization for gas chromatography-mass spectrometry based metabonomic profiling. *Anal. Chim. Acta*, 2007 583, 277-283.

[128] Kuhara, T. Diagnosis of inborn errors of metabolism using filter paper urine, urease treatment, isotope dilution and gas chromatography-mass spectrometry. *J. Chromatogr. B*, 2001 758, 3-25.

[129] Zhang, Q; Wang, G; Du, Y; Zhu, L; Jiye, A. GC/MS analysis of the rat urine for metabonomic research. *J. Chromatogr. B*, 2007 854, 20-25.

[130] Shoemaker, JD; Elliott, WH. Automated screening of urine samples for carbohydrates, organic and amino acids after treatment with urease. *J. Chromatogr.*, 1991 562, 125-138.

[131] Weckwerth, W; Wenzel, K; Fiehn, O. A comprehensive urinary metabolomic approach for identifying kidney cancer. *Proteomics*, 2004 4, 78-83.

[132] Gullberg, J; Jonsson, P; Nordstrom, A; Sjostrom, M; Moritz, T. Design of experiments: an efficient strategy to identify factors influencing extraction and derivatization of Arabidopsis thaliana samples in metabolomic studies with gas chromatography/mass spectrometry. *Anal. Biochem.*, 2004 331, 283-295.

[133] Fiehn, O; Kopka, J; Trethewey, RN; Willmitzer, L. Identification of uncommon plant metabolites based on calculation of elemental compositions using gas chromatography and quadrupole mass spectrometry. *Anal. Chem.*, 2000 72, 3573-3580.

[134] Jonsson, P; Gullberg, J; Nordström, A; Kusano, M; Kowalczyk, M; Sjöström, M; Moritz, T. A strategy for identifying differences in large series of metabolomic samples analyzed by GC/MS. *Anal. Chem.*, 2004 76, 1738-1745.

[135] Arbona, V; Iglesias, DJ; Talón, M; Gómez-Cadenas, A. Plant phenotype demarcation using nontargeted LC-MS and GC-MS metabolite profiling. *J. Agric. Food Chem.*, 2009 57, 7338-7347.

[136] Villas-Boas, SG; Hojer-Pedersen, J; Akesson, M; Smedsgaard, J; Nielsen, J. Global metabolite analysis of yeast: evaluation of sample preparation methods. *Yeast*, 2005 22, 1155-1169.

[137] Schaub, J; Schiesling, C; Reuss, M; Dauner, M. Integrated Sampling Procedure for Metabolome Analysis. *Biotechnol. Progr.*, 2006 22, 1434-1442.

[138] Buchholz, A; Hurlebaus, J; Wandrey, C; Takors, R. Metabolomics: quantification of intracellular metabolite dynamics. *Biomol. Eng.*, 2002 19, 5-15.

[139] Koek, MM; Muilwijk, B; van der Werf, MJ; Hankemeier, T. Microbial metabolomics with gas chromatography/mass spectrometry. *Anal. Chem.*, 2006 78, 1272-1281.

[140] Hiller, J; Franco-Lara, E; Weuster-Botz, D. Metabolic profiling of Escherichia coli cultivations: Evaluation of extraction and metabolite analysis procedures. *Biotechnol. Lett.*, 2007 29, 1169-1178.

[141] Rabinowitz, JD; Kimball, E. Acidic Acetonitrile for Cellular Metabolome Extraction from Escherichia coli. *Anal. Chem.*, 2007 79, 6167-6173.

[142] Canelas, AB; ten Pierick, A; Ras, C; Seifar, RM; van Dam, IC; van Gulik, WM; Heijnen, JJ. Quantitative Evaluation of Intracellular Metabolite Extraction Techniques for Yeast Metabolomics. *Anal. Chem.*, 2009 81, 7379-7389.

[143] Bedair, M; Sumner, LW. Current and emerging mass-spectrometry technologies for metabolomics. *Trends in Analytical Chemistry*, 2008 27, 238-250.

[144] Price, NPJ. Acylic Sugar Derivatives for GC/MS Analysis of ^{13}C-Enrichment during Carbohydrate Metabolism. *Anal. Chem.*, 2004 76, 6566-6574.

[145] Begley, P; Francis-McIntyre, S; Dunn, WB; Broadhurst, DI; Halsall, A; Tseng, A; Knowles, J; Goodacre, R; Kell, DB. Development and performance of a gas chromatography-time-of-flight mass spectrometry analysis for large-scale nontargeted metabolomic studies of human serum. *Anal. Chem.*, 2009 81, 7038-7046.

[146] Liu, Z; Phillips, JB. Comprehensive two-dimensional gas chromatography using an on-column thermal modulator interface. *J. Chromatogr. Sci.*, 1991 29, 227-231.

[147] Boutilier, K; Ross, M; Podtelejnikov, AV; Orsi, C; Taylor, R; Taylor, P; Figeys, D. Comparison of different search engines using validated MS/MS test datasets. *Anal. Chim. Acta*, 2005, 534, 11-20.

[148] Dalluge, J; Beens, J; Brinkman, UATh. Comprehensive two-dimensional gas chromatography: a powerful and versatile analytical tool. *J. Chromatogr. A*, 2003, 1000, 69-108.

[149] Bertsch, W. Two-dimensional gas chromatography. Concepts, instrumentation, and applications - part 1: fundamentals, conventional two-dimensional gas chromatography, selected applications. *J. High Resol. Chromatogr.*, 1999 22, 647-665.

[150] Gorecki, T.; Harynuk, J; Panic, O. The evolution of comprehensive two-dimensional gas chromatography. *J. Sep. Sci.*, 2004 27, 359-379.

[151] Welthagen, W; Shellie, RA; Spranger, J; Ristow, M; Zimmermann, R; Fiehn, O. Comprehensive two-dimensional gas chromatography-time-of-flight mass spectrometry (GC/GC-TOF) for high resolution metabolomics: biomarker discovery on spleen tissue extracts of obese NZO compared to lean C57BL/6 mice. *Metabolomics*, 2005 1, 65-73.

[152] Shellie, RA; Welthagen, W; Zrostlikova, J; Spranger, J; Ristow, M; Fiehn, O; Zimmermann, R. Statistical methods for comparing comprehensive two-dimensional gas chromatography-time-of-flight mass spectrometry results: Metabolomic analysis of mouse tissue extracts. *J. Chromatogr. A*, 2005 1086, 83-90.

[153] Almstetter, MF; Appel, IJ; Gruber, MA; Lottaz, C; Timischl, B; Spang, R; Dettmer, K; Oefner, PJ. Integrative normalization and comparative analysis for metabolic fingerprinting by comprehensive two-dimensional gas chromatography-time-of-flight mass spectrometry. *Anal. Chem.*, 2009 81, 5731-5739.

[154] Wilson, ID; Plumb, R; Granger, J; Major, H; Williams, R; Lenz, EM. HPLC-MS-based methods for the study of metabonomics. *J. Chromatogr. B*, 2005 817, 67-76.

[155] Lenz, EM; Wilson, ID. Analytical strategies in metabolomics. *J. Proteome Res.*, 2007 6, 443-458.

[156] Romani, A; Vignolini, P; Galardi, C; Araldi, C; Vazzana, C; Heimler, D. Polyphenolic content in different plant parts of soy cultivars grown under natural conditions. *J. Agric. Food Chem.*, 2003 51, 5301-5306.

[157] Cavaliere, C; Cucci, F; Foglia, P; Guarino, C; Samperi, R; Laganà, A. Flavonoid profile in soybeans by high-performance liquid chromatography/tandem mass spectrometry. *Rapid Commun. Mass Spectrom.*, 2007 21, 1–12.

[158] Cavaliere, C; Foglia, P; Pastorini, E; Samperi, R; Laganà, A. Identification and mass spectrometric characterization of glycosylated flavonoids in Triticum durum plants by high-performance liquid chromatography with tandem mass spectrometry. *Rapid Commun. in Mass Spectrom.*, 2005 19, 3143-3158.

[159] Nordström, A; Want, E; Northen, T; Lehtiö, J; Siuzdak, G. Multiple ionization mass spectrometry strategy used to reveal the complexity of metabolomics. *Anal. Chem.*, 2008 80, 421-429.

[160] Patterson, AD; Li, H; Eichler, G; Krausz, KW; Weinstein, JN; Fornace, AJ, Jr; Gonzalez, FJ; Idle, JR. UPLC-ESI-TOFMS-based metabolomics and gene expression dynamics inspector self-organizing metabolomic maps as tools for understanding the cellular response to ionizing radiation. *Anal. Chem.*, 2008 80, 665-674.

[161] Wilson, ID; Nicholson, JK; Castro-Perez, J; Granger, JH; Johnson, KA; Smith, BW; Plumb, RS. High resolution "ultra performance" liquid chromatography coupled to Q-TOF Mass Spectrometry as a tool for differential metabolic pathway profiling in functional genomic studies. *J. Proteome Res.*, 2005 4, 591-598.

[162] Hodson, MP; Dear, GJ; Griffin, JL; Haselden, JN. An approach for the development and selection of chromatographic methods for high-throughput metabolomic screening of urine by ultra pressure LC-ESI-ToF-MS. *Metabolomics*, 2009 5, 166-182.

[163] Moco, S; Bino, RJ; Vorst, O; Verhoeven, HA; de Groot, J; van Beek, TA; Vervoort, J; de Vos, CHR. A liquid chromatography-mass spectrometry-based metabolome database for tomato. *Plant Physiol.*, 2006 141, 1205-1218.

[164] De Vos, RCH; Moco, S; Lommen, A; Keurentjes, JJB; Bino, RJ; Hall, RD. Untargeted large-scale plant metabolomics using liquid chromatography coupled to mass spectrometry. *Nat. Protoc.*, 2007 2, 778-791.

[165] Rijk, JCW; Lommen, A; Essers, ML; Groot, MJ; Van Hende, JM; Doeswijk, TG; Nielen, MWF. Metabolomics approach to anabolic steroid urine profiling of bovines treated with prohormones. *Anal. Chem.*, 2009 81, 6879-6888.

[166] Harry, EL; Weston, DJ; Bristow, AWT; Wilson, ID; Creaser, CS. An approach to enhancing coverage of the urinary metabonome using liquid chromatography-ion mobility-mass spectrometry. *J. Chromatogr. B*, 2008 871, 357-361.

[167] Bruce, SJ; Tavazzi, I; Parisod, V; Rezzi, S; Kochhar, S; Guy, PA. Investigation of human blood plasma sample preparation for performing metabolomics using ultrahigh performance liquid chromatography/mass spectrometry. *Anal. Chem.*, 2009 81, 3285-3296.

[168] Evans, AM; DeHaven, CD; Barrett, T; Mitchell, M; Milgram, E. Integrated, nontargeted ultrahigh performance liquid chromatography/electrospray ionization tandem mass spectrometry platform for the identification and relative quantification of the small-molecule complement of biological systems. *Anal. Chem.*, 2009 81, 6656-6667.

[169] Flores-Valverde, AM; Hill, EM. Methodology for profiling the steroid metabolome in animal tissues using ultraperformance liquid chromatography-electrospray-time-of-flight mass spectrometry. *Anal. Chem.*, 2008 80, 8771-8779.

[170] Lutz, U; Lutz, RW; Lutz, WK. Metabolic profiling of glucuronides in human urine by LC-MS/MS and partial least-squares discriminant analysis for classification and prediction of gender. *Anal. Chem.*, 2006 78, 4564-4571.

[171] Thiocone, A; Farmer, EE; Wolfender, J-L. Screening for wound-induced oxylipins in Arabidopsis thaliana by differential HPLC-APCI/MS profiling of crude leaf extracts and subsequent characterisation by capillary-scale NMR. *Phytochem. Analysis*, 2008 19, 198-205.

[172] Iwasa, K; Cui, WH; Sugiura, M; Takeuchi, A; Moriyasu, M; Takeda, K. Structural analyses of metabolites of phenolic 1-benzyltetrahydroisoquinolines in plant cell cultures by LC/NMR, LC/MS, and LC/CD. *J. Nat. Prod.*, 2005 68, 992-1000.

[173] Tang, K; Smith, R. Physical/chemical separations in the break-up of highly charged droplets from electrosprays. *J. Am. Soc. Mass Spectrom.*, 2001 12, 343-347.

[174] Schmidt, A; Karas, M; Dulcks, T. Effect of different solution flow rates on analyte ion signals in nano-ESI MS, or: when does ESI turn into nano-ESI? *J. Am. Soc. Mass Spectrom.*, 2003 14, 492-1000.

[175] Tang, K; Page, J; Smith, R. Charge competition and the linear dynamic range of detection in electrospray ionization mass spectrometry. *J. Am. Soc. Mass Spectrom.*, 2004 15, 1416-1423.

[176] Cech, N; Enke, CG. Relating electrospray ionization response to nonpolar character of small peptides. *Anal. Chem.*, 2000 72, 2717-2723.

[177] Fisher, SM; Perkins, PD. Simultaneous multimode ion source for mass spectrometry. *Agilent technical note,* 2005.

[178] Giddings, J. (1991) *Unified Separation Science*. John Wiley & Sons Inc., New York, USA.

[179] McNair, GE; Lewis, KC; Jorgenson, JW. Ultrahigh-pressure reversed-phase liquid chromatography in packed capillary columns. *Anal. Chem.*, 1997 69, 983-989.

[180] Patel, KD; Jerkovic, AD; Link, JC; Jorgenson, JW. In-depth characterization of slurry packed capillary columns with 1.0-mm nonporous particles using reversed-phase isocratic ultrahigh-pressure liquid chromatography. *Anal. Chem.*, 2004 76, 5777-5786.

[181] Plumb, R; Rainville, P; Smith, B; Johnson, K; Castro-Perez, J; Wilson, I; Nicholson, J. Generation of ultrahigh peak capacity LC separations via elevated temperatures and high linear mobile-phase velocities. *Anal. Chem.*, 2006 78, 7278-7283.

[182] Cavaliere, C; Foglia, P; Gubbiotti, R; Sacchetti, P; Samperi, R; Laganà, A. Rapid-resolution liquid chromatography/mass spectrometry for determination and quantitation of polyphenols in grape berries. *Rapid Commun. Mass Spectrom.*, 2008 22, 3089-3099.

[183] Grata, E; Guillarme, D; Glauser, G; Boccard, J; Carrupt, P-A; Veuthey, JL; Rudaz, S; Wolfender, J-L. Metabolite profiling of plant extracts by ultra-high-pressure liquid chromatography at elevated temperature coupled to time-of-flight mass spectrometry. *J. Chromatogr. A*, 2009 1216, 5660-5668.

[184] Kirkland, JJ; Langlois, TJ; De Stefano, JJ. Fused core particles for HPLC columns. *American Laboratory (Shelton, CT, USA)*, 2007 39, 18-21.

[185] Hu, C; van Dommelen, J; van der Heijden, R; Spijksma, G; Reijmers, TH; Wang, M; Slee, E; Lu, X; Xu, G; van der Greef, J; Hankemeier, T. RPLC-ion-trap-FTMS Method for lipid profiling of plasma: method validation and application to p53 mutant mouse model. *J. Proteome Research*, 2008 7, 4982-4991.

[186] Shen, Y; Zhang, Y; Moore, RJ; Kim, J; Metz, TO; Hixon, KK; Zhao, R; Livesay, EA; Udseth, HR; Smith, RD. Automated 20 kpsi RPLC-MS and MS/MS with chromatographic peak capacities of 1000-1500 and capabilities in proteomics and metabolomics. *Anal. Chem.*, 2005 77, 3090-3100.

[187] Granger, J; Plumb, R; Castro-Perez, J; Wilson, ID. Metabonomic studies comparing capillary and conventional HPLC-Q-TOF MS for the analysis of urine from Zucker obese rats. *Chromatographia*, 2005 61, 375-380.

[188] Tolstikov, V; Lommen, A; Nakanishi, K; Tanaka, N; Fiehn, O. Monolithic silica-based capillary reversed-phase liquid chromatography/electrospray mass spectrometry for plant metabolomics. *Anal. Chem.*, 2003 75, 6737-6740.

[189] Tolstikov, V; Fiehn, O; Tanaka, N. (2007) Application of liquid chromatography-mass spectrometry analysis in metabolomics: Reversed-phase monolithic capillary chromatography and hydrophilic chromatography coupled to electrospray ionization-mass spectrometry. In: *Methods in Molecular Biology* 141-155. Humana Press Inc., Totowa, NJ, U SA

[190] Horie, K; Ikegami, T; Hosoya, K; Saad, N; Fiehn, O; Tanaka, N. Highly efficient monolithic silica capillary columns modified with poly(acrylic acid) for hydrophilic interaction chromatography. *J.Chromatogr.* A, 2007 1164 198-205.

[191] Idborg, H; Zamani, L; Edlund, P-O; Schuppe-Koistinen, I; Jacobsson, SP. Metabolic fingerprinting of rat urine by LC/MS. Part 1. Analysis by hydrophilic interaction liquid chromatography-electrospray ionization mass spectrometry. *J. Chromatogr. B*, 2005 828, 9-13.

[192] Cubbon, S; Bradbury, T; Wilson, J; Thomas-Oates, J. Hydrophilic interaction chromatography for mass spectrometric metabonomic studies of urine. *Anal. Chem.*, 2007 79, 8911-8918.

[193] Maxwell, EJ; Chen, DDY. Twenty years of interface development for capillary electrophoresis-electrospray ionization-mass spectrometry. *Anal. Chim. Acta*, 2008 627, 25-33.

[194] Schmitt-Kopplin, P; Frommberger, M. Capillary electrophoresis - mass spectrometry: 15 years of developments and applications. *Electrophoresis*, 2003 24, 3837-3867.

[195] Soga, T; Ohashi, Y; Ueno, Y; Naraoka, H; Tomita, M; Nishioka, T. Quantitative metabolome analysis using capillary electrophoresis mass spectrometry. *J. Proteome Res.*, 2003 2, 488-494.

[196] Edwards, JL; Chisolm, CN; Shackman, JG; Kennedy, RT. Negative mode sheathless capillary electrophoresis electrospray ionization-mass spectrometry for metabolite analysis of prokaryotes. *J. Chromatogr. A*, 2006 1106, 80-88.

[197] Soga, T; Baran, R; Suematsu, M; Ueno, Y; Ikeda, S; Sakurakawa, T; Kakazu, Y; Ishikawa, T; Robert, M; Nishioka, T; Tomita, M. Differential metabolomics reveals ophthalmic acid as an oxidative stress biomarker indicating hepatic glutathione consumption. *J. Biol. Chem.*, 2006 281, 16768-16776.

[198] Ishii, N; Nakahigashi, K; Baba, T; Robert, M; Soga, T; Kanai, A; Hirasawa, T; Naba, M; Hirai, K; Hoque, A; Ho, PY; Kakazu, Y; Sugawara, K; Igarashi, S; Harada, S; Masuda, T; Sugiyama, N; Togashi, T; Hasegawa, M; Takai, Y; Yugi, K; Arakawa, K; Iwata, N; Toya, Y; Nakayama, Y; Nishioka, T; Shimizu, K; Mori, H; Tomita, M. Multiple high-throughput analyses monitor the response of E. coli to perturbations. *Science*, 2007 316, 593-597.

[199] Ohashi, Y; Hirayama, A; Ishikawa, T; Nakamura, S; Shimizu, K; Ueno, Y; Tomita, M; Soga, T. Depiction of metabolome changes in histidine-starved Escherichia coli by CE-TOFMS. *Mol. Biosyst.*, 2008 4, 135-147.

[200] Yoshida, S; Imoto, J; Minato, T; Oouchi, R; Sugihara, M; Imai, T; Ishiguro, T; Mizutani, S; Tomita, M; Soga, T; Yoshimoto, H. Development of bottom-fermenting Saccharomyces strains that produce high SO_2 levels, using integrated metabolome and transcriptome analysis. *Appl. Environ. Microbiol.*, 2008 74, 2787-2796.

[201] Sato, S; Soga, T; Nishioka, T; Tomita, M. Simultaneous determination of the main metabolites in rice leaves using capillary electrophoresis mass spectrometry and capillary electrophoresis diode array detection. *Plant J.*, 2004 40, 151-163.

[202] Kinoshita, A; Tsukada, K; Soga, T; Hishiki, T; Ueno, Y; Nakayama, Y; Tomita, M; Suematsu, M. Roles of hemoglobin allostery in hypoxia-induced metabolic alterations in herythrocytes: simulation and its verification by metabolome analysis. *J. Biol. Chem.*, 2007 282, 10731-10741.

[203] Williams, BJ; Cameron, CJ; Workman, R; Broeckling, CD; Sumner, LW; Smith, JT. Amino acid profiling in plant cell cultures: an inter-laboratory comparison of CE-MS and GC-MS. *Electrophoresis*, 2007 28, 1371-1379.

[204] Soga, T; Igarashi, K; Ito, C; Mizobuchi, K; Zimmermann, H-P; Tomita, M. Metabolomic profiling of anionic metabolites by capillary electrophoresis mass spectrometry. *Anal. Chem.*, 2009 81, 6165-6174.

[205] Lapainis, T; Rubakhin, SS; Sweedler, JV. Capillary electrophoresis with electrospray ionization mass spectrometric detection for single-cell metabolomics. *Anal. Chem.*, 2009 81, 5858-5864.

[206] Kennedy, RT; Oates, MD; Cooper, BR; Nickerson, B; Jorgenson, JW. Microcolumn separations and the analysis of single cells. *Science*, 1989 246, 57-63.

[207] García-Pérez, I; Vallejoa, M; García A; Legido-Quigley, C; Barbasa, C. Metabolic fingerprinting with capillary electrophoresis. *J. Chromatogr.A*, 2008 1204, 130-139.

[208] Lafaye, A; Junot, C; Ramounet-le Gall, B; Fritsch, P; Tabet, J-C; Ezan, E. Metabolite profiling in rat urine by liquid chromatography/electrospray ion trap mass spectrometry. Application to the study of heavy metal toxicity. *Rapid Commun. in Mass Spectrom.*, 2003 17, 2541-2549.

[209] Sawada, Y; Akiyama, K; Sakata, A; Kuwahara, A; Otsuki, H; Sakurai, T; Saito, K; Hirai, MY. Widely targeted metabolomics based on large-scale MS/MS data for elucidating metabolite accumulation patterns in plants. *Plant and Cell Physiology*, 2009 50, 37-47.

[210] Tiller, PR; Yu, S; Castro-Perez, J; Fillgrove, KL; Baillie, TA. High-throughput, accurate mass liquid chromatography/tandem mass spectrometry on a quadrupole time-of-flight system as a 'first-line' approach for metabolite identification studies. *Rapid Commun. Mass Spectrom.*, 2008 22, 1053-1061.

[211] Stroh, JG; Petucci, CJ; Brecker, SJ; Huang, N; Lau, JM. Automated sub-ppm mass accuracy on an ESI-TOF for use with drug discovery compound libraries. *J. Am. Soc. Mass Spectrom.*, 2007 18, 1612-1616.

[212] Gika, HG; Theodoridis, GA; Wingate, JE; Wilson, ID. Within-day reproducibility of an HPLC-MS-based method for metabonomic analysis: application to human urine. *J. of Proteome Research*, 2007 6, 3291-3303.

[213] King, R; Fernandez-Metzler, C. The use of Q-trap technology in drug metabolism. *Curr. Drug Metab.*, 2006 7, 541-545.

[214] Giavalisco, P; Kohl, K; Hummel, J; Seiwert, B; Willmitzer, L. ^{13}C isotope-labeled metabolomes allowing for improved compound annotation and relative quantification in

liquid chromatography-mass spectrometry-based metabolomic research. *Anal. Chem.*, 2009 81, 6546-6551.

[215] Guo, K; Li, L. Differential ^{12}C-/^{13}C-Isotope dansylation labeling and fast liquid chromatography/mass spectrometry for absolute and relative quantification of the metabolome. *Anal. Chem.*, 2009 81, 3919-3932.

[216] Ding, J; Sorensen, CM; Zhang, Q; Jiang, H; Jaitly, N; Livesay, EA; Shen, Y; Smith, RD; Metz, TO. Capillary LC coupled with high-mass measurement accuracy mass spectrometry for metabolic profiling. *Anal. Chem.*, 2007 79, 6081-6093.

[217] Koulman, A; Woffendin, G; Narayana, VK; Welchman, H; Crone, C; Volmer, DA. High-resolution extracted ion chromatography, a new tool for metabolomics and lipidomics using a second-generation Orbitrap mass spectrometer. *Rapid Commun. Mass Spectrom.*, 2009 23, 1411-1418.

[218] Carr TW. (Ed.), (1984) *Plasma Chromatography*. Plenum Press, New York, USA.

[219] Guevremont, R; Siu, K; Wang, J; Ding, L. Combined ion mobility/time-of-flight mass spectrometry study of electrospray-generated Ions. *Anal. Chem.* 1997 69, 3959-3965.

[220] Van Pelt, CK; Zhang, S; Fung, E; Chu, I; Liu, T; Li, C; Korfmacher, WA; Henion, J. A fully automated nanoelectrospray tandem mass spectrometric method for analysis of Caco-2 samples. *Rapid Commun. Mass Spectrom.*, 2003 17, 1573-1578.

[221] Koulman, A; Cao, M; Faville, M; Lane, G; Mace, W; Rasmussen, S. Semi-quantitative and structural metabolic phenotyping by direct infusion ion trap mass spectrometry and its application in genetical metabolomics. *Rapid Commun. Mass Spectrom.*, 2009 23, 2253-2263.

[222] Breitling, R; Pitt, AR; Barrett, MP. Precision mapping of the metabolome. *Trends Biotech.*, 2006 24, 543-548.

[223] Stephen, CB; Kruppa, G; Dasseux, J-L. Metabolomics applications of FT-ICR mass spectrometry. *Mass Spectrom. Rev.*, 2005 24, 223-231.

[224] Aharoni, A; de Vos, CHR; Verhoeven, HA; Maliepaard, CA; Kruppa, G; Bino, R; Goodenowe, DB. Nontargeted metabolome analysis by use of Fourier transform ion cyclotron mass spectrometry. *Omics*, 2002 6, 217-234.

[225] Southam, AD; Payne, TG; Cooper, HJ; Arvanitis, TN; Viant, MR. Dynamic range and mass accuracy of wide-scan direct Infusion nanoelectrospray Fourier transform ion cyclotron resonance mass spectrometry-based metabolomics increased by the spectral stitching method. *Anal. Chem.*, 2007 79, 4595-4602.

[226] Payne, TG; Southam, AD; Arvanitis, TN; Viant, MR. A signal filtering method for improved quantification and noise discrimination in Fourier transform ion cyclotron resonance mass spectrometry-based metabolomics data. *J. Am. Soc Mass Spectrom.*, 2009 20, 1087-1095.

[227] Jones, JJ; Borgmann, S; Wilkins, CL; O'Brien, RM. Characterizing the phospholipid profiles in mammalian tissues by MALDI FTMS. *Anal. Chem.*, 2006 78, 3062-3071.

[228] Fraser, PD; Enfissi, EMA; Goodfellow, M; Eguchi, T; Bramley, PM. Metabolite profiling of plant carotenoids using the matrix-assisted laser desorption ionization time-of-flight mass spectrometry. *Plant J.*, 2007 49, 552-564.

[229] Shroff, R; Rulisek, L; Doubský, J; Svatoš, A. Acid-base-driven matrix-assisted mass spectrometry for targeted metabolomics. *PNAS*, 2009 106, 10092-10096.

[230] Edwards, JL; Kennedy RT. Metabolomic analysis of eukaryotic tissue and prokaryotes using negative mode MALDI time-of-flight mass spectrometry. *Anal. Chem.*, 2005 77, 2201-2209.

[231] Vaidyanathan, S; Jones, D; Ellis, J; Jenkins, T; Chong, C; Anderson, M; Goodacre, R. Laser desorption/ionization mass spectrometry on porous silicon for metabolome analyses: influence of surface oxidation. *Rapid Commun. Mass Spectrom.*, 2007 21, 2157-2166.

[232] Vaidyanathan, S; Goodacre, R. Quantitative detection of metabolites using matrix-assisted laser desorption/ionization mass spectrometry with 9-aminoacridine as the matrix. *Rapid Commun. Mass Spectrom.*, 2007 21, 2072-2078.

[233] Takats, Z; Wiseman, JM; Gologan, B; Cooks, RG. Mass spectrometry sampling under ambient conditions with desorption electrospray ionization. *Science*, 2004 306, 471-473.

[234] Cooks, RG; Ouyang, Z; Takats, Z; Wiseman, JM. Ambient Mass Spectrometry. *Science*, 2006, 311, 1566-1570.

[235] Chen, H.; Venter, A.; Cooks, R.G. Extractive electrospray ionization for direct analysis of undiluted urine, milk and other complex mixtures without sample preparation. *Chem. Commun.*, 2006, 2042-2044.

[236] Takats, Z; Wiseman, JM; Gologan, B; Cooks, RG. Electrosonic spray ionization. A gentle technique for generating folded proteins and protein complexes in the gas phase and for studying ion-molecule reactions at atmospheric pressure. *Anal. Chem.*, 2004 76, 4050-4058.

[237] Gu, H; Chen, H; Pan, Z; Jackson, AU; Talaty, N; Xi, B; Kissinger, C; Duda, C., Doug Mann, M; Raftery, D; Cooks, RG. Monitoring diet effects via biofluids and their implications for metabolomics studies. *Anal. Chem.*, 2007 79, 89-97.

[238] McDonnell, L.A; Heeren, RMA. Imaging mass spectrometry. *Mass Spectrometry Reviews*, 2007 26, 606-643.

[239] Castaing, R; Slodzian, G. Microanalysis by secondary ionic emission. *J. Microscopie*, 1962 1, 395-410.

[240] Liebl, HJ. Ion microprobe mass analyzer. *J. Appl. Phys.*, 1967 38, 5277-5283.

[241] Stoeckli, M; Farmer, TB; Caprioli, RB. Automated mass spectrometry imaging with a matrix-assisted laser desorption ionization time-of-flight instrument. *J. Am. Soc. Mass Spectrom.*, 1999 10, 67-71.

[242] Kleinfeld, AM; Kampf, JP; Lechene, C; Transport of 13C-oleate in adipocytes measured using multi imaging mass spectrometry. *J. Am, Soc. Mass Spectrom.*, 2004 15, 1572-1580.

[243] Seeley, EH; Caprioli, RM. Molecular imaging of proteins in tissues by mass spectrometry. *PNAS*, 2008 105, 18126-18131.

[244] Maharrey, S; Bastasz, R; Behrens, R; Highley, A; Hoffer, S; Kruppa, G; Whaley, J. High mass resolution SIMS. *Appl. Surf. Sci.*, 2004 231-232, 972-975.

[245] Taban, IM; Altelaar, AFM; Fuchser, J; van der Burgt, YE-M; McDonnell, LA; Baykut, G; Heeren, RMA. Imaging of peptides in the rat brain using MALDI-FTICR mass spectrometry. *J. Am. Soc. Mass Spectrom.*, 2006 18, 145-151.

[246] McDonnell, LA; Heeren, RMA; de Lange, RPJ; Fletcher, IW. Higher sensitivity secondary ion mass spectrometry of biological molecules for high resolution, chemically specific imaging. *J.Am. Soc. Mass Spectrom.*, 2006 17, 1195-1202.

[247] Altelaar, AFM; Klinkert, I; Jalink, K; de Lange, RPJ; Adan, RAH; Heeren, RMA; Piersma, SR. Gold-enhanced biomolecular surface imaging of cells and tissue by SIMS and MALDI mass spectrometry. *Anal. Chem.*, 2006 78, 734-742.

[248] Brunelle, A; Touboul, D; Laprévote, O. Biological tissue imaging with time-of-flight secondary ion mass spectrometry and cluster ion sources. *J. Mass Spectrom.*, 2005 40, 985-999.

[249] Wiseman, JM; Ifa, DR; Cooks, RG; Venter, A. Ambient molecular imaging by desorption electrospray ionization mass spectrometry. *Nat. Protoc.*, 2008 3, 517-524.

[250] Kertesz, V; Van Berkel, GJ; Vavrek, M; Koeplinger, KA; Schneider, BB; Covey, TR. Comparison of drug distribution images from whole-body thin tissue sections obtained using desorption electrospray ionization tandem mass spectrometry and autoradiography. *Anal. Chem.*, 2008 80, 5168-5177.

[251] Wiseman, JM; Puolitaival, SM; Takats, Z; Cooks, RG; Caprioli R. Mass spectrometric profiling of intact biological tissue by using desorption electrospray ionization. *Angew. Chem. Int. Ed.*, 2005 44, 7094-7097.

[252] Ifa, DR; Wiseman, JM; Song, QY; Cooks, RG. Development of capabilities for imaging mass spectrometry under ambient conditions with desorption electrospray ionization (DESI). *Int. J. Mass Spectrom.*, 2007 259, 8-15.

[253] Wiseman, JM; Ifa, DR; Song, Q; Cooks, RG. Tissue imaging at atmospheric pressure using desorption electrospray ionization (DESI) mass spectrometry. *Angew. Chem. Int. Ed.*, 2006 45, 7188-7192.

[254] Nemes, P; Vertes, A. Laser ablation electrospray ionization for atmospheric pressure, in vivo, and imaging mass spectrometry. *Anal. Chem.*, 2007 79, 8098-8106.

[255] Nemes, P; Barton, AA; Li, Y; Vertes, A. Ambient molecular imaging and depth profiling of live tissue by infrared laser ablation electrospray ionization mass spectrometry. *Anal. Chem.*, 2008 80, 4575-4582.

[256] Kovats, E. Gas chromatographic characterization of organic compounds. I. Retention indexes of aliphatic halides, alcohols, aldehydes, and ketones. *Helv. Chim. Acta*, 1958 41, 1915-1932.

[257] Gates, SC; Sweeley, CC. Quantitative metabolic profiling based on gas chromatography. *Clin. Chem.*, 1978 24, 1663-1673.

[258] Sumner, LW; Amberg, A; Barrett, D; Beger, R; Beale, MH; Daykin, C; Fan, T; Fiehn, O; Goodacre, R; Griffin, JL; Higashi, R; Kopka, J; Lindon, JC; Lane, AN; Marriott, P; Nicholls, AW; Reily, MD; Viant, M. Proposed minimum reporting standards for chemical analysis. Chemical Analysis Working Group (CAWG) Metabolomics Standards Initiative (MSI). *Metabolomics*, 2007 3, 211-221.

[259] Goodacre, R; Broadhurst, D; Smilde, AK; Kristal, BS; Baker, JD; Beger, R; Bessant, C; Connor, S; Capuani, G; Craig, A; Ebbels, T; Kell, DB; Manetti, C; Newton, J; Paternostro, G; Sjoestroem, M; Trygg, J; Wulfert, F. Proposed minimum reporting standards for data analysis in metabolomics. *Metabolomics*, 2007 3, 231-241.

[260] Lange, E; Tautenhahn, R; Nuemann, S; Gropl, C. Critical assessment of alignment procedures for LC-MS proteomics and metabolomics measurements. *BMC Bioinformatics*, 2008 9, 375-378.

[261] Issaq, HJ; Van, QN; Waybright, TJ; Muschik, GM; Veenstra, TD. Analytical and statistical approaches to metabolomics research. *J. Sep. Sci.*, 2009 32, 2183-2199.

[262] Strehmel, N; Hummel, J; Erban, A; Strassburg, K; Kopka, J. Retention index thresholds for compound matching in GC-MS metabolite profiling. *J. Chromatogr.B*, 2008 871, 182-190.

[263] Hoffmann, N; Stoye, J. ChromA: signal-based retention time alignment for chromatography-mass spectrometry. *Bioinformatics*, 2009 25, 2080-2081.

[264] Broeckling, CD; Reddy, IR; Duran, AL; Zhao, X; Sumner, LW. MET-IDEA: data extraction tool for mass spectrometry-based metabolomics. *Anal. Chem.*, 2006 78, 4334-4341.

[265] Sturm, M; Kohlbacher, O. TOPPView: An open-source viewer for mass spectrometry data. *J. of Proteome Research*, 2009 8, 3760-3763.

[266] Ott, MA; Vriend, G. Correcting ligands, metabolites, and pathways. *BMC Bioinformatics*, 2006 7, 517-532.

[267] Wishart, DS; Knox, C; Guo, AC; Eisner, R; Young, N; Gautam, B; Hau, DD; Psychogios, N; Dong, E; Bouatra, S; Mandal, R; Sinelnikov, R; Xia, I; Jia, J; Cruz, L; Lim, JA; Sobsey, E; Shrivastava, CA; Huang, S; Liu, P; Fang, P; Peng, L; Fradette, J; Cheng, D; Tzur, D; Clements, M; Lewis, A; De Souza, A; Zuniga, A; Dawe, M; Xiong, Y; Clive, D; Greiner, R; Nazyrova, A; Shaykhutdinov, L; Li, R; Vogel HJ; Forsythe, I. HMDB: a knowledge-base for the human metabolome. *Nucleic Acids Research*, 2009 37, D603-D610.

[268] Horai, H; Arita M; Nishioka, T. (2008) Comparison of ESI-MS spectra in MassBank data base. In: *Proceedings of the International Conference on Biomedical Engineering and Informatics, BMEI.* Hainan, China, Vol. 2, pp. 853-857, IEEE Computer society

[269] Kind, T; Fiehn, O. Metabolomic database annotations via query of elemental compositions: mass accuracy is insufficient even at less than 1 ppm. *BMC Bioinformatics*, 2006 7. No pp. given

[270] Kopka, J; Schauer, N; Krueger, S; Birkemeyer, C; Usadel, B; Bergmuller, E; Dormann, P; Weckwerth, W; Gibon, Y; Stitt, M; Willmitzer, L; Fernie, AR; Steinhauser D. GMD@CSB.DB: the Golm Metabolome Database. *Bioinformatics*, 2005 21, 1635-1638.

[271] *http://msi-workgroups.sourceforge.net/*

[272] Tohge, T; Fernie, AR. Web-based resources for mass-spectrometry-based metabolomics: A user's guide. *Phytochemistry*, 2009 70, 450–456.

In: Metabolomics: Metabolites, Metabonomics…
Editors: J.S. Knapp and W.L. Cabrera, pp. 163-180
ISBN: 978-1-61668-006-0

Chapter 4

PLANT ENVIRONMENTAL METABOLOMICS

Matthew P. Davey[*]
Department of Plant Sciences, University of Cambridge,
Downing Street, Cambridge, CB2 3EA, UK

Introduction

It was quoted in 1953 at the 'Changing flora of Britain' conference that 'we should mobilize a team which could tackle the problems, genetical, cytological, physiological, ecological and chemical, and see whether out of the available mass of material we can not only reach a settled nomenclature… but make a serious contribution to the problems of evolution' (Raven 1953). Nearly 60 years later, we are now starting to assemble such genomic and post-genomic teams with the appropriate infrastructure, technology and bioinformatic power to answer questions in plant ecology and evolution. Of course, the chemical component of the team can now be termed environmental metabolomics and is progression of the study of genes (genomics), mRNA (transcriptomics) and proteins (proteomics).

The main intention of plant metabolomics research is to provide an unbiased assessment of metabolism across multiple pathways. Ideally, all plant metabolites should be identified and quantified at a relevant temporal and spatial scale by untargeted metabolomic fingerprinting using mass spectrometry (Dunn WB 2005; Overy SA 2005) or NMR (Krishnan, Kruger et al. 2005; Colquhoun 2007) or by targeted, quantitative metabolite profiling (Shulaev, Cortes et al. 2008); to provide a comprehensive view of metabolism (Hurry, Strand et al. 2000; Last, Jones et al. 2007). Such global screening of the metabolites has been termed biochemical, or metabolic phenotyping (eg. Roessner, Willmitzer et al. 2002). This approach builds on the much valid work carried out by plant biologists such as Richard Dixon (Dixon 2001) and Jeffrey Harborne (Harborne 1999) to name but a very few. However, the ease of application and software to analyse results, alongside the increase in interdisciplinary science, has opened up such technology to more research fields to answer a

[*] E-mail address: mpd39@cam.ac.uk. (Corresponding author)

wider range of questions (Stitt and Fernie 2003; Davey, Bryant et al. 2004; Miller 2007; Bundy, Davey et al. 2009; Penuelas and Sardans 2009).

Many of the initial publications in metabolomics were on plant species, such as Fiehn, Kopka et al. (2000) and Roessner, Wagner et al. (2000) where the main aim was to identify the metabolic phenotype of different plant genotypes. Fiehn, Kopka et al. (2000) detected over 200 compounds in one sample run using gas-chromatography mass-spectrometry (GC-MS) and by using principal component analysis (PCA) managed to cluster groups of plants as to whether they are wild type or are genetic mutants. The approach was largely advanced for agronomical research (Kuiper 2001; Watkins, Hammock et al. 2001). However, it was not long before metabolomics approaches were being identified and used to help measure and predict a plants sensitivity or tolerance to environmental pressure and to better understand genetic variation and evolution within and between plant species (Trethewey, Krotzky et al. 1999; Jackson, Linder et al. 2002; Davey 2003; Gidman, Goodacre et al. 2003; Sumner, Mendes et al. 2003; Kunin, Vergeer et al. 2009).

Environmental metabolomics was finally defined as the application of metabolomics to the investigation of both free-living organisms obtained directly from the natural environment or laboratory conditions, where any laboratory experiments specifically serve to mimic scenarios encountered in the natural environment (Morrison N 2007). In the plant sciences, this includes a wide variety of environmental, ecological and evolutionary scenarios and questions. Naturally, the questions and research carried out under these headings are interchangeable. The environmental area would predominantly include research into natural abiotic and anthropogenic phytotoxic pollution effects on plants. There is an urgent need to assess the impact of largely anthropogenic pollutants entering and affecting plants from the atmosphere via stomata and/or by epithelial contact. Such conditions include carbon dioxide, methane, ozone, aerosols, sulphur and nitrogen oxides. Toxins, and other pressures such as nutrient availability, also affect the root (Lambers and Colmer 2005). Other abiotic factors that affect the whole plant include air or substrate temperature and water availability. The ecological research include areas of research largely based on biotic interactions such as herbivory, alleopathy, competition, pathogens, mychorrizal and fungal interactions and tissue decomposition. The evolutionary aspects would cover areas of research such as plant population spread and biogeography and identifying metabolic traits that have been selected for under a variety of environmental pressures. The identity of these metabolic traits may provide the assignment of function to genes and post-genomic processes. Together, these situations and questions apply to every ecosystem on every continent on Earth, and with up to 200,000 potential metabolites in the plant kingdom (Fiehn 2001), implies that there is much work to be carried out in plant environmental metabolomics (Shulaev, Cortes et al. 2008). This review will outline some of the advances made in such areas of plant environmental metabolomics.

Environment - Abiotic Interactions

Within the plant environmental physiology literature, there are many metabolic studies that target a certain class of compounds. These compounds are usually grouped into 'totals', such as 'total carbohydrates' or 'total phenolics' (Davey, Harmens et al. 2007). The advantage of a metabolic fingerprinting approach is that a range of metabolites are detected that cover diverse traits and pathways, such as defence, UV light, heat, cold and drought stress. This new information will be valuable in combination with genetic, transcript, protein and physiological data obtained for many species (Bohnert, Gong et al. 2006). Metabolites that can be identified as functional biomarkers may allow them to be targeted in future research in many more ecosystems. Some environmental responses will only result in temporal changes in metabolite concentrations (Sumner, Mendes et al. 2003). Therefore, as acclamatory, plastic and developmental metabolic changes occur in plants, trying to assess when the most appropriate time is to assess the metabolome of a plant to a certain environmental perturbation will need to be addressed by further basic research. Also, care should be taken when quantifying a plant's metabolome as results and conclusions could vary depending on whether the metabolite is recorded on a concentration or content basis (Koricheva 1999). Overall, this provides important information other than visual observations and gross traits such as biomass, which may take longer to quantify in long-lived plant species in the field and do not always indicate the internal stressors that plants are experiencing.

An overview of the use of metabolomics to assess how environmental pressures affect plants is given below:

Nitrogen Deposition

Some of the first publications on plant environmental metabolomics were carried out to assess how the global metabolic pool of shrub species was altered by increasing atmospheric N deposition. A quick screen approach was applied to assess whether Fourier-Transform-InfraRed (FT-IR) analysis followed by using Discriminant Function Analysis (DFA) of the FT-IR spectra could detect changes in Calluna vulgaris biochemistry caused by increased N deposition (Gidman 2004). The study nicely showed that increasing amount of N applied by wet deposition in misting units changed the global metabolism of the plants which was detectable. This work in open-top chambers was followed up by assessing the changes in FT-IR spectra in field sites that had controlled amounts of N, and additional watering, added to the experimental field plots (Gidman, Royston et al. 2005). This approach was taken further by successfully evaluating N deposition impacts on the landscape level where *Galium saxatile* (Heath bedstraw) samples were taken from sites with different levels of N deposition across the United Kingdom (Gidman, Stevens et al. 2006). Such techniques have shown that it is possible to detect the amount of N deposition, and the affect that it is likely to have on the plant, by using a quick, cheap diagnostic test for environmental pollution. Such approaches need to be tested in more plant communities, genotypes and pollutants to allow predictive modelling at the landscape level (Gidman et al. 2005).

Nutrient Deficiency

Converse to the problems of N deposition is the effect that nutrient deficiencies have on plants by altering biomass, yield and the allocation of resources to defence compounds against herbivores and pathogens (see Hermans, Hammond et al. 2006; Hoefgen and Nikiforova 2008 for a recent review on S deficiency). Although there have been no reported natural field studies, metabolomic information on the effect of N or S deficiencies have been obtained for *Arabidopsis thaliana* (Hirai, Yano et al. 2004) and tomato (Urbanczyk-Wochniak and Fernie 2005). Phosphorous (P) deficiency has also been studied in the roots of *Phaseolus vulgaris* which showed accumulation in carbohydrate and polyol concentrations (Hernandez, Ramirez et al. 2007) and in *Hordeum vulgare* (Barley) where slight P stress caused an accumulation of carbohydrates but severe P stress also increased metabolites that were related to ammonium metabolism (Huang, Roessner et al. 2008).

Salinity

Many plant species, such as coastal halophytic species living in estuaries, have adapted to growing in substrates that have a high saline concentration (Stewart GR 1979). Alongside the accumulation of compatible solutes, the wide variety of biochemical mechanisms that enable these species to survive in such saline environments has yet to be truly discovered and characterised. Taking a metabolite profiling approach, the very assumption that many of the metabolites that are traditionally termed compatible solutes can be questioned. For example, Gagneul, Ainouche et al. (2007) have shown in an in-depth study of the cellular compartmentalisation of metabolites in the halophyte *Limonium latifolium* that many compounds such as proline and betaine do not conform to the definition of compatible solutes, in that they maintain osmotic equilibrium under saline conditions. With an increase in salinisation of agricultural soils such an untapped metabolic resource is of huge economic value (Sanchez, Siahpoosh et al. 2008). Already, Johnson, Broadhurst et al. (2003) and Smith, Johnson et al. (2003) have shown global metabolic differences in tomato plants that were subjected to saline treatments. By using a FT-IR metabolic fingerprinting and chemometrics approach they were able to identify changes in metabolic fingerprints in varieties that were saline tolerant.

Drought

Completely understanding the metabolic basis of drought resistance in plants is of paramount important in today's, and the future, world as climate change, water availability and usage and land management continues to put pressure on successful plant growth and reproduction (Chaves, Maroco et al. 2003). Changes in metabolism of droughted and re-watered plants can be detected using NMR, where Pinheiro, Passarinho et al. (2004) found alterations in the abundance of carbohydrates and amino acids on an individual organ basis in *Lupinus albus* plants. Semel, Schauer et al. (2007) also studied the effect of drought on field-grown tomatoes and found that a hybrid between a commercial and wild type variety was more drought resistant then the commercial control. Also, the metabolite phenotype of the watered hybrid plants was similar to that measured in droughted plants to which Semel, Schauer et al. (2007) concluded that the wild type hybrid may be 'metabolically primed' for drought scenarios. On a field basis, Llusia, Penuelas et al. (2008) assessed whether manipulated water availability would affect emission rates of isoprenoids in Mediterranean shrublands where such compounds are important for pollination and anti-herbivory. There was a species-specific effect when under droughted conditions, as *Erica multiflora* decreased levels of isoprene emissions but *Globularia alypum* and *Pinus halepensis* increased terpene emissions.

Metals and Soil Pollution

There is also potential to use a metabolomics approach in identifying the mechanisms and selective factors that evolved traits involved in metal accumulation in plants such as *Thlaspi caerulescens* (Assuncao, Schat et al. 2003). Also, Bailey, Oven et al. (2003) have studied cadmium exposure in *Silene cucubalus* by NMR and PCA and found changes in abundance of a variety of metabolites, such as malic acid and glutamine. There is also a need to assess the impact that persistent organic pollutants have on plant metabolism. Phytoremediation is one way in which such pollutants could be removed from the environment, and by taking a metabolomics approach, it is possible to identify possible mechanisms in which phytoremediation could work (Narasimhan, Basheer et al. 2003; Ott, Aranibar et al. 2003; Van Aken 2008).

Atmospheric Carbon Dioxide (CO_2)

There has been much work on assessing how elevated concentrations of atmospheric CO_2 will alter plant growth and plant chemistry (Long, Ainsworth et al. 2004). In a world where there will be more C available to plants, changes in the concentration of phenolics, terpenes and structural polysaccharides may have knock-on affects at an ecosystem level as levels and rates of herbivory and litter decomposition are altered (Penuelas and Estiarte 1998). However, such changes in plant chemistry are dependent on the species inherent ability to acclimate to such new conditions, alongside the availability of other nutrients. Studies using LC-PDA-MS have identified changes in secondary metabolite concentrations and lignification in semi-natural Solardome experimental conditions (Davey, Bryant et al. 2004). The use of FACE (Free-Air CO_2 Enrichment) experiments allows plants to be grown in open-air fields at controlled elevated atmospheric CO_2 concentrations. Such large scale sites will allow more plant species, and material, to be obtained for metabolomics and other physiological measurements (Long, Ainsworth et al. 2004). Li, Sioson et al. (2006) have already studied the metabolic responses to elevated CO2 using FACE rings. They found differences in the metabolome and transcriptome of *Arabidopsis* grown at elevated CO_2. Not only were global metabolome differences observed, the metabolic response differed according to its ecotype implying some level of evolutionary adaptation at a species level to CO_2 concentrations.

Ozone

Low level tropospheric ozone (O_3) can severely damage, or even kill, plants by causing serious oxidative stress in the plant cells (Mittler 2002). Sensitivity to ozone is plant species specific, with some species in a community being resistant whilst others are not (Penuelas, Llusia et al. 1999). Such interspecific differences in resistance to ozone with the heightened ability to deal with oxidative stress, must rest with differences in genetic adaptation and the subsequent metabolic changes (Smirnoff 1998). One of the first publications on the effect of ozone pollution on plant tissues using a metabolomics approach was by Kontunen-Soppela, Ossipov et al. (2007). They studied the effect of ozone on white birch (*Betula pendula*) in field conditions and out of 339 metabolites identified by GC-MS and HPLC, 98 metabolites (such as increases in quercetins and decreases in triterpenoid concentrations) were associated with ozone treatments. More recently, (Cho 2008) investigated the effect that ozone had on rice plants and identified changes in amino acid concentrations in conjunction with alterations in gene transcript and protein expression.

Photoperiod

Day length differs across the globe and shifts annually so causing many phenological changes in nature. The subtle metabolic changes that occur when plants are exposed to different periods of light have started to be identified by Goodacre, York et al. (2003). They directly injected the leaf sap of *Pharbitis* into a mass spectrometer (Direct Injection Mass Spectrometry DIMS), and after DFA of the spectra were able to discriminate plants that were subjected to different photoperiods. Photoperiod also affects when buds burst. The subtle changes in C and N assimilation of developing leaves of Quaking Aspen (*Populus tremuloides*) have been analysed by GC-MS and PCA and HCA (Jeong, Jiang et al. 2004). This study is particularly nice as they incorporated other physiological measurements, such as leaf gas exchange. Identifying such changes in metabolism during photoperiod is important when trying to decipher the mechanisms involved in processes such as floral induction and the effect of climate change, especially warming.

UltraViolet Radiation

Ultra-Violet-B radiation (280-320nm) (UV-B) is an important abiotic stressor throughout the world, especially in polar regions where this has particularly adverse affects on plant growth (Day, Ruhland et al. 2001). UV-B does affect plant chemistry, in particular there has been much research on how this affects UV-absorbing compounds (mainly phenolics) (Lois 1994). Lake, Field et al. (2009) have successfully used a mixture of DIMS, HPLC and ms-ms to identify temporal changes in the metabolite fingerprints and phenylpropanoid and flavonoid metabolism in *Arabidopsis thaliana* exposed to elevated UV-B. However, there is still much work to be carried out to prove that increases in a variety of metabolites provide a protective function by blocking UV-B light before it reaches plant cells and to assess the role of metabolites in the consequences of UV-B exposure such as repairing damaged cells and other structures such as DNA (Lois 1994; Smirnoff 1998). Outside field conditions, cell cultures have also been used to study the changes in secondary metabolite profiles of the legume *Medicago truncatula* (Broeckling, Huhman et al. 2005) but there is also a call for metabolic research on UV-B to also be carried out in the field as the phenotype observed in controlled growth rooms may be different to those observed in field experiments (Kliebenstein 2004). There has also been discussion on whether UV-B induced changes in the metabolome of crop species may be beneficial to human health as many of the compounds that increase in abundance are phenolics and other antioxidants (Jansen 2008).

Temperature

Thermotolerance is an important, and somewhat expensive, trait for a plant and is likely to be a limiting factor in plant distribution (Woodward 1987; Browse and Lange 2004). There is significant variation in freezing tolerance even within a plant species (Zhen and Ungerer 2008) and changes in the metabolite content of the plant during cold temperatures may play an advantageous role in cell cryoprotection prior to freezing temperatures (Stitt and Hurry 2002). This process is known as cold acclimation (Thomashow 1999; Hurry, Strand et al. 2000). Such metabolic changes are likely to differ according to a plant species' inherent ability to adapt or acclimate to cold temperatures. Guy, Kaplan et al. (2008) have recently published a comprehensive review on the metabolomics of temperature stress. One of the first studies to assess the metabolome of temperature stressed plants was by Kaplan, Kopka et al. (2004). They used GC-MS, followed by PCA, to identify 143 and 311 metabolites in *Arabidopsis thaliana* that responded to heat or cold shock, respectively. They even identified changes in metabolite abundances that were previously not associated with temperature stress. Cook, Fowler et al. (2004) also reported the exploration of the metabolome of two contrasting ecotypes of *Arabidopsis thaliana* and Hannah, Wiese et al. (2006) have also analysed the metabolome, and transcriptome, of nine geographically diverse ecotypes of *Arabidopsis*. Again, both studies reported significant natural variation for freezing tolerances and the preceding acclamatory processes within the metabolome.

Outside the model species, we have identified significant metabolic changes, using DIMS, GC and HPLC in *Arabidopsis lyrata* spp. *petraea* grown from natural populations across Europe (Davey, Burrell et al. 2008; Davey, Woodward et al. 2009). In a similar study to assess the cold acclimated metabolome *Arabidopsis thaliana* using DIMS, Gray and Heath (2005) found 1187 masses (DIMS-Fourier Transform-Ion Cyclotron Resonance) of which about 8% significantly increased or decreased in intensity after seven days cold treatment. Such a large-scale assessment of the changes that plants make in metabolism to temperature enables insights into alterations in the different metabolic pools, the associated changes in gene transcripts and the possibility of identifying temperature related biomarkers (Browse and Lange 2004).

Ecology

Plant populations are affected by a variety of biotic interactions (Arany, de Jong et al. 2005) and metabolites, especially secondary metabolites, play a key role in surviving a multitude of ecological processes. The identification of such metabolites is likely to increase as metabolomics becomes a useable tool to assess ecological questions (Kliebenstein 2004; D'Auria and Gershenzon 2005). Most, if not all, plants are at risk of herbivory, pathogen and fungal attack and are in competition for resources from neighbouring plants. The effect that such biotic influences can have on the plants metabolome is reviewed below.

Herbivory

Herbivory is an important ecological and economical process. The induction and precise function of a wide variety of metabolites in non-commercial plant species remains to be obtained. The allocation of metabolites to either defence or growth functions has been the centre of physiological and ecological research and debate for a few decades (Herms and Mattson 1992; Hamilton, Zangerl et al. 2001). Allocation of carbon and nitrogen to classes of compounds such as phenolics for defence or amino acids for growth is complex, species-specific and resource limited (Jones and Hartley 1999; Davey, Harmens et al. 2007).

Metabolomics approaches have started to be used in assessing the metabolic alternations occurring in plants that were subjected to either a generalist or a specialist herbivore (Jansen, Allwood et al. 2009). Arany, de Jong et al. (2008) studied an inland and a coastal natural population of *Arabidopsis thaliana*. They detected differences in the metabolome of each species but more interestingly they detected that the differences in the metabolome, mainly in glucosinolate concentrations, affected the growth of specialist or generalist herbivores so implying chemical adaptation to herbivory type at a population level. Also, the study by Riipi, Haukioja et al. (2004) nicely show the within-season and between-year variation in leaf chemistry of mountain birch (*Betula pubescens*). They were able to detect temporal changes in leaf metabolites that may be used for anti-herbivory purposes, such as hydrolysable tannins and proanthocyanidins. This outlines the importance of assessing temporal scales when assessing metabolomic changes in the field, especially for ecological studies. Kant, Ament et al. (2004) have also used GC-MS screening techniques to detect volatile compounds that were emitted from tomato plants that were infested with spider mites and genetically-modified Aspen trees that over express sucrose phosphate synthetase in order to increase cell sucrose concentration and biomass was shown to change the metabolic composition of secondary metabolites associated with anti-herbivory (Hjalten, Lindau et al. 2007).

Competition

Plant competition in the field will be regulated by resources such as light, CO_2 and nutrients; allelopathy and the plants inherent capacity to compete. Such inter-specific competition is considered to influence the metabolome of plants, as shown by Gidman et al. (2003). They were able to identify chemical differences in the FT-IR spectra of *Brachypodium distachyion* when grown in competition with *Arabidopsis thaliana*. Interestingly, there were no detectable changes in the FT-IR spectra of *A. thaliana*, implying a species-specific response to competition.

Floral Scents

Floral scents are also important in agriculture, horticulture and in studying ecosystem function. Already, metabolomic approaches have been applied and it is hoped that the metabolomics approach will lead to a more detailed understanding of the underlying processes involved in floral scents, such as circadian rhythms, and the evolutionary ecology between plant and pollinator (Vainstein, Lewinsohn et al. 2001; Verdonk, de Vos et al. 2003; Fridman and Pichersky 2005).

Populations, Evolution and Genetics

Natural selection acts on variation in phenotypes, and understanding the origins and maintenance of this variation is the focus of ecological genetics (Jackson, Linder et al. 2002) and metabolic fingerprinting and profiling is currently being utilised for environmental genomics research to identify ecologically important genes and traits (Benfey and Mitchell-Olds 2008). To truly assign function to genes from metabolites, the best current approach is to use plant crossings and using single nucleotide polymorphism (SNP); quantitative trait loci (QTL) and amplified fragment length polymorphism (AFLP) techniques (Jansen and Nap 2001). Jansen and Nap (2001) have already successfully used the model tree *Populus* to identify candidate genes for regulated flavonoid biosynthesis, a class of compounds that have many ecological functions in plants (Morreel, Goeminne et al. 2006). Metabolomic approaches can also be used to link plant genotypes and phenotypes using either forward or reverse genetic approaches, however, in natural systems, the forward genetic approach is likely to take precedence over reverse genetics (Fiehn 2002). Therefore, metabolomics could help in assessing the evolutionary history of a plant species and it may also help to assess the evolution of the actual metabolites and pathways (Schwab 2003). For example, the metabolite profiles, mainly cyclitols by GC-MS, of eucalypts were obtained alongside related ecological data of each species, to assess the evolution of this genus in arid environments (Merchant, Richter et al. 2006). Also, the geographical and evolutionary diversity of glucosinolates, an anti-herbivore class of compounds in Brassicaceae (Windsor, Reichelt et al. 2005; Clauss, Dietel et al. 2006; Keurentjes, Fu et al. 2006). Intra and interspecific diversity in glucosinolates structures, or any other class of metabolites, may provide clues as to whether species and populations within a genus were evolved by parallel evolution or by a common ancestral phenotype (Windsor, Reichelt et al. 2005).

Genetic versus Environmental Influences on the Metabolome

The amount of influence that genetics or the environment has over metabolism is difficult to measure and interpret. A field trial to assess the influence of environment and genetics in metabolic variation in a number of Douglas-fir trees with a known genetical family history was carried out by Robinson, Ukrainetz et al. (2007). Here, metabolites in the tree xylem were examined by GC-MS and multivariate discriminant analysis. The metabolite phenotypes were largely associated with environmental site information rather than any associations with the known genetic family structure. However, more recently, Ossipova et al. (2008) set out to study how metabolomics and the associated chemometrics could be used to recognise two different genotypes and the metabolic phenotypes of field grown birch trees (*Betula pendula*) under elevated ozone treatments. From the GC-MS and LC-MS fingerprinting, they were able to discriminate the different genotypes and even able to discriminate which field the trees were grown in. However, there was less metabolic variation in the ozone treated plants when compared to the differences in genotype. Such interesting results showing the influence that genetics and environment have on the metabolome needs to be investigated further.

Populations

A major application of metabolomics in ecology is the understanding of why plants only grow in restricted areas. Plant adaptation to the local environment should result in traits that are relevant to the abiotic and biotic conditions of the plants realised niche (Hoffmann 2005). Plant environmental metabolomics can help us understand the adaptive significance of traits and gene functions as it is likely that many genes are expressed only in the realised niche, which may be difficult to replicate in the laboratory environment (Jackson, Linder et al. 2002). An example of the ecological application of metabolomics is the study of *Arabidopsis lyrata* ssp. *petraea*. Across Europe, this species occurs in small isolated populations in Wales, Scotland, Germany, Norway, Sweden and Iceland, usually growing on rocky or stony cliffs and shores. Genetic differences between the populations have been obtained, however, metabolomic fingerprinting using DIMS, HPLC and GC, followed by PCA, nicely differentiate the Welsh and Swedish populations, which also differ from the closely related *Arabidopsis thaliana* (Davey, Burrell et al. 2008). Along a similar vein, NMR has been used to characterise nine different ecotypes of *Arabidopsis thaliana* by NMR followed by PCA (Ward, Harris et al. 2003). Also, the different genotypes the same species of *Populus* trees can also be discriminated by GC-MS metabolite profiling (Robinson, Gheneim et al. 2005). Such approaches will ultimately aid the understanding of the complex genetic and environmental factors underlying plant metabolism and plant distribution.

A benefit of being able to identify a species ecotype, or genotype, by metabolic fingerprinting is that by using the correct statistical techniques it is possible to determine the geographical origin of the plant sample. The correct identification of a plants geographical origin can be made using techniques such as pyrolysis mass spectrometry and artificial neural networks (Salter, Lazzari et al. 1997). Here, olive oils were analysed by such methods and by assessing the spectra using training and test sets, oils were correctly assigned to its origin of growth in different regions of Italy. Such work on assessing and predicting the geographical origin of olive oils in Greece has also been carried out NMR fingerprinting (Petrakis, Agiomyrgianaki et al. 2008). Supervised modelling techniques, such as Partial Least Squares – Discriminant Analysis (PLS-DA) have already been successful in correctly classifying the Country of origin of wine samples using its chemical component (Capron, Smeyers-Verbeke et al. 2007). Another example of identifying metabolic differences between populations of the same plant species is in the medicinal herb *Ephedra sinica*. (Schaneberg, Crockett et al. 2003) showed that they were able to identify the geographical origin on a cross continental scale of plant extracts from this species using chemical fingerprinting. Even the origin of tea (*Camellia sinensis*) can be discriminated using metabolite profiling (Sultana, Stecher et al. 2008) and different types of tropical hardwoods can be characterised by FT-IR and NMR combined with processing data by PCA (Nuopponen, Wikberg et al. 2006).

Conclusion

It is becoming clear that plant environmental metabolomics will play an important part in the understanding and manipulation of the natural world (Wollenweber, Porter et al. 2005; Dixon, Gang et al. 2006; Schauer and Fernie 2006; Bundy, Davey et al. 2009). Such alterations in the metabolite fingerprints and phenotypes of plant populations in response to abiotic and biotic interactions need to be investigated. This will ultimately aid our understanding of the complex genetic and environmental factors underlying plant metabolism and plant distribution which is important for assessing how plants may respond to climatic change (Thomas, Cameron et al. 2004; Jump and Penuelas 2005). There is recognition that field-based measurements, alongside other traditional measurements in plant physiology, is required (Blanchard 2004). It is also clear that plant metabolomics will increasingly become integrated into the other omic approaches, after which the true power of the technology will become apparent (Fridman and Pichersky 2005; Usadel, Blasing et al. 2008).

Abbreviations

DFA : Discriminant Function Analysis;
DIMS : Direct Injection Mass Spectrometry;
FT-IR : FourierTransform-InfraRed;
GC : Gas Chromatography;
GC-MS : Gas Chromatography-Mass Spectrometry;
HPLC : High Performance Liquid Chromatography;
MS : Mass Spectrometry;
NMR : Nuclear Magnetic Resonance;
PCA : Principal Component Analysis;
PDA : Photo Diode Array;
PLS-DA : Partial Least Squares-Discriminant Analysis

Some text within this article has been reproduced with kind permission from Springer Science+Business Media from the article by the author: Bundy, J., M. Davey, et al. (2009). "Environmental metabolomics: a critical review and future perspectives." Metabolomics 5(1): 3-21.

References

Arany, A. M., de Jong, T. J., et al. (2008). "Glucosinolates and other metabolites in the leaves of Arabidopsis thaliana from natural populations and their effects on a generalist and a specialist herbivore." *Chemoecology*, **18**(2), 65-71.

Arany, A. M., de Jong, T. J., et al. (2005). "Herbivory and abiotic factors affect population dynamics of Arabidopsis thaliana in a sand dune area." *Plant Biology*, **7**(5), 549-555.

Assuncao, A. G. L., Schat, H., et al. (2003). "Thlaspi caerulescens, an attractive model species to study heavy metal hyperaccumulation in plants." *New Phytologist*, **159**(2), 351-360.

Bailey, N. J. C., Oven, M., et al. (2003). "Metabolomic analysis of the consequences of cadmium exposure in Silene cucubalus cell cultures via H-1 NMR spectroscopy and chemometrics." *Phytochemistry*, **62**(6), 851-858.

Benfey, P. N. & T. Mitchell-Olds (2008). "Perspective - From genotype to phenotype: Systems biology meets natural variation." Science, **320***(***5875***)*, 495-497.

Blanchard, J. L. (2004). "Bioinformatics and Systems Biology, rapidly evolving tools for interpreting plant response to global change." *Field Crops Research*, **90**(1), 117-131.

Bohnert, H. J., Gong, Q. Q., et al. (2006). "Unraveling abiotic stress tolerance mechanisms - getting genomics going." *Current Opinion in Plant Biology*, **9**(2), 180-188.

Broeckling, C. D., Huhman, D. V., et al. (2005). "Metabolic profiling of Medicago truncatula cell cultures reveals the effects of biotic and abiotic elicitors on metabolism." *Journal of Experimental Botany*, **56**(410), 323-336.

Browse, J. & Lange, B. M. (2004). "Counting the cost of a cold-blooded life: Metabolomics of cold acclimation." *Proceedings of the National Academy of Sciences of the United States of America*, **101**(42), 14996-14997.

Bundy, J., Davey, M., et al. (2009). "Environmental metabolomics: a critical review and future perspectives." *Metabolomics*, **5**(1), 3-21.

Capron, X., Smeyers-Verbeke, J., et al. (2007). "Multivariate determination of the geographical origin of wines from four different countries." *Food Chemistry*, **101**(4), 1585.

Chaves, M. M., Maroco, J. P., et al. (2003). "Understanding plant responses to drought - from genes to the whole plant." *Functional Plant Biology*, **30**(3), 239-264.

Cho, K., Shibato, J., Agrawal, G. K., Jung, Y., Kubo, A., Jwa, N., et al (2008). "Integrated Transcriptomics, Proteomics, and Metabolomics Analyses To Survey Ozone Responses in the Leaves of Rice Seedling." *Journal of Proteome Research*, **7**, 2980-2998.

Clauss, M. J., Dietel, S., et al. (2006). "Glucosinolate and trichome defenses in a natural Arabidopsis lyrata population." *Journal Of Chemical Ecology*, **32**(11), 2351-2373.

Colquhoun, I. J. (2007). "Use of NMR for metabolic profiling in plant systems." *Journal of Pesticide Science*, **32**(3), 200-212.

Cook, D., Fowler, S., et al. (2004). "A prominent role for the CBF cold response pathway in configuring the low-temperature metabolome of Arabidopsis." *Proceedings of the National Academy of Sciences of the United States of America*, **101**(42), 15243-15248.

D'Auria, J. C. & Gershenzon, J. (2005). "The secondary metabolism of Arabidopsis thaliana: growing like a weed." *Current Opinion in Plant Biology*, **8**(3), 308-316.

Davey, M., Woodward, F. I., et al. (2009). "Intraspecfic variation in cold-temperature metabolic phenotypes of Arabidopsis lyrata ssp. petraea." *Metabolomics*, **5**(1), 138-149.

Davey, M. P. (2003). The effect of an elevated atmospheric CO2 concentration on secondary metabolism and resource allocation in Plantago maritima and Armeria maritima, Durham University, UK. Ph.D. Thesis.

Davey, M. P., Bryant, D. N., et al. (2004). "Effects of elevated CO2 on the vasculature and phenolic secondary metabolism of Plantago maritima." *Phytochemistry*, **65**(15), 2197-2204.

Davey, M. P., Burrell, M. M., et al. (2008). "Population-specific metabolic phenotypes of Arabidopsis lyrata ssp petraea." *New Phytologist*, **177**(2), 380-388.

Davey, M. P., Harmens, H., et al. (2007). "Species-specific effects of elevated Co-2 on resource allocation in Plantago maritima and Armeria maritima." *Biochemical Systematics And Ecology*, **35**(3), 121-129.

Day, T. A., Ruhland, C. T., et al. (2001). "Influence of solar ultraviolet-B radiation on Antarctic terrestrial plants: results from a 4-year field study." *Journal of Photochemistry and Photobiology B-Biology*, **62**(**1-2**), 78-87.

Dixon, R. A. (2001). "Phytochemistry in the genomics and post-genomics eras." *Phytochemistry*, **57**(2), 145-148.

Dixon, R. A., Gang, D. R., et al. (2006). "Perspective - Applications of metabolomics in agriculture." *Journal of Agricultural and Food Chemistry*, **54**(24), 8984-8994.

Dunn, W. B., Quick, O. S. W. P. (2005). "Evaluation of automated eletrospray-TOF mass spectrometry for metabolic fingerprinting of the plant metabolome." *Metabolomics*, **1**, 137-148.

Fiehn, O. (2001). "Combining genomics, metabolome analysis, and biochemical modelling to understand metabolic networks." *Comparative and Functional Genomics*, **2**(3), 155-168.

Fiehn, O. (2002). "Metabolomics - the link between genotypes and phenotypes." *Plant Molecular Biology*, **48**(**1-2**), 155-171.

Fiehn, O., Kopka, J., et al. (2000). "Metabolite profiling for plant functional genomics." *Nature Biotechnology*, **18**(11), 1157-1161.

Fridman, E. & Pichersky, E. (2005). "Metabolomics, genomics, proteomics, and the identification of enzymes and their substrates and products." *Current Opinion in Plant Biology*, **8**(3), 242-248.

Gagneul, D., Ainouche, A., et al. (2007). "A reassessment of the function of the so-called compatible solutes in the halophytic Plumbaginaceae Limonium latifolium." *Plant Physiology*, **144**(3), 1598-1611.

Gidman, E., Goodacre, R., et al. (2003). "Investigating plant-plant interference by metabolic fingerprinting." *Phytochemistry*, **63**(6), 705-710.

Gidman, E., Goodacre, R., Emmett, B., Sheppard, L., Leith, Ian. & Gwynn-Jones, D. (2004). " Applying Metabolic Fingerprinting to Ecology: The Use of Fourier-Transform Infrared Spectroscopy for the Rapid Screening of Plant Responses to N Deposition." Water, *Air and Soil Pollution*, **4**(6), 251-258.

Gidman, E. A., Royston, G., et al. (2005). "Metabolic fingerprinting for bio-indication of nitrogen responses in Calluna vulgaris heath communities " *Metabolomics*, **1**(3), 1573-3882.

Gidman, E. A., Stevens, C. J., et al. (2006). "Using metabolic fingerprinting of plants for evaluating nitrogen deposition impacts on the landscape level." *Global Change Biology*, **12**(8), 1460-1465.

Goodacre, R., York, E. V., et al. (2003). "Chemometric discrimination of unfractionated plant extracts analyzed by electrospray mass spectrometry." *Phytochemistry*, **62**(6), 859-863.

Gray, G. R. & Heath, D. (2005). "A global reorganization of the metabolome in Arabidopsis during cold acclimation is revealed by metabolic fingerprinting." *Physiologia Plantarum*, **124**(2), 236-248.

Guy, C., Kaplan, F., et al. (2008). "Metabolomics of temperature stress." *Physiologia Plantarum*, **132**(2), 220-235.

Hamilton, J. G., Zangerl, A. R., et al. (2001). "The carbon-nutrient balance hypothesis: its rise and fall." *Ecology Letters*, **4**(1), 86-95.

Hannah, M. A., Wiese, D,. et al. (2006). "Natural genetic variation of freezing tolerance in arabidopsis." *Plant Physiology*, **142**(1), 98-112.

Harborne, J. B. (1999). "Recent advances in chemical ecology." *Natural Product Reports*, **16**(4), 509-523.

Hermans, C., Hammond, J. P., et al. (2006). "How do plants respond to nutrient shortage by biomass allocation?" *Trends in Plant Science*, **11**(12), 610-617.

Herms, D. A. & Mattson, W. J. (1992). "The dilemma of plants - to grow or defend." *Quarterly Review of Biology*, **67**(4), 478-478.

Hernandez, G., Ramirez, M., et al. (2007). "Phosphorus stress in common bean: Root transcript and metabolic responses." *Plant Physiology*, **144**(2), 752-767.

Hirai, M. Y., Yano, M., et al. (2004). "Integration of transcriptomics and metabolomics for understanding of global responses to nutritional stresses in Arabidopsis thaliana." *Proceedings of the National Academy of Sciences of the United States of America*, **101**(27), 10205-10210.

Hjalten, J., Lindau, A. et al. (2007). "Unintentional changes of defence traits in GM trees can influence plant-herbivore interactions." *Basic and Applied Ecology*, **8**(5), 434-443.

Hoefgen, R. & Nikiforova, V. J. (2008). "Metabolomics integrated with transcriptomics: assessing systems response to sulfur-deficiency stress." *Physiologia Plantarum*, **132**(2), 190-198.

Hoffmann, M. H. (2005). "Evolution of the realized climatic niche in the genus Arabidopsis (Brassicaceae)." *Evolution*, **59**(7), 1425-1436.

Huang, C. Y., Roessner, U., et al. (2008). "Metabolite profiling reveals distinct changes in carbon and nitrogen metabolism in phosphate-deficient barley plants (Hordeum vulgare L.)." *Plant and Cell Physiology*, **49**(5), 691-703.

Hurry, V., Strand, A., et al. (2000). "The role of inorganic phosphate in the development of freezing tolerance and the acclimatization of photosynthesis to low temperature is revealed by the pho mutants of Arabidopsis thaliana." *Plant Journal*, **24**(3), 383-396.

Jackson, R. B., Linder, C. R., et al. (2002). "Linking molecular insight and ecological research." *Trends in Ecology & Evolution*, **17**(9), 409-414.

Jansen, J., Allwood, J., et al. (2009). "Metabolomic analysis of the interaction between plants and herbivores." *Metabolomics*, **5**(1), 150-161.

Jansen, M. A. K., Hectors, K., O'Brien, N. M., Guisez, Y. & Potters, G. (2008). "Plant stress and human health: Do human consumers benefit from UV-B acclimated crops?" Plant Science.

Jansen, R. C. & Nap, J. P. (2001). "Genetical genomics: the added value from segregation." *Trends in Genetics*, **17**(7), 388-391.

Jeong, M. L., Jiang, H. Y., et al. (2004). "Metabolic profiling of the sink-to-source transition in developing leaves of quaking aspen." *Plant Physiology*, **136**(2), 3364-3375.

Johnson, H. E., Broadhurst, D., et al. (2003). "Metabolic fingerprinting of salt-stressed tomatoes." *Phytochemistry*, **62**(6), 919-928.

Jones, C. G. & Hartley, S. E. (1999). "A protein competition model of phenolic allocation." *Oikos*, **86**(1), 27-44.

Jump, A. S. & Penuelas, J. (2005). "Running to stand still: adaptation and the response of plants to rapid climate change." *Ecology Letters*, **8**(9), 1010-1020.

Kant, M. R., Ament, K., et al. (2004). "Differential timing of spider mite-induced direct and indirect defenses in tomato plants." *Plant Physiology*, **135**(1), 483-495.

Kaplan, F., Kopka, J., et al. (2004). "Exploring the temperature-stress metabolome of Arabidopsis." *Plant Physiology*, **136**(4), 4159-4168.

Keurentjes, J. J. B., Fu, J. Y., et al. (2006). "The genetics of plant metabolism." *Nature Genetics*, **38**(7), 842-849.

Kliebenstein, D. J. (2004). "Secondary metabolites and plant/environment interactions: a view through Arabidopsis thaliana tinged glasses." *Plant Cell and Environment*, **27**(6), 675-684.

Kontunen-Soppela, S., Ossipov, V., et al. (2007). "Shift in birch leaf metabolome and carbon allocation during long-term open-field ozone exposure." *Global Change Biology*, **13**(5), 1053-1067.

Koricheva, J. (1999). "Interpreting phenotypic variation in plant allelochemistry: problems with the use of concentrations." *Oecologia*, **119**(4), 467-473.

Krishnan, P., Kruger, N. J., et al. (2005). "Metabolite fingerprinting and profiling in plants using NMR." *Journal of Experimental Botany*, **56**(410), 255-265.

Kuiper, H. A. (2001). "Environmental and food safety issues of genetically modified crops." *J Environ Monit*, **3**(2), 26N-32N.

Kunin, W. E., Vergeer, P., et al. (2009). "Variation at range margins across multiple spatial scales: environmental temperature, population genetics and metabolomic phenotype." Proceedings of the Royal Society B: *Biological Sciences*, **276***(***1661***)*, 1495-1506.

Lake, J. A., Field, K. J., et al. (2009). "Metabolomic and physiological responses reveal multi-phasic acclimation of Arabidopsis thaliana to chronic UV radiation." *Plant, Cell & Environment*, **32**, 1377-1389.

Lambers, H. & Colmer, T. D. (2005). "Root physiology - from gene to function - Preface." *Plant and Soil*, **274***(***1-2***)*, VII-XV.

Last, R. L., Jones, A. D., et al. (2007). "Towards the plant metabolome and beyond." *Nature Reviews Molecular Cell Biology*, **8**(2), 167-174.

Li, P. H., Sioson, A., et al. (2006). "Response diversity of Arabidopsis thaliana ecotypes in elevated [CO2] in the field." *Plant Molecular Biology*, **62***(***4-5***)*, 593-609.

Llusia, J., Penuelas, J., et al. (2008). "Contrasting species-specific, compound-specific, seasonal, and interannual responses of foliar isoprenoid emissions to experimental drought in a mediterranean shrubland." *International Journal of Plant Sciences*, **169**(5), 637-645.

Lois, R. (1994). "Accumulation of UV-absorbing flavonoids induced by UV-B radiation in Arabidopsis thaliana L. Mechanisms of UV-resistance in Arabidopsis." *Planta*, **194**(4), 498-503.

Long, S. P., E. Ainsworth, A., et al. (2004). "Rising atmospheric carbon dioxide: Plants face the future." *Annual Review of Plant Biology*, **55**, 591-628.

Merchant, A., Richter, A., et al. (2006). "Targeted metabolite profiling provides a functional link among eucalypt taxonomy, physiology and evolution." *Phytochemistry*, **67**(4), 402-+.

Miller, M. G. (2007). "Environmental metabolomics: A SWOT analysis (strengths, weaknesses, opportunities, and threats)." *Journal of Proteome Research*, **6**(2), 540-545.

Mittler, R. (2002). "Oxidative stress, antioxidants and stress tolerance." *Trends in Plant Science*, **7**(9), 405-410.

Morreel, K., Goeminne, G., et al. (2006). "Genetical metabolomics of flavonoid biosynthesis in Populus: a case study." *Plant Journal*, **47**(2), 224-237.

Morrison, N. B. D., Bundy, J., Collette, T., Currie, F., Davey, M. P., Haigh, N. S., Hancock, D., Jones, O., Rochfort, S., Sansone, S. A., Štys, D., Teng, Q., Field, D. & Viant, M. (2007). "Standard Reporting Requirements for Biological Samples in Metabolomics Experiments: Environmental Context." *Metabolomics*, **3**(3), 203-210.

Narasimhan, K., Basheer, C., et al. (2003). "Enhancement of plant-microbe interactions using a rhizosphere metabolomics-driven approach and its application in the removal of polychlorinated biphenyls." *Plant Physiology*, **132**(1), 146-153.

Nuopponen, M. H., Wikberg, H. I., et al. (2006). "Characterization of 25 tropical hardwoods with Fourier transform infrared, ultraviolet resonance Raman, and C-13-NMR cross-polarization/magic-angle spinning spectroscopy." *Journal of Applied Polymer Science*, **102**(1), 810-819.

Ossipov, V., Ossipova, S., et al. (2008). "Application of metabolomics to genotype and phenotype discrimination of birch trees grown in a long-term open-field experiment." *Metabolomics*, **4**(1), 39-51.

Ott, K. H. & Aranibar, N., et al. (2003). "Metabonomics classifies pathways affected by bioactive compounds. Artificial neural network classification of NMR spectra of plant extracts." *Phytochemistry*, **62**(6), 971-985.

Overy, S. A., Malone, W. H., Howard, S., Baxter, T. P., Sweetlove, C. J., Hill, L. J. & Quick, S. A., (2005). "Application of metabolite profiling to the identification of traits in a population of tomato introgression lines." *Journal of Experimental Botany*, **56**, 287-296.

Penuelas, J. & Estiarte, M. (1998). "Can elevated CO2 affect secondary metabolism and ecosystem function?" *Trends in Ecology & Evolution*, **13**(1), 20-24.

Penuelas, J., Llusia, J., et al. (1999). "Effects of ozone concentrations on biogenic volatile organic compounds emission in the Mediterranean region." *Environmental Pollution*, **105**(1), 17-23.

Penuelas, J. & Sardans, J. (2009). "Ecology: Elementary factors." *Nature*, **460***(***7257***)*, 803-804.

Petrakis, P. V., Agiomyrgianaki, A., et al. (2008). "Geographical characterization of Greek virgin olive oils (cv. Koroneiki) using H-1 and P-31 NMR fingerprinting with canonical dascriminant analysis and classification binary trees." *Journal of Agricultural and Food Chemistry*, **56**(9), 3200-3207.

Pinheiro, C., Passarinho, J. A., et al. (2004). "Effect of drought and rewatering on the metabolism of Lupinus albus organs." *Journal of Plant Physiology*, **161**(11), 1203-1210.

Raven, C. E. (1953). The significance of a changing flora. The changing flora of Britain. J. E. Lousley. Arbroath, UK, *Botanical Society of the British Isles*; Buncle and Co. Ltd.

Riipi, M., Haukioja, E., et al. (2004). "Ranking of individual mountain birch trees in terms of leaf chemistry: seasonal and annual variation." *Chemoecology*, **14**(1), 31-43.

Robinson, A. R., Gheneim, R., et al. (2005). "The potential of metabolite profiling as a selection tool for genotype discrimination in Populus." *Journal of Experimental Botany*, **56**(421), 2807-2819.

Robinson, A. R., Ukrainetz, N. K., et al. (2007). "Metabolite profiling of Douglas-fir (Pseudotsuga menziesii) field trials reveals strong environmental and weak genetic variation." *New Phytologist*, **174**(4), 762-773.

Roessner, U., Wagner, C., et al. (2000). "Simultaneous analysis of metabolites in potato tuber by gas chromatography-mass spectrometry." *Plant Journal*, **23**(1), 131-142.

Roessner, U., Willmitzer, L., et al. (2002). "Metabolic profiling and biochemical phenotyping of plant systems." *Plant Cell Reports*, **21**(3), 189-196.

Salter, G. J., Lazzari, M., et al. (1997). "Determination of the geographical origin of Italian extra virgin olive oil using pyrolysis mass spectrometry and artificial neural networks." *Journal of Analytical and Applied Pyrolysis*, 40-1: 159-170.

Sanchez, D. H., Siahpoosh, M. R., et al. (2008). "Plant metabolomics reveals conserved and divergent metabolic responses to salinity." *Physiologia Plantarum*, **132**(2), 209-219.

Schaneberg, B. T., Crockett, S., et al. (2003). "The role of chemical fingerprinting: application to Ephedra." *Phytochemistry*, **62**(6), 911-918.

Schauer, N. & Fernie, A. R. (2006). "Plant metabolomics: towards biological function and mechanism." *Trends in Plant Science*, **11**(10), 508-516.

Schwab, W. (2003). "Metabolome diversity: too few genes, too many metabolites?" *Phytochemistry*, **62**(6), 837-849.

Semel, Y., Schauer, N., et al. (2007). "Metabolite analysis for the comparison of irrigated and non-irrigated field grown tomato of varying genotype." *Metabolomics*, **3**(3), 289-295.

Shulaev, V., Cortes, D., et al. (2008). "Metabolomics for plant stress response." *Physiologia Plantarum*, **132**(2), 199-208.

Smirnoff, N. (1998). "Plant resistance to environmental stress." *Current Opinion in Biotechnology*, **9**(2), 214-219.

Smith, A. R., Johnson, H. E., et al. (2003). "Metabolic fingerprinting of salt-stressed tomatoes." Bulgarian Journal of Plant Physiology(Special Issue): 153-163.

Stewart, G. R., Ahmed, L. F. & Lee, I. J. A. (1979). Nitrogen Metabolism and Salt-Tolerance in Higher Plant Halophytes. Ecological Processes in Coastal Environments. R. Jefferies and A. Davy. Oxford, *Blackwell Scientific Publications*, 211-227.

Stitt, M. & Fernie, A. R. (2003). "From measurements of metabolites to metabolomics: an 'on the fly' perspective illustrated by recent studies of carbon-nitrogen interactions." *Current Opinion in Biotechnology*, **14**(2), 136-144.

Stitt, M. & Hurry, V. (2002). "A plant for all seasons: alterations in photosynthetic carbon metabolism during cold acclimation in Arabidopsis." *Current Opinion in Plant Biology*, **5**(3), 199-206.

Sultana, T., Stecher, G., et al. (2008). "Quality assessment and quantitative analysis of flavonoids from tea samples of different origins by HPLC-DAD-ESI-MS." *Journal of Agricultural and Food Chemistry*, **56**(10), 3444-3453.

Sumner, L. W., Mendes, P., et al. (2003). "Plant metabolomics: large-scale phytochemistry in the functional genomics era." Phytochemistry, **62**(6), 817-836.

Thomas, C. D., Cameron, A., et al. (2004). "Extinction risk from climate change." *Nature*, ***427(6970)***, 145-148.

Thomashow, M. F. (1999). " Plant cold acclimation: freezing tolerance genes and regulatory mechanisms." *Annual Review of Plant Physiology and Molecular Biology*, **50**, 571-599.

Trethewey, R. N., Krotzky, A. J., et al. (1999). "Metabolic profiling: a Rosetta Stone for genomics?" *Current Opinion in Plant Biology*, **2**(2), 83-85.

Urbanczyk-Wochniak, E. & Fernie, A. R. (2005). "Metabolic profiling reveals altered nitrogen nutrient regimes have diverse effects on the metabolism of hydroponically-grown tomato (Solanum lycopersicum) plants." *Journal of Experimental Botany*, **56**(410), 309-321.

Usadel, B., Blasing, O. E., et al. (2008). "Multilevel genomic analysis of the response of transcripts, enzyme activities and metabolites in Arabidopsis rosettes to a progressive decrease of temperature in the non-freezing range." *Plant Cell and Environment*, **31**(4), 518-547.

Vainstein, A., Lewinsohn, E., et al. (2001). "Floral fragrance. New inroads into an old commodity." *Plant Physiology*, **127**(4), 1383-1389.

Van Aken, B. (2008). "Transgenic plants for phytoremediation: helping nature to clean up environmental pollution." *Trends in Biotechnology*, **26**(5), 225-227.

Verdonk, J. C., de Vos, C. H. R., et al. (2003). "Regulation of floral scent production in petunia revealed by targeted metabolomics." *Phytochemistry*, **62**(6), 997-1008.

Ward, J. L., Harris, C., et al. (2003). "Assessment of H-1 NMR spectroscopy and multivariate analysis as a technique for metabolite fingerprinting of Arabidopsis thaliana." Phytochemistry, **62**(6), 949-957.

Watkins, S. M., Hammock, B. D., et al. (2001). "Individual metabolism should guide agriculture toward foods for improved health and nutrition." *American Journal of Clinical Nutrition*, **74**(3), 283-286.

Windsor, A. J., Reichelt, M., et al. (2005). "Geographic and evolutionary diversification of glucosinolates among near relatives of Arabidopsis thaliana (Brassicaceae)." Phytochemistry, **66**(11), 1321-1333.

Wollenweber, B., Porter, J. R., et al. (2005). "Need for multidisciplinary research towards a second green revolution - Commentary." Current Opinion in Plant Biology, **8**(3), 337-341.

Woodward, F. I. (1987). Climate and plant distribution. Cambridge, Cambridge University Press.

Zhen, Y. & Ungerer, M. C. (2008). "Clinal variation in freezing tolerance among natural accessions of Arabidopsis thaliana." *New Phytologist*, **177**(2), 419-427.

In: Metabolomics: Metabolites, Metabonomics...
Editors: J.S. Knapp and W.L. Cabrera, pp. 181-200
ISBN: 978-1-61668-006-0

Chapter 5

MICROBIAL METAGENOMICS: CONCEPT, METHODOLOGY AND PROSPECTS FOR NOVEL BIOCATALYSTS AND THERAPEUTICS FROM THE MAMMALIAN GUT MICROBIOME

B. Singh*[*], *T.K. Bhat, O.P. Sharma and N.P. Kurade
Indian Veterinary Research Institute, Regional Station Palampur-176 061, India

Abstract

Despite enormous advancements in microbial culturing methods, more than 95% of the global microbial diversity still remains cryptic. Microbial metagenomics- the applications of modern genomics techniques to the study of communities of microbes directly in their diverse natural environments, bypassing the need for isolation, is changing our comprehension of the biosphere. Advances in technologies designed to access this wealth of genetic information through environmental nucleic acids extraction and analysis have provided the means of overcoming the limitations of conventional culture-dependent microbial exploitation. Further developments and applications of these methods promise to provide opportunities to link distribution and identity of gut microbes in their natural habitats, and explore their use for promoting livestock health and industrial biotechnological applications.

Introduction

The microbial diversity exhibits a ubiquitous presence ranging from fossils that are about 3.5 billion years old and gastrointestinal (GI) tract of animals to the extremophiles. The total number of prokaryotic cells on the earth has been estimated at $4x10^{30}$ to $6x10^{30}$ (Whitman et al., 1998), comprising of 10^6 to 10^8 separate genospecies (distinct taxonomic groups based on gene sequence analysis) (Amann et al., 1995). The microbial populations that account for a major proportion of Earth's biological diversity are of enormous practical significance in

[*] E-mail address: bsbpalampur@yahoo.co.in; Fax; +91 1894 233063; Phone +91 1894 230526. (Corresponding author)

medicine, industry, engineering and agriculture. It is impossible to imagine life without association of microbes. The global microbial diversity, therefore, presents an enormous, largely untapped genetic and biological pool that could be exploited for the recovery of novel genes, enzymes and biomolecules for metabolic engineering and industrial development.

Trillions of microbes inhabit mammalian digestive tract and influence the host in profound and diverse ways. These microbes are indispensable to nutrition, immunity and health of the host. Recent gene- and genome-based analyses of the gut ecosystem have revealed novel insights into many microbial-mediated important symbiotic functions. The system-wide gene analysis of a microbial community specialized in plant lignocellulose degradation and detoxification of phytometabolites and xenobiotics, has both basic and applied implications. This chapter presents an overview of the concept and methodology of microbial metagenomics. Applications of genomics and metagenomic tools for exploring the rumen microbiome for identification of novel biocatalysts and therapeutically relevant products are discussed. As the metagenomics tools were originally validated in various environmental microbial niches, examples of these systems are also cited.

The Concept of Genomics in Microbial Ecology

Certain habitats like deep sea water, soil and compost, and gut of the animals are inhabited by a range **of** microbial populations. The mammalian gut contains a dense, complex, and diverse microbial community whose genome is called as gut microbiome. Conventionally, the gut microbes are studied by classical microbiological approaches involving culturing the microorganisms in synthetic culture media depending on their nutritional and physiological requirements. However, the general in vitro culture conditions tend to impose a selective pressure, thereby inhibiting the growth of a number of important microorganisms, and thus, provide only a few identification clues regarding gut microbial ecology and metabolism (Pace et al., 1986).

Concerning microbial taxonomy, one of the first and most successful applications of molecular phylogeny was the recognition of the Archaea and building of a tripartite tree of life by C. R. Woose and collaborators from the late 1970s (cited in Lopez-Garcia and Moreira, 2008). Since then, microbiology is under dynamic revolution and has emerged as fast-moving scientific discipline. During the past decade there was a remarkable evolution in the development and applications of traditional and DNA-based molecular tools that allowed the microbiologists to characterize and understand microbial communities in unprecedented ways. By creatively leveraging these newly emerging data sources, microbial ecology has potential to have a transition from a purely descriptive to a predictive framework in which ecological principles are integrated and exploited to engineer the systems that are biologically optimized for a desired goal. Molecular genomics enables the microbiologists to have a look at a more complete scenario of environmental microbial communities, and thus, to better understand the microbe–environment interaction. DNA–DNA hybridization, which has been used for many years, is still considered to be the standard protocol for bacterial species identification. However, PCR-based methods employing 16S rRNA gene sequences along with other approaches such as bioinformatics, are being increasingly applied to study complex microbial niches and to identify novel microbial genes with potential pharmaceutical and biotechnological applications. The ability to obtain whole or partial genome sequences from microbial community samples has opened the door for other system level studies of microbial communities such as community proteomics or metaproteomics. Hence, in view of the advancements in exploring the microbial species, there is a growing belief that the term "unculturable" is inappropriate and that in reality we have yet to discover the appropriate and new microbial culture methods.

What is Metagenomics?

The term metagenomics ('*meta*' Greek, for transcending; more comprehensive), which constitutes a challenging domain to discover and exploit novel enzymes from diverse niches, was first coined by Handelsman et al. (1998) to study the genomes from all microbes in a particular environment as opposed to the genome from organism isolated and cultured in vitro. The concept was based on earlier report by Schmidt et al. (1991) on the construction of a lambda phage library from 16S rRNA gene cloning and sequencing of a marine planktonic community. Metagenomics presents the greatest opportunity perhaps since the invention of the microscope to revolutionize the understanding of microbial world. It is aimed at elucidating the genomes of nonculturable microbes, and to better understand the global microbial ecology on one side, and on the other side driven by industrial biotechnological demands for novel enzymes and biomolecules. Thus, metagenomics has emerged as a promising tool for exploiting the diverse microbial ecosystems including extremophiles, termites' hind gut and mammalian GI flora.

The sequencing of ribosomal RNA (rRNA) and the genes encoding them pioneered a new era of microbial ecology. The early studies were technically challenging, relying on direct sequencing of RNA or sequencing of reverse transcription-generated DNA copies. The next technical breakthrough was made with establishment of PCR technology, purification of DNA polymerase and designing primers that revolutionized the amplification of almost entire gene. Thus, for genomic analysis of microbial populations the metagenomics has emerged as a powerful tool to gain insights into physiology and genetics of uncultured organisms. Initially, noncultured microorganisms and ancient DNA analysis had been the prime targets of metagenomic studies, but at present, the technology is being applied to study diverse microbial niches like deep-sea aquatic microflora, various extremophiles, soil and compost microbes and GI microbiome of humans and animals (Lu et al., 2007; Shanks et al., 2006; Singh et al., 2008a). Technical advances in construction of high efficiency cloning vectors like cosmids, phosmids, bacterial artificial chromosomes, BACs or yeast artificial chromosomes, YACs (Babcock et al., 2007; Xu 2006), which allow cloning and functional expression of larger and complex genes, and powerful algorithms and statistical methods for analysis of huge data have completely transformed the concept of microbial metagenomics to a practical reality.

Microbial Metagenomics: Major Procedural Steps

Microbial metagenomics comprises of a series of technical steps and analytical methods. The basic steps are described here, though depending on the microbial communities to be studied, the basic protocols can be modified.

Sampling and Microbial Nucleic Acids Extraction

In a typical metagenomic analysis, it is necessary that DNA extracted should represent all the microbes within a community. The samples could be from any environment or habitat, including GI ecosystem. Several procedural refinements like freeze-thawing, ultrasonication, glass-bead mediated homogenization have been made in extraction and recovery of high purity intact open reading frames (ORFs) from the complex microbial environments. The physical methods of DNA isolation have certain limitations like uncontrolled shearing of DNA, and increased risk of formation of chimeric DNA molecules during downstream PCR amplification. Chemical methods of nucleic acid extraction using SDS are gentle and efficient, and yield high purity genomic DNA. However, combination of physical and chemical methods that suits different types of the environmental samples may offer an ideal option.

The rare or less represented microbes in an environmental sample need to be enriched by applying suitable in vitro enrichment methods. Among these methods, differential centrifugation of a microbial community could be a simple enrichment protocol. The microbial-enrichment using a selective culture medium could also favor the growth of target microbes. Although culture-enrichment will inevitably result in the loss of a large proportion of the microbial diversity by promoting the fast-growing cultivable species, this can be partially minimized by reducing the selection pressure to a mild level after a short period of stringent treatment. Nevertheless, in vitro microbial-enrichment results in efficient isolation of large DNA fragments for the cloning of the operons and intact larger size genes for precise characterization and purification of the end product. A recent paper (Singh et al., 2008b) have reviewed the strategies for in vitro enrichment of various types of microbial cultures for isolating high purity genomic DNA.

The purity and recovery of larger size genomic DNA or intact ORFs from microbes is a critical step as DNA extracted is to be cloned for constructing metagenomic libraries. However, owing to the physiochemical diversity in matrices serving as microbial habitats, instead of a universal method different nucleic acids extraction protocols are used.

Since some microbial species are likely to be overshadowed by dominant or fast-growing microbial populations, therefore, genomes of rare organisms contribute a relatively low proportion of the extracted nuclear material (Bohannan and Hughes 2003). This leads to a selective bias in downstream analyses such as PCR amplification of the nucleic acids, sequencing of the cloned genes and subsequent data analysis. The problem could be partially resolved by means of experimental normalization (Short and Mathur, 1999). Normalization of the genomic materials can also be achieved by denaturing the extracted genomic DNA fragments, and re-annealing the single stranded DNA (ssDNA) under stringent conditions (e.g. 68.8°C for 12–36 h). Abundant ssDNAs anneal more rapidly to generate double stranded nucleic acids compared to DNA from rare species. The remaining single-stranded sequences are then separated from the double-stranded nucleic acids, resulting in an enrichment of rarer sequences within an environmental sample. Methods have been described for extracting high quality microbial genomic DNA from the vertebrate fecal samples (Nordgard et al., 2005) or from rumen digesta for phylogenetic analysis of metabolically active members of microbial communities (Sharma et al., 2003; Kang et al., 2009). The technologies for recovering RNA from environmental samples are largely similar to those used for DNA isolation, but modified to optimize the yield of intact mRNA by minimizing single-stranded polynucleotide degradation.

Microbial Genome- and Gene-Enrichment

Genome-enrichment strategies are aimed at targeting the active components of a specific microbial population. With the advent of genome enrichment and amplification techniques in metagenomics, overcoming the limitations in DNA purity and yield have become easier. A method known as stable-isotope probing (SIP) was developed by Radajewski et al. (2000) to identify the organisms involved in metabolism of specific substrates without the prerequisite for their in vitro cultivation. Modifications of the methods like nucleic acids-SIP involve labeling the nonculturable microbes in environmental samples using a substrate enriched with certain stable isotopes (^{13}C and/or ^{15}N, etc.), which are assimilated by the microbes and subsequently incorporated into their organelles and genomes. The isotopically labeled DNA is then retrieved by density gradient centrifugation, and the target microorganisms are identified by molecular analysis of their genomes (Friedrich, 2006). Other labeled biomarkers, such as phospholipid-derived fatty acid (PLFA), ribosomal RNA, and DNA can also be probed using a range of molecular analytical techniques, and used to identify the organisms that have incorporated the labeled substrates. Another method, termed suppression subtractive hybridization (SSH) identifies the genetic differences between different microorganisms and is therefore, a powerful tool for specific gene enrichment and detection in microorganisms. The technique has also been used to identify differences between complex DNA samples from the rumen of steers (Galbraith et al., 2004), identifying the unique genes encoding plant cell wall hydrolytic enzymes and some novel molecular features of the GI bacterium *Fibrobacter intestinalis* DR7, not shared with *F. succinigenes* (Qi et al., 2005). To selectively enrich a specific target gene within a metagenome, a more practical approach would be to use differential expression analysis (DEA) technologies that rely on the isolation of mRNA (transcriptome) to target transcriptional differences in gene expression.

Metagenome Cloning and Targeting

The cloning strategies depend strongly on suitability of a gene cloning vector and overall goal of the study. In many cases, the generation of large insert libraries is required to analyze the size, complexity and diversity of environmental metagenome. Large insert libraries can be generated using cosmids, BACs, YACs or phosmids. Small insert libraries may be more suitable for generating large amounts of DNA sequence information rather than functional analysis per se (Venter et al., 2004; Banfield et al., 2005).

Gene targeting approaches have been used in understanding the key community regulators in intestinal bacteria in diseases like Crohn's disease (Kobayashi et al., 2005), and developing new antibiotics targeting pathogenic bacterial genes whose expression is essential for their in vivo viability (Clatworthy et al., 2007). The microorganisms with specific metabolic traits can be probed using gene-specific PCR applications. However, as a tool for biocatalyst investigations, gene-specific PCR has some limitations. First, the design of primers is dependent on existing microbial gene sequence information which skews the search in favor of already known DNA sequence types. Functionally similar genes resulting from convergent evolution are not likely to be detected by a single gene-family-specific set of PCR primers. Second, only a fragment of a structural gene will typically be amplified by gene-specific PCR, thus requiring additional steps to access full-length genes in new microbial groups.

Amplicons could be labeled as probes to identify the putative full-length gene (s) in conventional metagenomic libraries. Alternatively, PCR-based strategies for the recovery of either the up- or down-stream flanking regions can be used to access the full length gene. For example, universal ''fast walking'' (Myrick and Gelbart, 2002), panhandle PCR (to amplify known sequence flanked by unknown sequence) (Myrick and Gelbart, 2002), inverse PCR and adaptor-ligation PCR (Ochman et al., 1993) are some important tools in use in recent microbial genomic analyses. These techniques are likely to revolutionize the current approaches to study microbial ecology in the GI tract and to provide, not simply a refinement or increased understanding, but a complete description of the gut ecosystem.

Screening and Analysis of Metagenomic Libraries

Hundreds of clones are generated with only a small fraction of colonies containing the target(s) of interest. Efficient screening methodologies are, therefore, needed to allow a targeted clone selection. Two approaches, namely, the function-driven analysis (screening the metagenomic libraries for an expressed and detectable trait) and sequence-driven analysis (metagenomic libraries screened for particular DNA sequences) are used to analyze metagenomic data.

The Function-Driven Analysis

The technique involves screening of clones expressing a desired trait or molecules of interest. This approach is based on identification of the constructed clones that express a desired trait in surrogate host, followed by characterization of the active clones based on their biochemical and molecular (gene sequence) features. This helps in identification of entirely new classes of genes for known functional applications like pharmaceutical, agricultural or industrial applications. Though being a highly preferred approach, this method has certain limitations including low expression of the cloned genes, hence, requires additional strategies to improve the gene expression and detection of the functional product in the host cell. The process may also require clustering of all the genes encoding a single product. Furthermore, it depends much on availability of assays for the function of interest that can be performed efficiently on vast metagenomic libraries. The functional metagenomics approach provides a unique tool for dissecting the metabolic contribution of human gut microbiota and moreover, since it employs culture-independent techniques, it has potential to generate testable scientific hypotheses concerning the functional and ecological role of bacteria till date far recognizable only as entries in 16S rRNA gene sequence database or completely new to science (Tuohy et al., 2009).

Table 1. Microbial metagenomics in the microbiological and biotechnological interventions in animals' GI ecosystem.

A. Ruminants/ herbivores	
Targets	**Future prospects**
Novel hydrolytic enzymes	Identification and characterization of enzymes for use in animals feeds, pulp and paper, and textile industry Enhanced utility of plant biomass for improving rumen productivity
Novel microbial Species	Establishing new in vitro culture conditions for therapeutically or nutritionally relevant novel gut microbes for use as probiotics or direct-fed microbials (DFMs)
	Developing strategies for biomonitoring of the transinoculated gut-based DFMs or probiotics
	Studying the rumen ecosystem of the animals exhibiting natural adaptation of diets containing antinutritional PSMs Identification of novel gut microbes and microbial enzymes and using them as prebiotics in susceptible animals for overcoming toxicity due to dietary phytometabolites
	Studying the interactions among different microbial consortia and between host and the gut symbionts
Novel genes, enzymes, and antimicrobials	Exploiting gut microbiome as resource of novel genes, restriction enzymes, and plasmids as tools for genetic engineering of the resident or normal flora Identification of the gut bacteriocins for use as rumen modulators or suppression of spoilage and opportunistic GI pathogens
Methanogenesis	Identification of the novel methanogens in the GI tract, manipulation of rumen for lowering methane emissions Inducing animal immune system-mediated antibodies against rumen methanogens, developing vaccines against the selected methanogens
B. Monogastrics (poultry, swine etc.)	
Novel microbial species	Identifying the potentially useful novel gut microbes, and establishing strategies for culturing them *in vitro* Studying host-gut microbe interactions and their symbiotic significance Identifying the novel gut flora producing bacteriocins and other antimicrobial peptides for use against opportunistic pathogens and spoilage bacteria Identification of species-specific molecular markers in selected elite microbes for their biomonitoring in new host

The functional screening is limited by the fact that metagenomic genes must be expressed in a heterologous background. Improved systems for heterologous gene-expression are being developed with shuttle vectors that facilitate screening of the metagenomic DNA in selected broad range hosts. As a host *Escherichia coli* alone cannot fulfill the requirements for functional activity of the gene product, *Streptomyces lividans* and *Pseudomonas putida* have been developed as alternative hosts (Martinez et al., 2004). Through functional screening of metagenomic libraries, several novel and previously described antibiotics (Amann et al., 1995; Gillespie et al., 2002), antibiotics-resistance genes in the human oral and infant fecal microorganisms (Diaz-Torres et al., 2003) and some commercially important novel enzymes with valuable hydrolytic activities (Lammle et al., 2007; Henne et al., 2000) have been identified. Also, from the mammalian GI tract some novel hydrolases (Ferrer et al., 2005; Lammle et al., 2007; Feng et al., 2007; Duan et al., 2006; 2009) and polyphenol oxidases (Beloqui et al., 2006) have been documented (Table 2).

Table 2. Some novel enzymes and microbes identified using the metagenomic tools from the animals GI tract.

Enzymes/ microbes	Source/ host	Salient findings/ remarks studied
Rumen microbial ecosystem	Bovine rumens (SSH)	Molecular complexity of rumen Archaeal communities revealed at molecular levels (Lammle et al., 2007)
Acetylxylan esterase (R.4) family carbohydrate esterase (CE 6)	Rumen	Identification and characterization of novel hydrolases (Beloqui et al., 2006)
Hybrid glycosyl hydrolase	Bovine rumen	Identification and characterization of enzyme, and their industrial importance (Lopez-Cortes et al., 2007)
RA.04 (α-amylase Family)	Bovine rumen	Identification and characterization of the enzyme (Lan et al., 2006 ; Palackal et al., 2007)
RL-5, gene encoding polyphenol oxidase	Bovine rumen	Purification and characterization of enzymes for industrial applications (Feng et al., 2007)
umcel3G, a gene encoding beta-gluconase	Buffalo rumen	Fermentatitive production of ethanol by simultaneous saccharification and co-fermentation (SSCF) of lignocellulose (Guo et al., 2008)
Novel cellulases	Buffalo rumen	Characterization, and purification of enzyme expressed in *E. coli*, for future industrial applications (Duan et al., 2009)
Low G+C bacteria and *Cytophaga-Flexibacter Bactyeroides* phyla	Guangxi buffao rumen	Cellulose hydrolysis, and similar abundance of the microbes revealed in rumens of yak, cattle and sheep (Liu et al., 2009)
umbgl3B (β-glycosidase)	Rabbit caecum	Characterization of the enzymes (Feng et al., 2009)
Cel A, *Xyl A* genes and their produts	Cow rumen	Purification and characterization of enzymes cel5 A and xyl A from the metagenome library (Shedova et al., 2009)
RlipE1 and *RlipE2* genes and their products	Cow rumen	Purification and characterization of recombinant lipases, and their possible applications in rumen lipid metabolism (Liu et al., 2009)
Novel methanogens	Cattle, sheep rumens (16S/18S rRNA rDNA, TTGE)	New opportunities for identification of methanogens in rumens (Ferrer et al., 2007)
Fungal texa	Murine GI tract	Elucidation of diverse fungal texa and their role in the GI tract. (Toyoda et al., 2009)

The Sequence-Driven Analysis

Identification of potential enzymes in metagenomes based on sequence-similarity is a viable and rewarding approach. This involves the complete sequencing of clones containing phyogenetic anchors, such as 16S rRNA genes and the archaeal DNA repair genes, which indicate the taxonomic group and functional information about the organisms from which these clones were derived. Sequence-driven analysis relies on the conserved DNA sequences to design hybridization probes or PCR primers for screening the metagenomic libraries for clones that are expected to contain nucleotide sequences of interest. The sequencing and analysis of genomic DNA/RNA from the uncultured environmental microorganisms are well-established technologies, and the massive sequencing of nucleic acids as a way to establish global inventory of metagenomic DNA from environmental sources, is technically feasible (Venter et al., 2004). Highly advanced sequencing technologies, independent of gene cloning, are available now (Margulies et al., 2005; Hall 2007) and the elaborate algorithms subsequently assist identifying the ORFs in silico and detect the related sequence entries in databases. For instance, the DOTUR software was developed and used to determine whether a genomic library contains sufficient genes for it to be considered representative of the original microbial diversity (Schloss and Handelsman 2005). Another software, called MetaGene, utilizes besides other various measures, two sets of codon frequency interpolations, one for bacteria and one for archaea, estimated by the guanine-cytosine (GC) content of a given sequence. The software was applied to metagenomic sequences of Sargasso Sea dataset, almost all annotated genes were predicted by MetaGene and in addition 0.4 million novel genes were also detected (Noguchi et al., 2006). MEGAN (MetaGenome ANalyzers), another computer program that generates specific profiles from sequencing data by assigning the reads to NCBI taxonomy using a straight-forward assigned algorithm, is used for the analysis of various metagenomic data (Huson et al., 2007). The MEGAN approach has been found applicable to several data sets including subset of the Sargasso Sea data set (obtained by Sanger's sequencing method), data obtained from mammoth (*Mammuthus primigenius*) bone (obtained by ''sequencing-by-synthesis'' approach), and identifying the microbial species based on already available microbial (*E. coli* and *Bdelovibrio bacteriovorus*) genome sequence information (Huson et al., 2007). To study mobile genetic elements including plasmids in gut bacteria, a culture-independent ''transposon aided capture'' (TRACA) method, independent of plasmid-encoded traits was developed to study the plasmids of bacteria in gut metagenome (Jones and Marchesi, 2007). The application of TRACA to further study plasmids resident in the gut and other bacteria, is likely to identify new plasmids encoding diverse functions important for adaptation, survival, interaction between bacteria within a microbial ecosystem, and interactions between gut symbionts and their host species.

Genomics in Mammalian Gut Microbial Diversity

The mammalian gut ecosystem is one of the most complex microbial ecosystems. Also called as ''normal flora'', the gut microbes have adapted in such a manner that they have no adverse effects on the host's overall health, and often they are beneficial or even obligatory to the host, especially in the herbivores. The commensal microbiota helps maintain immune homeostasis within the gut-associated lymphoid tissues, provides developmental cues, and supplements nutritional intake by the host. Certain regions of the mammalian GI tract, notably the rumen and large intestine, harbor extremely dense microbial communities, in which bacterial number can exceed as much as 10^{11} per gram of rumen fluid (Flint et al., 2008). These regions are active sites of the microbial metabolism of the dietary plant polysaccharides, which are resistant to host gastric enzymes. The bacteria in the large intestine are also involved in a range of metabolic transformations and complex interactions with the host and its immune system.

The current global drive to promote the white (industrial) biotechnology as a central feature of the sustainable economic future of modern industrialized societies requires development of novel enzymes, processes and biomolecules for industrial applications. Gut ecosystem offers an inexhaustible source of enzymes, biotherapeutics, genes and novel products for applications in health, nutrition and industrial development (Selinger et al., 1996; Singh et al., 2001; Flint et al., 2008; Morrison et al., 2009). Microbial biotechnological applications (Table 2) from GI microbiome will be fostered by the pursuit of fundamental ecological studies (Table 3) and focused screening for bioprospecting, just as both basic and applied approaches have contributed to the discovery of antibiotics and enzymes from other nonculturable microbes.

Metagenomics in Rumen Microbiome-Motives and Applications

The herbivores retain within their gastrointestinal tract a microbiome that specializes in the rapid hydrolysis and fermentation of lignocellulosisc plant biomass (Morrison et al., 2009). The rumen is the fermentative forestomach of the ruminant animals and is densely populated by the microbes which are classified into three main domains, namely **Bacteria** (bacteria) **Archaea** (methanogens) and **Eucarya** (fungi and protozoa). The symbiosis of this extended genome plays a pivotal role in host homeostasis, nutrient and energy derivation from the crude dietary resources. Collectively, these symbionts are responsible for the digestion of roughage diets, detoxification of a number of plant metabolites and synthesis of volatile fatty acids (VFAs) and microbial proteins which are utilized by the host. Due to the presence of unique obligate anaerobes (fungi, protozoa, bacteria and archaea) and continuous formation of microbial products, the rumen has been regarded as fountain head of valuable fibrolytic enzymes (hemicellulases, xylanases, cellulases, endoglucanases and acetyl xylan esterases, etc.) that could be exploited in feed (plant biomass saccharification for supplying critical nutrients to the animals from low quality dietary forages), textile, and pulp and paper processing (Selinger et al., 1996; Palackal et al., 2007; Flint et al., 2008; Singh et al., 2009).

However, despite enormous potential (discussed below), the rumen microbiome has not been completely studied. This is primarily due to survival of rumen microbes only in obligate anaerobic environment in vivo, and inability of most of these microbes to grow in vitro. Culture-independent genomics and metagenomics methods, therefore, may provide unique insights into this complex ecosystem.

Table 3. List of innovations leading to improvements in metagenomic analysis.

1.	In vitro enrichment of rare microbial species within a microbial population/ community
2.	Extraction of high purity intact DNA fragments/ ORFs or operons
3.	Minimized mechanical shearing of the DNA during extraction
4.	Exclusion of predominately present contaminating impurities from the environmental samples
5.	Direct in situ extraction of DNA from microbial communities
6.	Use of pre-cultivation step to improve quality of microbial environmental DNA
7.	Innovations in developing high-capacity gene cloning vectors
8.	Technical innovations in sequencing the genes of interest, and development of high accuracy computer programs and software to analyze the data
9.	Availability of data in data banks and their online accessibility

1. Rumen Microbes as Sources of Valuable Hydrolytic Enzymes

Industrial or white biotechnology is currently a buzzword in the biobusiness community, and requires development of enzymes, processes and products with diverse functions. The industries are interested in tapping the elite microbial resources, particularly the uncultured environmental microorganisms that are identified through large scale environmental genomics. Rumen fibrolytic enzymes could be of enormous significance in livestock feed processing (e.g. plant biomass saccharification for deriving critical nutrients from low quality dietary forages, detoxification of antinutritional PSMs), food and beverages, and textile and pulp industries.

Review of the literature reveals that metagenomics have made remarkable advances in studying the rumen ecosystem which may have important applications in future. For instance, sequence analysis of the metagenomic expression library from cow rumen revealed that 36% (8/22) gene sequences were entirely from new phylogenetic lineages (Ferrer et al., 2007). In another study, *RL5*, a gene responsible for a novel polyphenol oxidase was identified from a metagenome expression library from the bovine rumen microbes (Beloqui et al., 2006). Multifunctional glycosyl hydrolases from a microbial consortium from cow rumen have been shown to have potential industrial applications in plant biomass processing, and applications of the identified enzymes in textile and paper processing (Palackal et al., 2007). Similarly, novel genes encoding acidic cellulases have been identified from the rumen of buffalo, and these enzymes have potential industrial applications (Duan et al., 2009). The β-glycosidases from the metagenome of buffalo rumen have been shown to have applications in the fermentative production of ethanol by simultaneous saccharification and co-transformation of indigestible lignocellulases (Guo et al., 2008).

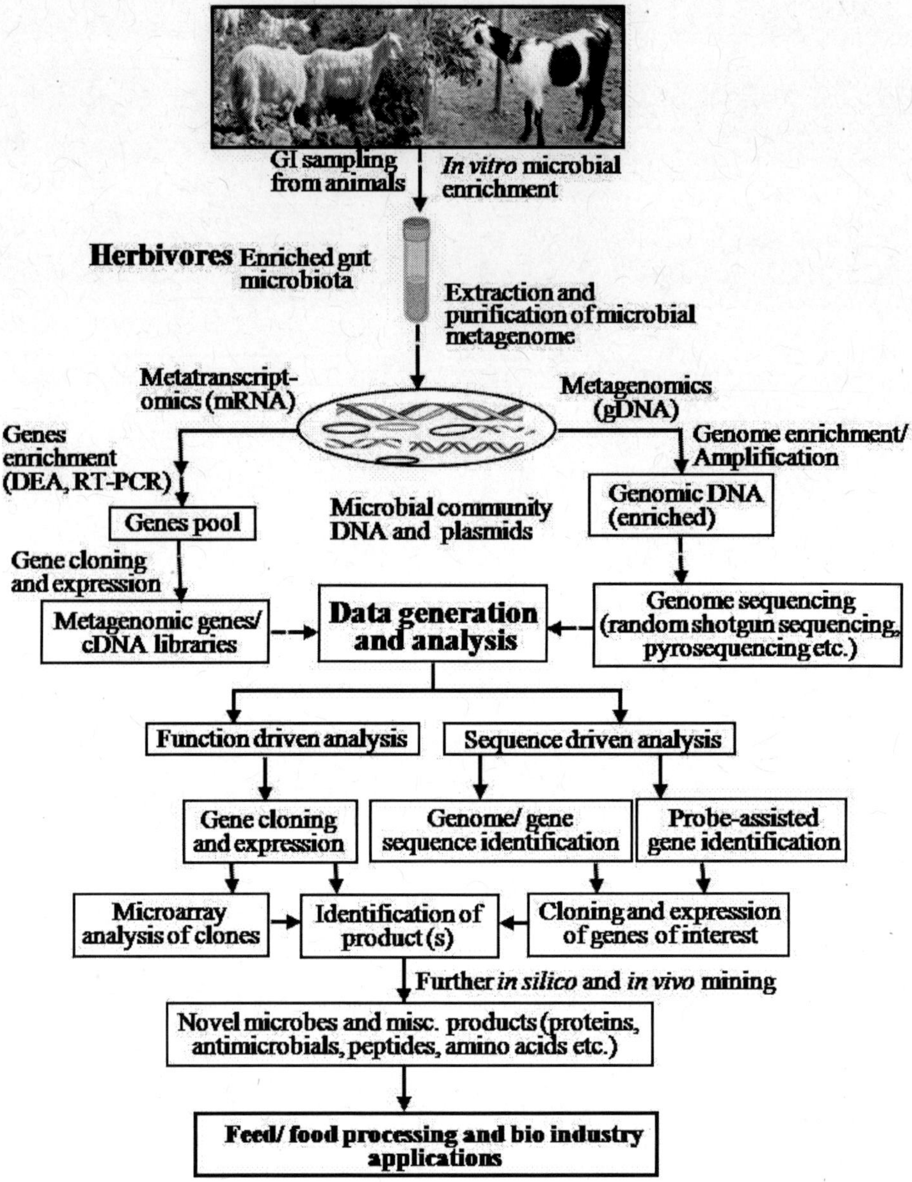

Figure 1. An overview metagenomic analysis of the gut microbiome. The animals adapted to diets containing high fiber and antinutritional PSMs for their nutrition, could harbor a wealth of novel microbes, biocatalysts and therapeutically important biomolecules for promoting livestock production and industrial development. The abbreviations used here have been discussed in the text.

Two novel lipase genes *RlipE1* and *RlipE2* which encoded 361- and 265-amino acid peptides, respectively, were recovered from metagenomic library of the rumen microbes of Chinese Holstein cow (Liu et al., 2009). Characterization of these enzymes, phylogenetic affiliation and high specificity for long chain fatty acids may make these enzymes interesting targets for manipulation of rumen lipid metabolism (Liu et al., 2009). A metagenomic expression library of bulk DNA extracted from the rumen content of dairy cattle was established in a phage vector, and the activity-based screening was employed to explore the functional activity of rumen microbes. Sequence analysis of retrieved enzymes revealed that 36% (8/22) sequences were entirely new and formed deep-branched phylogenetic lineages with no close relatives among the known esterases and glycosyl hydrolases (Ferrer et al., 2007). Some other studies have also demonstrated the usefulness of the metagenomic approach to identify novel hydrolytic enzymes from the ruminants (Ferrer et al., 2005; Ferrer et al., 2007). The rumen bacteria with relevance to fiber degradation, for which genome sequences are available, are *F. succinogenes*, *Ruminococcus albus* and *Prevotella ruminicolla* strain 23. These sequences are likely to be used in future for comparing the sequences from newly identified isolates of the rumen bacteria with similar traits. *F. succinogenes* has been highlighted as a potent rumen bacterium for biodegradation of lignocellulose in anaerobic biogas reactors. The metagenome analysis of this bacterium has yielded significant insights into an unexplored GI microbial niche, as from the gene-list at least 24 genes encoding endoglucanases and cellodextrinases have been identified compared to six genes identified by conventional recombinant DNA strategies (Nelson et al., 2003; Lissens et al., 2004). Table-2 presents an account of the novel gut microbes and microbial products which are elucidated using the microbial metagenomics.

2. Direct Fed Microbials (DFM) from the Rumen

One of the important applications of rumen metagenomics in livestock nutrition would be the identification of genetically superior microbial species from gut of the ruminants exhibiting a natural adaptation to the diets containing high lignocellulose contents and/ or anti-nutritional PSMs such as tannin-polyphenols, non-protein amino acids and oxalates etc. This is because the rumen-originated microbes used as DFM may carry a connotation of being "natural" and safe. Feral herbivores or browsing animals like goats and sheep can consume these forages without apparent adverse effects and may, therefore, be the sources of valuable gut microbes that could be used as DFM or probiotics to enhance rumen fermentation and overcome dietary toxicity in certain susceptible animals. High producing cows in early lactation would be the best candidates for feeding the rumen-based DFM because these animals are in negative energy balance. Identification and dietary supplementation of novel lactate-utilizing bacteria as DFM may have important implications when animals are offered high grain diets. At present *Megaspahera elsdenni* is the major species known to utilize lactate in the rumen. Similarly, supplementation of gut-based elite lactobacilli may be useful in the close-up dry period of lactation when intake is depressed and animals are stressed. Purified rumen hydrolases and phytases can be used as prebiotics in the diets of poultry and swine for promoting utilization of certain dietary nutrients and reducing environmental pollution due to release of mineral nutrients in feces of these animals.

3. Rumen-Originated Antimicrobial Products

The recent progress in molecular biology and microbial genome analysis has an enormous impact on antibacterial drug research. The bacteria with abilities to produce antimicrobial compounds (organic acids, hydrogen peroxide, diacetyl and antibiotics or antibiotic-like compounds), are ubiquitously distributed in all habitats. A family of microbial proteins or peptides, called bacteriocins, is in high demand for livestock health and food industrial applications. Rumen bacteriocins can be used for manipulating rumen ecosystem. For instance, bovicin HC5 purified from *Streptococcus bovis* HC5, can target hyper ammonia-producing bacteria, thus inhibit wasteful ruminal amino acid-degradation (Lima et al., 2009).

At present, a number of bacteriocins have been purified from the rumen and intestinal bacteria. Bacteriocins are proteins that are digestible by the host gastric enzymes, hence, leave no adverse residual effects in milk or meat products. When used as alternatives for ionophores in the feedlot the bacteriocins can improve environmental sustainability of milk and meat production. Metagenomic tools need to be applied to identify new bacterial species for production of bacteriocins for use as dietary supplements to reduce fecal pathogenic load, and as a feed additive to promote growth in milk and meat producing animals. Applications of bacteriocins in food industries as inhibitory agents against spoilage and pathogenic microbes in processed milk or meat products are well documented.

4. Lowering Methane Emissions

Rumen fermentation produces VFAs and methane at faster rates. Microbial genomics can be used for identification of rumen methanogens, a majority of which are still unidentified. The metagenomics has proved to be a promising tool for identification of some new methanogens in the rumen. A temporal temperature gradient gel electrophoresis (TTGE) method developed to determine the diversity of methanogens in cattle and sheep rumens, showed that uncultured methanogens account for the majority of methanogenic archaea in these species (Nicholson et al., 2007). Understanding the adaptation of methanogenic archaea to dietary ingredients in the rumen, and cellular and molecular mechanisms of association between rumen archaea and protozoa is another topic of thorough investigation to minimize methane emissions by the ruminants. Once ecology of the methanogens, and the methane production pathways are identified, novel strategies to manipulate the rumen for lowering methane emission or development of vaccine against the rumen methanogens may be developed. This is of great concern for developing countries where majority of the livestock populations feed on high roughage diets, which favor higher enteric methane emissions.

5. Determining the Protozoal Ecology of the Rumen

An important application of microbial metagenomics in animal nutrition is the quantitative determination of total rumen microbial biomass and differentiating the bacterial and protozoal biomass. This is because the ciliate protozoa are present in most ruminants (10^5–10^6 cells/ml of rumen fluid) and can represent up to half of the total microbial nitrogen. Despite the importance of protozoal ecology, there is no widely applicable marker to measure and differentiate protozoal mass from the overall bacterial populations.

The current knowledge of rumen functioning, therefore, needs to be integrated with a future perspective regarding how the metagenomics could be used to correlate rumen microbiology with animal nutrition. A better understanding of mechanistic processes altering the production and uptake of amino nitrogen will help the livestock nutritionists to improve overall conversion of dietary nitrogen into microbial protein. It will provide key information needed to further improve mechanistic models describing rumen function and evaluating dietary conditions that influence the efficiency of conversion of dietary nitrogen into milk protein (Firkins et al., 2007).

6. Genes and Genetic Engineering Tools from Rumen Microbiome

The rumen bacteria and fungi can be promising sources of genes encoding hydrolytic enzymes. Also, the rumen bacteria are found to contain a range of plasmids with antibiotic resistance markers. A few bacterial species are reported to produce restriction endunucleases which can be used to genetically engineer the resident bacteria in the rumen (Singh et al., 2001). This may possibly increase the establishment and survival of genetically engineered gut bacteria when they are transferred into other ruminants.

Bottlenecks of the Technology

The technology has certain bottlenecks that limit its wider implementation to explore the mammalian gut microbiome. The gut metagenomics is still in initial phases of experimental validation. Only a few genes and genes-encoded products obtained using the metagenomic tools are practically in use in biotechnology process. Within many novel DNA sequences, though new enzymatic functions are identified, but none of them has been practically isolated (Schmeisser et al., 2007). Furthermore, the emerging technologies and revival in culturing techniques may make metagenomic approaches less attractive for microbial physiologists (Kowalchuk et al., 2007).

Conclusion

In conclusion, the genomic studies have made great advances in understanding the complex microbial ecosystems. Metagenomic and metaproteomic analyses have further established the promising potential of the gut ecosystem for biotechnological and pharmaceutical applications. In rumen ecosystem, these techniques need to be focused on identifying the microbes and microbial mechanisms for deriving nutrients from low quality forages, enhanced dietary fiber digestion by the selected elite rumen microbes, and studying the nutrient–host tissue interactions. The long-term goal of metagenomics is to reconstruct the genomes of unculturable important gut microorganisms by identifying overlapping fragments in metagenomic libraries and ''walking'', clone to clone, to assemble each chromosome. To exploit the potential of biotechnological applications of the gut flora, it is essential that both basic biology and utility streams be pursued as a part of the new field of metagenomics of mammalian gut micrbiome.

References

Amann, R. I., Lusdwig, W. & Schleifer, K. H. (1995). Phylogenetic identification and *in situ* detection of individual microbial cells without cultivation. *Microbiol Rev.*, **59**, 143-169.

Babcock, D. A., Wawrik, B., Paul, J. H., McGuinness, L. & Kerkhof, L. J. (2007). Rapid screening of a large insert BAC library for specific 16S rRNA genes using TRFLP. *J Microbiol Methods.*, **71**, 156-161.

Banfield, J. F., Verberkmoes, N. C., Heittich, R. L. & Thelen M. P. (2005). Proteogenomic approaches for the molecular characterization of natural microbial communities. *OMICS.*, **9**, 301-333.

Beloqui, A., Pita, M., Polaina, J., Martinez-Arias, A., Golyshina, O. V. & Zumarraga, M. (2006). Novel polyphenol oxidase mined from a metagenome expression library of bovine rumen: biochemical properties, structural analysis, and phylogenetic relationship. *J Biol Chem.*, **281**, 22933-22942

Bohannan, B. J. & Hughes, J. (2003). New approaches to analyzing microbial biodiversity data. Review. *Curr Opin Microbiol.*, **6**, 282-287.

Clatworthy, A. E., Pierson, E. & Hung D. T. (2007). Targeting virulence: a new paradigm for antimicrobial therapy. Review. *Nature Chem Biol.*, **3**, 541-548.

Diaz-Torres, M. L., McNab, R., Spratt, D. A., Villedieu, A., Hunt, N., Wilson, M. & Mullany, P. (2003). Novel tetracycline resistance determined from the oral metagenome. *Antimicrob Agents Chemother.*, **47**, 1430-1432.

Duan, Z. Y., Guo, Y. Q. & Liu, J. X. (2006). Applications of modern molecular biology techniques to study micro-ecosystem in the rumen. *Wei Sheng Wu Xue Bao.*, **46**, 166-169. (Article in Chinese, abstract in English)

Duan, C. J., Xian, L., Zhao, G. C., Feng, Y., Pang, H., Bai, X. L., Tang, J. L., Ma, Q. S. & Feng, J. X. (2009). Isolation and partial characterization of novel genes encoding acidic cellulases from metagenome of buffalo rumens. *J Applied Microbiol.*, **107**, 245-256.

Feng, Y., Duan, C. J., Liu, L., Tang, J. & Feng J. (2009). Properties of a metagenome-derived β-glucosidase from the contents of rabbit cecum. *Biosci Biotechnol Biochem.*, **73**, 1470-1473.

Feng, Y., Duan, C. J., Pang, H., Mo, X. C., Wu, C. F., Yu, Y., Hu, Y. L., Wei, J., Tang, J. L. & Feng, J. X. (2007). Cloning and identification of novel cellulase genes from uncultured microorganisms in rabbit cecum and characterization of the expressed cellulases. *Appl Microbiol Biotechnol.,* **75**, 319-328.

Ferrer, M., Beloqui, A., Golyshina, O. V., Plou, F. J., Neef, A. & Chernikova, T. N. (2007). Biochemical structure features of a novel cyclodextrinase from cow rumen metagenome. *Biotechnol J.,* **2**, 207-213.

Ferrer, M., Golyshina, O. V., Chernikova, T., Khachane, A. N., Reyes-Durate, D., Santos, V. A., Strompl, C., Elborough, K., Jarvis, G., Neef, A., Yakimov, M. M., Timmis, K. N. & Golyshin, P. N. (2005). Novel hydolase diversity retrieved from a metagenome library of bovine rumen microflora. *Environ Microbiol.,* **7**, 1996-2010.

Firkins, J. L. Yu, Z. &Morrison, M. (2007). Ruminal nitrogen metabolism: perspectives for integration of microbiology and nutrition for dairy. *J Dairy Sci.*, **90** *(Suppl* **1***)*, E1-E16.

Flint, H. J., Bayer, E. A., Rincon, M. T., Lamed, R. & White, B. A. (2008). Polysaccharide utilization by gut bacteria: potential for new insights from genomic analysis. *Nature Rev Microbiol.,* **6**, 121-131.

Friedrich, M. W. (2006). Stable-isotope probing of DNA-insights into the function of uncultivated microorganisms from isotopically labeled metagenomes. *Curr Opin Biotechnol.,* **17**, 59-66.

Galbraith, E. A., Antonopoulos, D. A. & White, B. A. (2004). Suppressive subtractive hybridization as a tool for identifying genetic diversity in an environmental metagenome: the rumen as a model. *Environ Microbiol.*, **6**, 928-937.

Gillespie, D. E., Brady, S. F., Bettermann, A. D., Cianciotto, N. P., Liles, M. R. & Rondon, M. R. (2002). Isolation of antibiotics turbomycin A and B from a metagenome library of soil microbial DNA. *Appl Environ Microbiol.*, **68**, 4301-4306.

Guo, H., Feng, Y., Mo, X., Duan, C., Tang, J. & Feng, J. (2008). Cloning and expression of beta-glucosidase gene umcel3G from metagenome of buffalo rumen and characterization of the translated product. *Sheng Wu Gong Cheng Xue Bao.* 24, 232-38. (Article in Chinese, abstract in English)

Hall, N. (2007). Advanced sequencing technologies and their wider impact in microbiology. *J Exp Biol.*, **210**, 1518-1525.

Handelsman, J., Rondon, M. R., Brady, S. F., Clardy, J. & Goodman, R. M. (1998). Molecular biological access to the chemistry of unknown soil microbes: a new frontier for natural products. *Chem Biol.,* **5**, R245-R249.

Henne, A., Schmitz, R. A., Bomeke, M., Gottschalk, G. & Daniel R. (2000). Screening of environmental DNA libraries for the presence of genes conferring lipolytic activity on *Escherichia coli. Appl Environ Microbiol.*, **66**, 3113-3116.

Huson, D. H., Auch, A. F., Qi J. & Schuster, S. C. (2007). MEGAN analysis of metagenomic data. *Genome Res.*, **17**, 377-386.

Jones, B. V. & Marchesi, J. R. (2007). Transposon-aided capture (TRACA) of plasmids resident in the human gut mobile metagenome. *Nature Methods.*, **4**, 55-61.

Kang, S., Denman, S. E., Morrison, M., Yu, Z. & McSweeny, C. S. (2009). An efficient RNA extraction method for estimating gut microbial diversity by polymerase chain reaction. *Curr Microbiol.*, **58**, 464-471.

Kobayashi, K. S., Chamaillard, M., Ogura, Y., Henegariu, O., Inohara, N. & Nunez, G. (2005). Nod2-dependent regulation of innate and adaptive immunity in the intestinal tract. *Science.*, **307**, 731-734.

Kowalchuk, G. A., Speksnijder, A. G. C., Zhang, K., Goodman, R. M. & van Veen, J. A. (2007) Finding the needles in metagenome haystack. *Microbial Ecol.*, **53**, 475-485

Lammle, K., Zipper, H., Breuer, M., Hauer, B., Buta , C., Brunner, H. & Rupp, S. (2007). Identification of novel enzymes with different hydrolytic activities by metagenome expression cloning. *J Biotechnol.*, **127**, 575-592.

Lan, P. T., Sakamoto, M., Sakata, S. & Benno, Y. (2006). *Bacteroides barnesiae* sp. nov., *Bacteroides salanitronis* sp. nov., and *Bacteroides gallinarum* sp. nov., isolated from chicken caecum. *Int J Syst Evol Microbiol.*, **56**, 2853-9.

Lima, J. R., Ribin, A. O., Russell, J. B. & Mantovani, H. C. (2009). Bovicin HC5 inhibits wasteful amino acid degradation by mixed ruminal bacteria in vitro. *FEMS Microbiol Lett.*, **292**, 78-84.

Lissens, G., Verstraete, W., Albrecht, T., Brunner, G., Creuly, C., Seon, J., Dussup, G. & Lasseur C. (2004). Advanced anaerobic bioconversion of lignocellulosic waste for bioregenerative life support following thermal water treatment and biodegradation by *Fibrobacter succinogenes*. *Biodegradation.*, **15**, 173-83.

Liu, K., Wang, J., Bu, D., Zhao, S., McSweeny, C. S., Yu, P. & Li, D. (2009). Isolation and biochemical characterization of two lipases from a metagenomic library of China Holstein cow rumen. Biochem. *Biophys Res Commun.*, **385**, 605-611.

Liu, L., Tang, J. & Feng, J. (2009). Bacterial diversity in Guangxi buffalo rumen. *Wei Sheng Wu Xue Bao.*, **49**, 251-256. (article in Chinese, abstract in English)

Lopez-Cortes, N., Reyes-Duarte, D., Beloqui, A., Polaina, J., Ghazi, I. & Golyshina, O. V. (2007). Catalytic role of conserved HQGE motif in the CE6 carbohydrate esterase family. *FEBS Lett.*, **581**, 4657-4662.

Lopez-Garcia, P. & Moreira D. (2008). Tracking microbial diversity through molecular and genomic ecology. Review. *Res Micrbiol.*, **159**, 67-73.

Lu, J., Santo Domingo, J. & Shanks, O. C. (2007). Identification of chicken-specific fecal microbial sequences using a metagenomic approach. *Water Res.*, **41**, 3561-3574.

Margulies, M., Egholm, M., Altman, W. E., Attiya, S., Bader, J. S. & Bemben, L. A., (2005). Genome sequencing in microfabricated high-density picolitre reactions. *Nature.*, **437**, 376-380.

Martinez, A., Kolvek, S. J., Yip, C. L., Hopke, J., Brown, K. A., MacNeil, I. A. & Osburne, M. S. (2004). Genetically modified bacterial strains and novel bacterial artificial chromosome shuttle vectors for constructing environmental libraries and detecting heterologous natural products in multiple expression hosts. *Appl Environ Microbiol.*, **70**, 2452-2463.

Morrison, M., Pope, P. B., Denman, S. E. & McSweeney, C. S. (2009). Plant biomass degradation by gut microbiomes: more of the same or something new? *Curr Opin Biotechnol.*, **20**, 358-363.

Myrick, K. V. & Gelbart, W. M. (2002). Universal fast walking for direct and versatile determination of flanking sequence. *Gene.*, **284**, 125-131.

Nelson, K. E, Zinder, S. H., Hance, I., Burr. P., Odongo, D., Wasawo, D., Odenyo, A. & Bishop R. (2003). Phylogenetic analysis of the microbial populations in the wild herbivore gastrointestinal tract: insights into an unexplored niche. *Environ Microbiol.*, **5**, 1212-1220.

Nicholson, M. J., Evans, P. N. & Joblin, K. N. (2007). Analysis of methanogens diversity in the rumen using temporal temperature gradient gel electrophoresis: identification of uncultured methanogens. *Microb Ecol.*, **54**, 141-50.

Noguchi, H., Park, J. & Takagi, T. (2006). MetaGene: prokaryotic gene finding from environmental genome shotgun sequences. *Nucleic Acid Res.*, **34**, 5623-5630.

Nordgard, L., Traavik, T. & Nielsen, K. M. (2005). Nucleic acid isolation from ecological samples—vertebrate gut flora. *Methods Enzymol.*, **395**, 38-48.

Ochman, H., Ayala, F. J. & Hartl, D. L. (1993). Use of polymerase chain reaction to amplify segments outside boundaries of known sequences. *Methods Enzymol.*, **218**, 309-321.

Pace, N. R., Stahal, D. A., Lane, D. J. & Olsen, G. J. (1986). The analysis of natural microbial populations by ribosomal RNA sequences. *Adv Microb Ecol.*, **9**, 1-55.

Palackal, N., Lyon, C. S., Zaidi, S., Luginbuhl, P., Dupree, P. & Goubet, F. (2007). A multifunctional hybrid glycosyl hydrolase discovered in an uncultured microbial consortium from ruminant gut. *Appl Micobiol Biotechnol.*, **74**, 113-124.

Qi, M., Nelson, K. E., Daugherty, S. C., Nelson, W. C., Hance, I. R , Morrison, M. & Forsberg, C. W. (2005). Novel molecular features of the fibrolytic intestinal bacterium *Fibrobacter intestinalis* not shared with *Fibrobacter succinogens* as determined by suppressive subtractive hybridization. *J Bacteriol.,* **187**, 3739-3751.

Radajewski, S., Ineson, P., Parekh, N. R. & Murell, J. C. (2000). Stable-isotope probing as a tool in microbial ecology. *Nature.*, **403**, 646-649.

Schloss, P. D. & Handelsman, J. (2005). Introducing DOTUR, a computer program for defining operational taxonomic units and estimating species richness. *Appl Environ Microbiol.*,1501-1506.

Schmeisser, C., Steele, H. & Streit, W. R. (2007) Metagenomics, biotechnology with non-culturable microbes. *Appl Microbiol Biotechnol.*, **75**, 955-962

Schmidt, T. M., DeLong, E. F. & Pace, N. R. (1991). Analysis of marine plankton community by 16S rRNA gene cloning and sequencing. *J Bacteriol*, **173**, 4371-4378.

Selinger, L. B., Forsberg, C. W. & Cheng. K. J. (1996). The rumen: a unique source of enzymes for enhancing livestock production. *Anaerobe.*, **2**, 263-284.

Shanks, O. C., Santo Domingo, J. W., Lamendella, R., Kelty, C. A. & Graham, J. E. (2006). Competitive metagenomic DNA hybridization identifies host-specific microbial genetic markers in cow fecal samples. *Appl Environ Microbiol.,* **72**, 4054-4060.

Sharma, R., John, S. J., Damgaard, M. & McAllister, T. A. (2003). Extraction of PCR quality plant and microbial DNA from total rumen contents. *Biotechniques.*, **34**, 92-94, 96-97.

Shedova, E. N., Lunina, N. A., Berezina, O. V., Zverlov, V. V., Schwarz, V. & Velikodvorskaia, G. A. (2009). Expression of the genes *celA* and *XylA* isolated from a fragment of metagenomic DNA in *Escherichia coli. Mol Gen Mikrobiol Virol.*, **2**, 28-32 (Article in Russian, abstract in English)

Short, J. M. & Mathur, E. J. (1999). Production and use of normalized DNA libraries. *US Patent No.* **6001574**.

Singh, B., Bhat, T. K. & Singh, B. (2001). Exploiting gastrointestinal microbes for livestock and industrial development. *Asian-Aust J Anim Sci.*, **14**, 567-586.

Singh, B., Gautam S. K. & Mukesh, M. (2009). Rumen ecosystem to boost productivity- a metagenomic overview. *Indian Dairyman.* **61** (9), 50-55.

Singh, B., Gautam, S. K., Verma, V., Kumar, M. & Singh, B. (2008a). Metagenomics in animals gastrointestinal tract- potential biotechnological prospects. *Anaerobe.*, **14**, 138-144.

Singh, B., Bhat, T. K., Sharma, O. P. & Kurade, N. P. (2008b). Metagenomics in animal gastrointestinal tract- a microbiological and biotechnological perspective. *Indian J Microbiol.*, **48**, 216-227.

Toyoda, A., Iio, W., Mitsumori, M. & Minato, H. (2009). Isolation and identification of cellulose binding proteins from sheep rumen contents. *Appl Environ Microbiol.*, **75**, 1667-1673.

Tuohy, K. M., Gougoulias, C., Shen, Q., Walton, G., Fava, F. & Ramani, P. (2009). Studying the human gut microbiota in the trans-omics era- focus on metagenomics and metabolomics. *Curr Pharmaceut Des.,* **15**, 1415-1427.

Whitman, W. B., Coleman, C. D. & Wiebe, W. J. (1998). Prokaryotes: the unseen majority. *Proc Natl Acad Sci., USA*. **95**, 6578-6583.

Xu, J. (2006). Microbial ecology in the age of genomics and metagenomics: concepts, tools, and recent advances. *Mol Ecol*., **15**, 1713-1731.

Venter, J. C., Remington, K., Heidelberg, J. F., Halpern, A. L., Rusch, D. & Eisen, J. A. (2004). Environmental genome shotgun sequencing of the Sargasso Sea. *Science*., **304**, 66-74.

In: Metabolomics: Metabolites, Metabonomics…
Editors: J.S. Knapp and W.L. Cabrera, pp. 201-213
ISBN: 978-1-61668-006-0

Chapter 6

NUTRIGENOMICS, METABOLOMICS AND METABONOMICS: EMERGING FACES OF MOLECULAR GENOMICS AND NUTRITION

B. Singh[1,*], M. Mukesh[2], M. Sodhi[2], S.K. Gautam[3], M. Kumar[4] and P.S. Yadav[5]

[1]Regional Station, IVRI, Palampur-176 061, India
[2]Animal Genetics Division, NBAGR Karnal-132 001, India
[3]Department of Biotechnology, Kurukshetra University, Kurukshetra-136 119, India
[4]Dairy Microbiology Division, NDRI Karnal-132 001, India.
[5]Department of Buffalo Physiology and Reproduction, CIRB, Hisar-125 001, India

Abstract

Nutrition exhibits the most important life-long environmental impact on health. Nutrients, gut microbial metabolites and other bioactive food constituents interact with body at system, organ, cellular and molecular levels, and affect the expression of genome at several levels, and subsequently, the overall production of metabolites. Direct measurement of cellular metabolites is essential for the study of biological processes, and may allow causes of disease, toxicological progression, and novel disease-biomarkers to be identified. Advances in analytical techniques and the algorithms for management of the data has allowed a precise and global analysis of biological substances such as DNA (genomics), RNA (transcriptomics), proteins (proteomics) and smaller molecules (metabolomics). Holistic "**omics**" approaches are indispensable to cover the complex nutrient-cell and gut microbial-host interactions. This chapter presents an overview of nutrigenomics and metabolomics tools with reference to their perspective in livestock health and production.

* E-mail address: bsbpalampur@yahoo.co.in; Fax; +91 1894 233063; Phone +91 1894 230526. (Corresponding author)

Introduction

Imbalanced nutrition in terms of deficits of critical nutrients or excessive intake of certain anti-nutritional compounds may cause a number of diseases and metabolic disorders. The classic or traditional research related to human or animal nutrition, deals mainly with studying the interactions of host-dietary components directly or using biomarker approaches (van Ommen, 2004) or aims to study either deficiency or excess of nutrients in relation to health ailments. Recent advances in genomics have propelled the development of new technologies that have provided the researchers with methods to quickly analyze genes and their products *en masse*.

The advent of modern analytical tools has led to realization that not only are certain nutrients essential, but also that specific quantities of each were necessary for optimal health, thereby leading to such notions as dietary recommendations, nutritional epidemiology, and realization that food can directly contribute to disease onset (Mutch et al., 2005). Further, the availability of human genome sequence information and large set of single nucleotide polymorphisms (SNPs) in candidate genes and their correlation with metabolic imbalances have pioneered a new era in modern genomics, and added new parameters to the molecular nutrition panel. There is wide support to the theory that genetic variation in selected SNPs, haplotypes, and copy number variants can have remarkable effect not only in an individual's response to dietary components, but also for their optimal utilization (Ferguson, 2006).

Nowadays it is recognized that understanding the effect of diet on health requires the study of mechanisms of nutrients and other bioactive food constituents at cellar as well as molecular levels. This is supported by the increasingly growing number of studies in humans, animals and cell culture studies revealing that nutrients and other bioactive dietary ingredients have crucial role in regulating gene-expression in diverse ways (Mead, 2007). Also, there are enough scientific evidences illustrating that micronutrients and certain plant secondary metabolites (PSMs) can interact with the genome, modify gene-expression and regulation, alter protein and metabolite composition within the cells and tissues, and even participate in genome repair (Fig. 1) (Singh et al., 2003; Marambaud et al., 2005; Zheng and Chen, 2006; Fenech 2008). This concept puts nutritional genomic area at the food/ gene interphase creating opportunities for the industry to develop commercially viable supplements or nutraceuticals that can modify expression of the genes of interest (Subbiah, 2008).

Nutrigenomics Concept

Nutritional genomics is a recent offshoot of genetic revolution that was experienced over the past decade. The term nutritional genomics, or nutrigenomics, appears to have its origin in the context of plant biology, wherein it refers to work in interface of plant biochemistry (specifically, secondary metabolism), genomics, and human nutrition (DellaPenna, 1999). Muller and Kersten (2003) define nutrigenomics as "the applications of high throughput genomics in nutrition research, and studying the genome wide influences of nutrition". From a nutrigenomics perspective, nutrients and dietary signals are detected by cellular sensor or receptor systems that in turn, influence gene and protein expression, and subsequently the metabolite production by the cell (Muller and Kersten, 2003; Mariman, 2006; Fenech, 2008). Hence, pattern of gene expression, protein synthesis and modification and production of metabolites in response to the nutrients or nutritional regimes can be viewed as "dietary signatures". The nutrigenomics is primarily aimed at examining these "dietary signatures" in a target cell, tissue or even entire organism, and thereby elucidating the impact of nutrition on host homeostasis. According to Kaput et al. (2005) nutrigenomics is the study of how constituents of the diet interact with genes, and their products, to alter phenotype and, conversely, how genes and their products metabolize these constituents into nutrients, antinutrients, and bioactive compounds. This new era of nutrition recognizes the complex relation and interaction between the health of individual, its genome, and the life-long dietary exposure, and has led to realization that nutrition is essentially a gene-environment interaction science.

The nutritional genomics encompasses two broad areas namely, nutrigenomics, which deals with interaction between dietary components and the genome as well as the resulting changes in proteins and other metabolites, and the nutrigenetics, which aims at understanding gene-based differences in response to dietary components and to develop novel nutraceuticals that are most compatible with the health status of individuals based on their genetic makeup (Kaput, 2008). Nutrigenetics and nutrigenomics have emerged as nascent areas that are evolving quickly and riding on the wave of "personalized medicine" that is providing opportunities in the discovery and development of nutraceutical compounds (Panagiotou and Nielsen, 2009).

Nutrient-Gene Interaction

It is evident that diet is a complex mixture of substances and gut microbial metabolites (Fig. 2) that supply both energy and building blocks to develop and sustain the cells or organisms. The nutrients exhibit a variety of biological activities, ranging from protection against diseases and acting as signaling molecules (Muller and Kersten, 2003). At molecular levels, the nutrients relay signals and communicate a specific cell about the dietary components. Therefore, the cellular processes including every step of genetic information from gene-expression to the protein synthesis and degradation might be affected by the diet and environmental factors.

In some ways, the nutrigenomics can be compared with pharmacogenomics (personalizing drug therapy based on individual SNPs) which has made tremendous headway in recent years as a tool to reduce individual drug toxicity and designing personalized medicine (Giacomini et al., 2007; Relling and Hoffman, 2007). However, the important difference is that pharmacogenomics is concerned with the effects of drugs that are pure compounds administered in precise and small doses, whereas nutrigenomics encompasses complexity and variability of nutrients (Muller and Kersten, 2003).

Advanced technologies for genomic analysis have led to identification of genes or markers associated with genes of interest. Microarray technology for high-throughput screening of changes in gene-expression has enormously advanced the nutrigenomic studies. At the moment, the gene-expression microarrays have become a de facto golden standard to evaluate changes in genome wide gene-expression under different conditions.

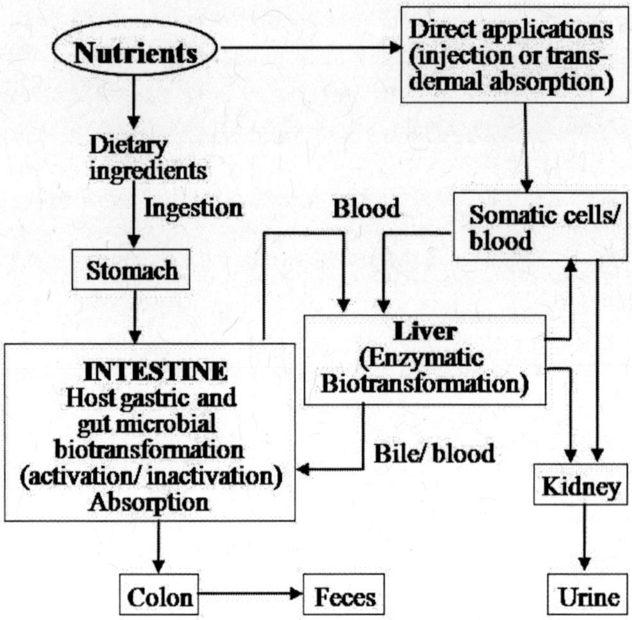

Figure 1. A diagrammatic illustration of applications and effects of nutrients and bioactive PSMs. In addition to oral intake, certain compounds can reach blood circulation via absorption or inhalation. Gut and liver are the major sites for metabolism of these components. The gut microbes have ability to detoxify as well as to generate bioactive compounds from dietary components.

Nutrigenomics in Livestock Perspective

Application of modern molecular biological techniques has potential to revolutionize the animal nutrition. Presently, many of these technologies are used in research on searching and identifying candidate genes and disease diagnosis. Veterinary nutritionists have begun applying animal genomics to the field of nutrition. Integrating the information encoded in the genome to applied nutrition and ultimately augmenting livestock production is the goal. Further, when genomics is combined with metabolomics (discussed below), the whole animal assessment may be achieved and may provide the opportunity for corrective interaction *via* specific nutrients like, retinoic acid, fatty acids, vitamins and other compounds. The ability to thoroughly understand the role of nutrients will be significantly enhanced by using nutrigenomics and metabolomic approaches. In addition, nutrigenomics tools are being used to link expression data with gene-function in the bovines, such as in vitro models of bovine adipogenesis and bioinformatics tools to map gene network (Lehnert et al., 2006).

In the changing scenario of ruminant nutrition, the nutritional genomics has several applications (Zdunczyk and Pareek, 2008). However, at present the nutrigenomics studies in livestock sciences are rare, but they are likely to become more important, as we develop an understanding of the relationship between nutrition, genetics, fertility and tissue growth (Dawson, 2006). Molecular nutrition will serve as a new tool for nutritional research in mitigating the problems related to animal health and production (Table 1).

Table 1. A summary of challenging areas in augmenting livestock production using nutrigenomics.

1. Gaining insight into the mechanisms or pathways by which dietary components affect animal growth, tissue structure and overall performance by "up- or down-regulation" of target genes
2. Identifying the novel strategies for controlling the key metabolic processes by managing gene expression rather than looking at animal performance based on traditional nutritional responses
3. Evaluating the diet-mediated differential gene expression, identifying diet-induced alterations in gene expression profile (Reverter et al., 2003)
4. Using proteomics tools (e. g. 2-D electrophoresis) to reveal information concerning composition of egg and poultry meat proteins, and effect of dietary methionine on breast-meat accretion
5. Evaluating the effects of use of transgenic crops on animal nutrition and health
6. Identification of the genes involved in carcinogenesis and anti-carcinogenesis process in response to dietary toxins or PSMs
7. Elucidating the role of gut microbiota in host immunity and nutrient utilization
8. Analysis of regulation of myogenesis and its regulatory pathways in meat producing animals

What is Metabolomics?

With the advent of functional genomics during the last two decades and recent advances in sequencing technologies, a substantial progress has been made in biological investigations. With the evolution of second generation sequencing it is possible to sequence the entire genome of an organism, and precisely analyze the huge data produced. The "omics" applications now enable us to understand various aspects of cellular physiology and/ or biology of an organism as affected or influenced by environmental stimuli or genetic perturbations.

The current rise in diet-related diseases continues to be one of the most significant health problems. The technologies are developed that have enormous impact on disease investigation by studying the metabolic profile of a cell or an organism. The ^{1}H-nuclear magnetic resonance (NMR) and mass spectrometry (MS)-based technologies to generate profiles of metabolites in biofluids permit profiling of the entire metabolome, which provides a sensitive intermediate phenotype linking the genotype, gut microbial composition and personal health status (Oresic, 2009). The metabolomics combines strategies to identify and quantify the cellular metabolites using sophisticated analytical technologies along with applications of statistical and multivariate methods for deriving information and data interpretation (Roessner and Bowne, 2009). The assessment of both essential nutrient status and the more comprehensive systemic metabolic response to dietary, lifestyle and environmental influences are necessary for the evaluation of physiological status in individuals that can identify multiple targets of interventions needed to address metabolic diseases (Zivkovik and German, 2009). As the cellular metabolites are considered to act as "spoken language, or broadcasting signals" from the genetic architecture and the environment, the metabolomics is considered to provide a direct "functional readout of the physiological state" of an organisms (Gieger et al., 2008). Salient applications of metabolomics have been summarized in Table 2.

Techniques and Data Analysis in Metabolomics

Unlike *transctriptomics* and *proteomics*, which intend to determine a single or unique class of end products (mRNA and proteins, respectively), the metabolomics has to deal with components of very diverse physiochemical properties. Moreover, concentration of these metabolites in biofluids varies from millimoler level (or higher) to picomoler, making it to exceed the linear range of conventially employed analytical techniques (Garcia-Canas et al., 2010). Hence, metabolomics relies on additional technologies to isolate and characterize biological metabolites which can combine automation and miniaturization as for nucleic acids. This includes techniques for tissue sampling, extraction of specific classes of molecules, their storage, sample preparation and analyses. A combination of methods (Table 3) based on gas chromatography/mass spectrometry (GC/MS) and liquid chromatography/ mass spectrometry (LC/MS) has attained a high technical robustness, which makes them more comparable to microarrays used for nucleic acids or protein studies (Hocquette, 2005).

The process of metabolomics comprises of four broad conceptual approaches, namely, i) target analysis, ii) metabolic profiling, iii) metabolomics, and iv) metabolic fingerprinting. Therefore, the specific application depends on the subject and requirement. Innovative experimental designs combined with novel computational tools (Table 4) for handling metabolomics data offer new opportunities for early disease detection as well as characterization of dietary and therapeutic interventions in the context of human physiology (Oresic, 2009). MS-based small molecular metabolite analysis is rapidly becoming a method of choice and enables multiple biological paths discovering and validating functional assignments (Baran et al., 2009). The combination of MS-based metabolic profiling with genome-scale models of metabolisms and other –'omics' approaches provide opportunities to expand our understanding of microbial metabolic-networks, stress-responses, and to identify genes associated with specific enzymatic and regulatory activities (Baran et al., 2009). NMR however, still remains most important instrument in metabolomic studies, though initially it had certain limitations which were later overcome by incorporating software called as Eclipse Version 3.0.

Table 2. Applications of metabolomic tools.

1. Predicting the physiological status of a cell or organism, detection of drug residues through global profiling of metabolites in blood or body fluids
2. Developing actionable metabolic diagnostics and more comprehensive systemic metabolic response to dietary, lifestyle and environmental influences (Zivkovik and German, 2009)
3. Using 'metabolome fingerprints' for predicting embryonic development through dynamic changes in its metabolome (Hayashi et al., 2009)
4. Rapidly assessing the disturbances in metabolic profiles following administration of drugs, and identifying the biomarkers of toxicity to assess the health risk of specific toxins

Comprehensive multidimensional techniques, such as GCxGC or LCxLC, are also a revolutionary improvement in separation techniques that will be applicable in nutritional metabolomics in near future. They provide enhanced resolution and a huge increase in selectivity and sensitivity in comparison with conventional separation techniques (Garcia-Canas et al., 2010). A number of commercially available software can be used for quantitative analysis of the desired markers (Issaq et al., 2009). As metabolomics is an emerging technology, so new analytical techniques and method are being developed and will continue in future in order to achieve its goals.

Table 3. List of major equipments used in nutrigenomic investigations and the software for multivariate statistical analysis.

Technique	Salient features	Major limitations
Gas chromatography (GC)	Simple, quick and cheaper analytical tool	Analysis is limited to small compounds that are thermally stable and volatile Derivatization of samples is required Detection is limited to certain compounds only unless MS is the method of choice
Gas chromatography mass spectrometry (GC/MS)	Applicable for metabolic profiling of body fluids Separation of large number of compounds in a mixture can be accomplished using multidimensional separation	Some compounds may not ionize sufficiently to be detected at low levels Needs derivatization with ionizable moiety
High pressure chromatography (HPLC)	Broad range of separation Much wider range of applications than GC/MS Well suited for analysis of global metabolomics and the disease markers Offers various modes of separation and purification	Conventional HPLC systems use pumps that are limited to 6000 psi, and reverse phase (RP) columns are packed with 3-5 mm particles which limit the applications HPLC as an efficient system
Ultra high HPLC	Coupled with MS, it is a powerful tool that allows separation and characterization of various compounds Well suited for global metabolomic studies Can resolve large number of metabolites, thus, suitable for elucidating several disease markers	Highly expensive Cost of the sample analysis is high which limit its use
Capillary electerophoresis (CE)	High resolution, suitable for charged, neutral, polar, and hydrophobic compounds for targeted as well global metabolomics Requires minute organic solvent consumption, and minimal solvent waste production	The sensitivity is poor because of very less (nanoliters) injection volume and low optical path when UV/ absorbance is the mode of action
Capillary electrochromato-graphy (CEC)	Efficient than HPLC and CE It has ability to separate complex mixtures with large number of metabolites than are possible with HPLC. Higher sample loading may make it a method of choice in future	Major problem is formation of bubbles in the column The information on applications of CEC in metabolomics is scarce
NMR Spectroscopy	Ideal instrumental platform for metabolic analysis of biofluids Useful in structure and conformational analysis of a number of compounds/ metabolites It is non-invasive method, and offers high reproducibility and non-selectivity in metabolite detection Has ability to simultaneously quantify multiple classes of metabolites	Sophisticated software needed for precise analysis of data It has lower sensitivity compared to other techniques (Baxan et al., 2008)
Microarray analysis	Highly useful tool in functional genomics A number of gene transcripts can be analyzed at a time	High cost of instrument and the arrays

Table 4. Useful websites and database in nutrigenomic studies to aid identification of unknown metabolites.

Name	Web addresses	Database
Biological Magnetic Resonance Data Bank	www.bmbr.wisc.edu	NMR
CyberCell Database	redpoll.pharmacy.ulberta.ca.CCDB	-
ExPASy-GlycolMod tool	www.expasy.org/tools/glycomod	MS
Functional Glycomics Gateway	www.functionalglycomics.org	MS
HORA suit	www.paternostroblab.org	-
Human metabolome database	www.hmdb.ca	NMR & MS
Lipid Maps	www.lipmaps.org	MS
Madison Metabolomic Consortium Database	mmcd.nmrfam.wisc.edu	MS& NMR
MassBank	www.massbank.jp	MS
METLIN	metlin.scrips.edu	MS
NIST Chemistry Web book	webbook.nist.gov/chemistry	MS
NMRShiftDB	nmrshiftdb.ice.mpg.de	NMR
Spectral Database for Organic Compounds	riod01.ibase.aist.go.jp/sdbs/cgi-bin/cre_index.cgi	MS & NMR
The Magnetic Resonance Metabolomic Database	www.ilu.se/hu/md1/main	NMR

Adapted from Issaq et al. (2009). The abbreviations are explained in the text

What Is Metabonomics?

Metabonomics is a relatively new term, and a post genomic research field having been coined by Nicholson et al., (1999), and is concerned with developing methods for high throughput analysis of low molecular weight compounds in the metabolome. The term was used to describe quantitative measurement of the dynamic multiparametric metabolic responses of a cell or an individual to pathophysiologic stimuli or genetic alterations. During the past few years the metabonomics has emerged as a rapidly expanding area of scientific research and is one of the new "omics" methods joining genomics, transcriptomics and proteomics in the field of biological sciences. Metabonomics, a variant of metabolomics, thus, examines the changes in hundreds or thousands of metabolites in an intact tissue or biofluid. After going through the published research it is clear that metabolomics is now moving in an exciting direction.

It is evident that cellular metabolites display cell-specific concentrations of metabolites, therefore, co-vary with gene-expression signatures for individual cell-types. Through metabonomics it is possible to quantitatively understand the metabolite complement of integrated living systems and its dynamic responses to the changes in both endogenous factors (e.g. physiology and development) and exogenous factors like environmental factors and xenobiotics.

Table 5. Metabolic pathways websites.

BioCyc	www.biocyc.org
ExPASy	www.expasy.org
KEGG	www.genome.jp
NuGO	www.nugowiki.org
PMN	www.arabidopsis.org
Reactome	www.reactome.org
SGD	www.yeastgenome.org

Adapted from Issaq et al. (2009).

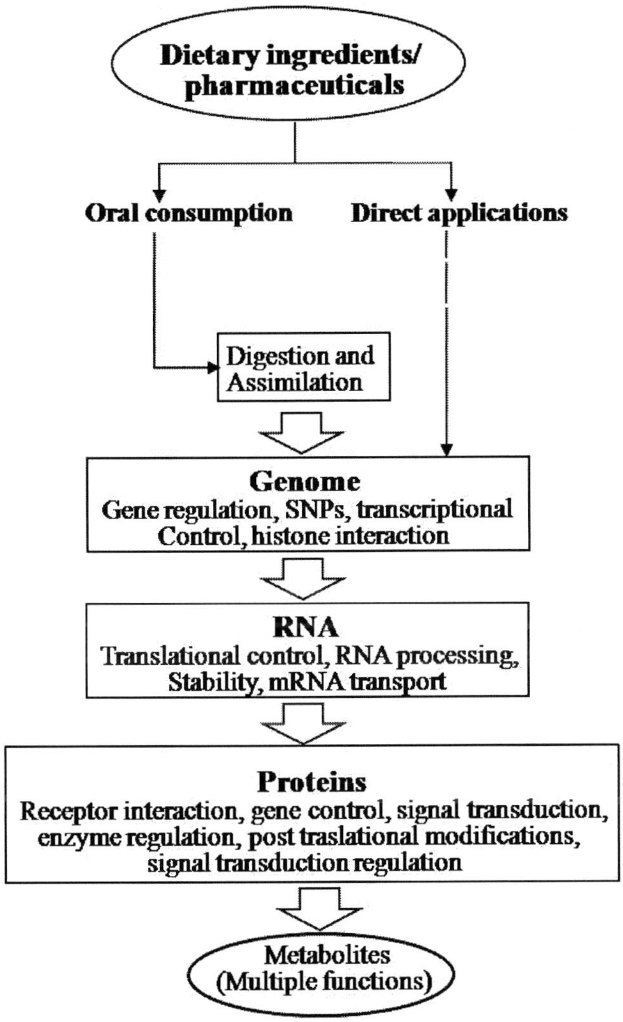

Figure 2. Health effects of food components. Though diet is the most important environmental factor affecting the host's health, other bioactive molecules also have profound effect on metabolism and determining the phenotype of the cells, organs or organisms.

As a holistic approach, the metabonomics aims to detect cellular metabolites, quantifies and catalogues the temporal metabolic processes of an integrated biological system, and ultimately correlates such processes to the physiological or pathophysiological status of a cell or organism. By measuring the metabolites simultaneously a picture referred to as a "fingerprint" of the current metabolic status of the organism is generated. It is then possible to compare this metabolic profile in the same organism at different times or else in different organisms.

The profiling of metabolites is known to initiate in 1950s, though the subsequent progress was slow, and it is only since the beginning of the new millennium, the metabonomics has emerged as a fast-growing area with several medical applications. The advancement can be attributed to the innovations in techniques like NMR and MS to study metabolic composition of biological fluids, cells and tissues (Rezzi et al., 2007), and the data handling and processing.

NMR offers a number of advantages (Table-3) including very high reproducibility as indicted by co-efficient of variation for replicate measures of the same sample that are in the range 0.5-2% across the NMR spectrum, which compares with 3-10% for techniques such as ELISA. The data on metabolic profiles are processed by multivariate statistics to maximize recovery of information to be correlated with well-defined stimuli such as dietary intervention or with any phenotypic data. From the profile or “spectral fingerprint”, it is possible to uncover information about organisms’ disease or physiological state, diet, biological age or nutritional regime or drug treatment. The ability to detect cellular metabolites from body fluids makes the metebonomics uniquely suitable to access metabolic responses to deficiencies or excess of nutrients and bioactive components.

Sample Analysis and Data Processing in Metabonomics

Basically, the metabonomics consists of two parts. In the first part, the experimental technique must be used to collect the input dataset - the concentration of multiple metabolites within the sample under study. Secondly, a data processing technique must be applied to the dataset in order to sift out patterns of interest. However, there is no single instrument that can correctly detect and analyze all cellular metabolites (Dettmer et al., 2007), nor there is a standard complete metabolic database for analysis of the data (Goldsmith et al., 2009). The need for a method to be applied depends on the question being asked. For global metabolomics, a comprehensive procedure should be used that might employ more than one chromatographic technique or separation mode, whereas for targeted metabolomics, the decision depends on group of metabolites of interest, and which separation technique is particularly well suited for its analysis (Issaq et al., 2009).

A machine, the Metabolic Profiler, combines NMR and time-of-flight mass spectrometry (TOF-MS) with Brukers Biospin (www.bruker-biospin.com) analysis software. This platform is applicable to toxicity and efficacy studies in preclinical and clinical development, as well as to clinical research in disease screening and patient satisfaction. Another application of the system is the discovery of new small molecular diagnostic markers. NMR-based metabonomics is non-destructive, non selective, fast, cost effective and needs a minimal amount of the sample, hence remains as a prioritized choice.

The data processing challenges in metabolomics are quite unique and require specialized data-analysis programs, and a detailed knowledge of cheminformatics, bioinformatics and statistics. Due to huge amount of data generated it is necessary to develop strategies to convert the complex raw data into useful information. For obtaining metabolic profile of an organism, multivariate methods need to be applied to derive latent information.

In order to classify and sharpen separation between groups of observations, projection methods such as principal component analysis (PCA), SIMCA, and PLS-discriminant analysis (PLS-DA) are suited. In order to quickly analyze and identify the molecules, certain commercial companies have developed high throughput platforms (Tables-4 and 5) combining high-end LC/MS, GC/MS with proprietary software. Once identified the detected molecules can be related back to biochemical pathways. Certain commercial companies are now using metabonomics for preclinical work with animals and in early-stage clinical trials.

In an effort to simplify metabolomic data analysis while at the same time improving user accessibility, a web server for metabolomic data analysis called MetaboAnalyst has been developed (Xia et al., 2009). MetaboAnalyst accepts a variety of input data (NMR peak lists, binned NMR or mass spectra, MS peak lists, compound/ concentration data) in a wide varity of formats, and supports such techniques as fold change analysis, *t*-tests, PCA, PLS-DA, hierarchical clustering and a number of more sophisticated statistical or machine learning methods (Xia et al., 2009).

Whilst metabonomics is at the endpoint of the ''omics cascade'' and closest to the individual phenotype, at present, there is no single-instrument that can be used to analyze all metabolites simultaneously (Dettmer et al., 2007). For comprehensive details of the instrumentation in metabonomics and metabolomics, the readers may refer to recent publications (Lindon et al., 2004; Dettmar et al., 2007; Goldsmith et al., 2009; Issaq et al., 2007; 2009).

Conclusion

In mammals the impact of nutrients on gene-expression and nutritional interventions to manage health has emerged as thrust of post genomic area. The current evidence based on nutrigenomics has begun to identify subgroups of individuals who benefit more from dietary interventions. Although still in infancy, the nutrigenomics has shown immense promise in areas as diverse as toxicology studies to discovery of biomarkers of the disease. It is likely that during next decade the nutritional supplementation and functional food industries will experience robust growth in response to advances in nutritional genomics research and applications. Metabolomics forms a useful platform for further biomarker development, and in the field of medicine. The continuous progress in the field will allow in future to provide targeted gene-based dietary advice.

References

Baran, R., Reindl, W. & Northen, T. R. (2009). Mass spectrometry based metabolomics and enzymatic assays for functional genomics. *Curr Opin Microbiol.,* **12**, 547-552.

Baxan, N., Rabeson, H., Pasquet, G., Chateaux J., Briguet A., Morin, P., Graveron-Demilly, D. & Fakri-Bouchet, L. (2008). Limit of detection of cerebral metabolites by localized NMR spectroscopy using microcoils. *Comptes Rendus Chimie.*, **11**, 448-456.

Dawson, K. A. (2006). Nutigenomics: feeding the genes for improved fertility. *Anim Reprod Sci.,* **96**, 312-322.

DellaPenna, D. (1999). Nutritional genomics: manipulating plant micronutrients to improve human health. Review. *Science.*, **285**, 375-379.

Dettmer, K., Aronov, P. A. & Hammock, B. D. (2007). Mass-psectrometry based metabolomics. *Mass Spectrom Rev.*, **26**, 51-78

Fenech, M. (2008). Genome health nutrigenomics and nutrigenetics: diagnosis and nutritional treatment of genome damage on an individual basis. *Food Chem Toxicol.,* **46**, 1365-1370.

Ferguson, L. R. (2006). Nutrigenomics: integrating genomic approaches into nutrition research. Review. *Mol Diagn Ther.,* **10**, 101-108.

Garcia-Canas, V., Somo, C., Leon, C. & Cifuentes A. (2010). Advances in nutrigenomics research: novel and future analytical approaches to investigate the biological activity of natural compounds and food functions. Review. *J Pharmaceut Biomed Anal.,* **51**, 290-304.

Giacomini, K. M., Brett C. M., Altman, R. B., Benowitz, N. L., Dolan, M. E., Flockhart, D. A., Johnson, J. A., et al., (2007). The pharmacogenetics research network: from SNP discovery to clinical drug response. Review. *Clin Pharmacol Ther.*, **81**, 328-345.

Gieger, C., Geistlinger, L., Altmaier, E., Hrabe de Angelis, M., Kronenberg, F., Meitinger, T., Mewes, H. W., Wichmann, H. E., Weinberger, K. M., Adamski, J., Illig, & T. Suhre, K. (2008). Genetic meets metabolomics: a genome-wide association of metabolite profiles in human serum. *PLoS Genet.*, **4**, e1000282.

Goldsmith, P., Fenton, H., Morris-Stiff, G., Ahmed, N., Fisher, J. & Prasad, R. (2009). Metabonomics: a useful tool for the future surgeon. *J Surg Res*. (in press).

Hayashi, S., Akiyama, S., Tamaru, Y., Takeda, Y., Fujiwara, T., Inoue, K., Kobayashi, A., Maegawa, S. & Fukusaki, E. (2009). A novel application of metabolomics in vertebrate development. *Biochem Biophys Res Commun.,* **386**, 268-272.

Hocquette, J. F. (2005). Where are we in genomics? *Annu Rev Physiol Pharmacol (Suppl.* **3**), 37-70.

Issaq, H. J., Abbott, E. & Veenstra, T. D. (2008). Utility of separation science in metabolomic studies. *J Sep Sci.,* **31**, 1936-1947.

Issaq, H. J., Van, Q. N., Waybright, T. J., Muschik, G. M. & Veenstra, T. D. (2009). Analytical and statistical approaches to metabolomics research. Review. *J Sep Sci.,* **32**, 2183-2199.

Kaput, J. (2008). Nutrigeneomics research for personalized nutrition and medicine. *Cur Opin Biotechnol.,* **19**, 110-120.

Kaput, J., Ordovas, J. M., Ferguson. L., van Ommen, B., Rodriguez, R. L., Allen, L., Ames, B. N., Dawson, K., et al., (2005). The case for strategic international alliance to harness nutritional genomics for public and personal health. *Br J Nutr.,* **94**, 623-632.

Lehnert, S. A., Wang,Y. H., Tan, S. H. & Reverter, A. (2006). Gene expression-based approaches to beef quality research. *Aust J Exp Agric.,* **46**, 165-172.

Lindon, J. C., Holmes, E., Bollard, M. E., Stanley, E. G. & Nicholson, J. K. (2004). Metabolomics technologies and their applications in physiological monitoring, drug safety assessment and disease diagnosis. Review. *Biomarkers.*, **9**, 1-31.

Marambaud, P., Zhao, H. & Davies P. (2005). Resveratol promotes clearance of Alzheimer's disase amylid-beta peptides. *J Biol Chem.*, **280**, 37377-37382.

Mariman, E. C. (2006). Nutrigenomics and nutrigenetics: the 'omics' revolution in nutritional science. Review. *Biotechnol Appl Biochem.,* **44** (Pt.3), 119-128.

Mead, M. N. (2007). Nutrigenomics: the genome-food interface. *Environ. Health Perspect.*, **115**, A582-A589.

Mulller, M. & Kersten, S. 2003. Nutrigenomics: goals and strategies. *Nature Rev Genet.*, **4**, 315-322.

Mutch, D. M., Wahli, W. & Williamson, G. (2005). Nutrigenomics and nutrigenetics: the emerging faces of nutrition. Review. *FASEB J.*, **19**, 1602-1616.

Nicholson, J. K., Lindon, J. C. & Holme, E. (1999). Metabonomics: understanding the metabolic responses of living systems to pathophysiologiacl stimuli via multivariate statistical analysis of biological NMR spectroscopic data. *Xenobiotica.*, **29**, 1181-1189.

Oresic, M. (2009). Metabolomics, a novel tool for studies of nutrition, metabolisms and lipid dysfunction. *Nutr Metab Cardiovasc Dis.*, **19**, 816-824.

Panagiotou, G. & Nielsen, J. (2009). Nutritional system biology: definitions and approaches. *Annu Rev Nutr.*, **29**, 329-339.

Relling, M. V. & Hoffman J. M. (2007). Should pharmacogenomic studies be required for new drug approval? Review. *Clin Pharmacol Ther.,* **81**, 425-428.

Reverter, A., Byrne, K. A., Brucet, H. L., Wang, Y. H., Dalrymple, B. P. & Lehnert, S. A. (2003). A mixture model-based cluster analysis of DNA microarray gene expression data on Brahman and Brahman composite steers fed high-, medium-, and low-quality diets. *J Anim Sci.,* **81**, 1900-1910.

Rezzi, S., Ramadan, Z., Fay, L. B. & Kochhar, S. (2007). Nutritional metabonomics: applications and perspectives. Review. *J Proteome Res.,* **6**, 513-525.

Roessner, U. & Browne J. (2009). What is metabolomics all about? *Biotechniques.,* **46**, 363-365.

Singh, B., Bhat, T. K. & Singh, B. (2003). Potential therapeutic applications of some antinutritional plant secondary metabolites. *J Agric Food Chem.* 51, 5579-5597.

Subbiah, M. T. (2008). Understanding the nutrigenomics definitions and concepts at the food-genome junction. *OMICS.*, **12**, 229-235.

van Ommen, B. (2004). Nutrigenomics: exploiting systems biology in the nutrition and health arenas. Review. *Nutrition.,* **20**, 4-8.

Xia, J., Psychogios N., Young, N. & Wishart, D. S. (2009). MetaboAnalyst: a web server for metabolomic data analysis and interpretation. *Nucleic Acid Res.*, **37**, W652-W660.

Zdunczyk, Z. & Pareek, C. S. (2008). Applications of nutrigenomics tools in animals feeding and nutritional research. *J Anim Feed Sci.*, **17**, 3-16.

Zheng, S. & Chen, A. (2006). Curcumin suppresses the expression of extracellular matrix genes in activated hepatic stellate cells by inhibiting gene expression of connective tissue growth factor. *Am J Physisol Gastrointest Liver Physiol.*, **290**, G883-G893.

Zivkovik, A. M. & German, J. B. (2009). Metabolomics for assessment of nutritional status. Review. *Curr Opin Clin Nutr Metab Care.*, **12**, 501-507.

In: Metabolomics: Metabolites, Metabonomics ISBN 978-1-61668-006-0
Editors: J.S. Knapp and W.L. Cabrera, pp. 215-228

Chapter 7

Machine Reconstruction of Metabolic Networks from Metabolomic Data through Symbolic-Statistical Learning

Marenglen Biba[1,2,*]**, *Stefano Ferilli***[1,†] ***and Floriana Esposito***[1,‡]
[1]Department of Computer Science, University of Bari, Italy
[2]Department of Computer Science, University of New York Tirana, Albania

Abstract

Metabolomics is a rapidly growing field with the goal of measuring and interpreting the complex time and condition dependent concentration, activity or flux of metabolites in cells, tissues and other biosamples. On the other side, the integrated approach to studying biological systems in Systems Biology has led to significant improvement of our understanding of such systems. Since biological circuits are hard to model and simulate, many efforts are being made to develop computational models that can handle their intrinsic complexity. However, a large part of the biological networks remains unknown and hard to understand and Metabolomics technology that allows simultaneous acquisition of many metabolite measurements can lead to further analysis for discovering novel pathway components and unknown network relationships. Metabolic networks are structurally complex and behave in a stochastic fashion. In this paper we describe how symbolic-statistical machine learning techniques can be used to reconstruct metabolic networks from metabolic profiling data. We show that symbolic machine learning methods have the power to model structural and relational complexity while statistical machine learning ones provide principled approaches to uncertainty modeling. We apply a symbolic-statistical learning framework to analyze sequences of reactions for biologically active paths in metabolic networks. We show through experiments that our approach provides a robust methodology for machine reconstruction of metabolic networks from metabolomic data.

*E-mail address: biba@di.uniba.it
†E-mail address: ferilli@di.uniba.it
‡E-mail address: esposito@di.uniba.it

1. Introduction

Metabolomics [1] is a rapidly growing field. Analytical techniques and instruments such as Mass Spectrometry (MS) and Nuclear Magnetic Resonance (NMR) for gathering and analyzing voluminous metabolic data are being intensively refined. MS is now able to detect molecules at concentrations as low as 10^{-18} molar, and high-field NMR can efficiently differentiate between molecules that are highly similar in structure. The main problem in this research area is the study of the metabolome [2] which represents the collection of all the metabolites in a biological organism. This set of molecules consists of metabolic intermediates, hormones and other signalling molecules, and secondary metabolites. All these represent the chemical fingerprints that every specific cellular process leaves behind. Thus, in order to understand how cells work it is important to explore the metabolome in a principled and robust manner. However, the separate study of the metabolome would not give a deep comprehension of the organism, because biological systems' behavior is determined by complex interactions between their building components. Therefore, an integrated approach to studying biological systems is necessary. This has given rise to the Systems Biology [3] approach to modeling biological phenomena. In Systems Biology the main problem is to uncover and model how function and behavior of the biological machinery are implemented through complex interactions among its building blocks. Metabolomics data provide precious traces of the cell's circuits functioning, hence it is highly important for the Systems Biology approach to integrate metabolomics for a deeper understanding of biological systems [4].

Since biological circuits are hard to model and simulate, many efforts [5] have been made to develop computational models that can handle their intrinsic complexity. In this paper we focus on a particular problem of Systems Biology that concerns the modeling of metabolic pathways and the possibility to discover biologically active paths. A metabolic pathway is a sequence of chemical reactions occurring within the cell. These reactions are catalyzed by enzymes which are particular proteins that convert metabolites (input molecules) in other molecules that represent the products of the reaction. These products can be stored in the cell under certain forms or can cause the initiation of another metabolic pathway. A metabolic network of a cell is formed by the metabolic pathways occurring in the cell. It is through the metabolic networks that every single living organism carries out all its activities. Thus, pathway analysis is crucial to understand cell's behavior and machine learning methods, that are not limited to only simulate biological networks, are essential to infer knowledge from exponentially growing observation data gathered by high-throughput instruments.

Since a reaction can happen if the input molecules are available to the catalytic enzyme, a modeling framework must be able to model relations among entities. Symbolic approaches such as logic-based techniques have the potential to model relations in structural complex domains. First-order logic representations have also the advantage that models are easily comprehensible to humans. Moreover, since most part of biological systems performs its activity remaining hidden to the human modeler, machine learning techniques can play an important role in discovering latent phenomena. However, symbolic-only approaches suffer from the incapability of handling uncertainty. In models built with symbolic-only approaches, the learned rules are deterministic and do not incorporate any kind of mechanism for uncertainty modeling. On the other side, biological systems intrinsically behave in a stochastic fashion with many interactions probable to happen. Since cell's life is determined by the most probable interactions, handling uncertainty is crucial when the cell's machinery must be modeled. Statistical approaches based on the probability theory represent a valuable mechanism to govern uncertainty. However, observations of biological systems rarely reflect exactly what happens inside them. Therefore, estimation techniques are precious in order to model what we cannot observe. Statistical machine learning methods have the ability to learn probability distributions from observations and hence are suitable for modeling biological systems. On the other side, statistical-only approaches rarely are able to reason about relations and/or interactions among biological circuits as symbolic approaches do. Hence, there is strong motivation on developing and applying hybrid approaches to modeling biological systems.

Machine learning and data mining communities have traditionally focused their attention on vector data which is mainly independent and identically distributed. However, since in the real-world, data is mainly stored in relational databases and involves interactions among entities and their attributes, relational databases pose for machine learning the serious problem of learning from relational and non i.i.d data. Moreover, a critical problem in knowledge discovery tasks is that most relational real-world databases are noisy and present a lot of missing data. This characteristic of real-world data greatly affects the performance of standard machine learning algorithms making them very unsuccessful for real tasks.

Recently, to deal with both aspects, relational structure and noisy data, statistical relational models [6] are being developed in order discover knowledge from noisy relational databases. These models exploit statistics to properly handle uncertainty in the data due to missing values and logic-based formalisms to represent relations among entities. Combining both formalisms has a long history in artificial intelligence and machine learning and starts with the works in [7, 8, 9]. Later, several authors began exploiting logic programs to define compact Bayesian networks. This approach was known as knowledge-based model construction [10]. Recently, different approaches for combining logic and statistics have been proposed such as Probabilistic Relational Models [11], First-order Probabilistic Models with Combining Rules [12], Relational Dependency Networks [13], Relational Bayesian Networks [14], and others. The advantage of these models is that they are able to represent probabilistic dependencies between attributes of related different objects in a certain domain.

The contribution of this paper is at the intersection of Systems Biology, Metabolomics and Machine Learning. We apply a hybrid symbolic-statistical framework to the problem of modeling metabolic pathways and mining active paths from time-series data. We show through experiments the feasibility of mining significant paths from metabolomics data in the form of traces of sequences of reactions.

The paper is organized as follows. Section 2 describes the problem of modeling metabolic pathways and the necessity for symbolic-statistical machine learning. Section 3 describes the hybrid framework PRISM. Section 4 describes modeling in PRISM of the Bisphenol A Degradation pathway of Dechloromonas aromatica. Section 5 presents experiments on mining stochastically generated sequences of reactions for biologically active paths. Section 6 concludes discussing related and future work.

2. Metabolic Pathways

Metabolic pathways can be represented as graphs where each node represents a chemical compound and a chemical reaction corresponds to a directed edge labeled by a protein that catalyzes the reaction. Thus, there is an edge from one compound (metabolite) to another compound (product) if there is an enzyme that transforms the metabolite into product. Figure 1. shows part of the pathway of Bisphenol A Degradation in Dechloromonas aromatica extracted from KEGG database. We have chosen this pathway from the KEGG because, as we can see from Figure 1, starting from one point in the pathway there are multiple paths that can be explored. Therefore, the task of mining biologically active paths is harder because more paths should be explored in order to discover the active ones.

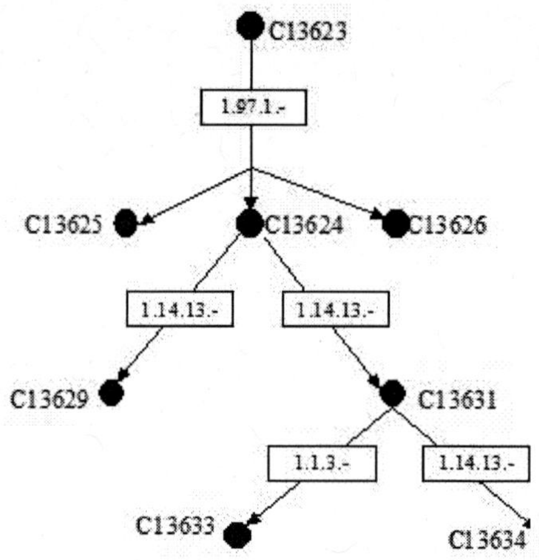

Figure 1. Part of the pathway of Bisphenol A Degradation.

In order to model a metabolic pathway, a suitable framework for their simulation and mining must be able to handle relations. First-order logic representations have the expressive power to model structural and relational problems. The metabolic pathway in Figure 1 can be easily represented in a first-order logic formalism as follows:

$$enzyme(1.97.1.-, reaction\_1\_97\_1\_, [c13623], [c13625, c13624, c13626]).$$
$$enzyme(1.14.13.-, reaction\_1\_14\_13\_a, [c13624], [c13629).$$
$$enzyme(1.14.13.-, reaction\_1\_14\_13\_b, [c13624], [c13631]).$$
$$enzyme(1.1.3.-, reaction\_1\_1\_3, [c13631], [c13633]).$$
$$enzyme(1.14.13.-, reaction\_1\_14\_13\_c, [c13631], [c13634]).$$

However, this representation does not incorporate any further information about the reactions. For example, as we can see there are two competing reactions because the enzyme 1.14.13.- catalyzes two different reactions with the same chemical compound c13624 in input. Subsequently, two enzymes, 1.14.13.- and 1.1.3.-, can elaborate the same input metabolites and thus two reactions compete among them. The occurring of any of the reactions determines a certain sequence of successive reactions instead of another. Hence, it is important to know which reaction among the two is more probable to happen. The most probable reaction determines the biologically active path under certain conditions. This means that under certain conditions, a biological path becomes inactive or useless and another path may become active and yield different overall products in the whole pathway. The conditions under which the reactions happen, may change stochastically due to the random behavior of the biological environment. For example, some input metabolites can suddenly be not available. Their absence can cause a certain reaction not to occur and give rise to another sequence in the metabolic pathway. Therefore, it is crucial to know how probable a certain reaction is. This situation can be modeled by attaching to each reaction the probability that it happens. This requires a first-order representation framework that can handle for each predicate that expresses a reaction the probability that the predicate is true.

The simple incorporation of probabilities is not enough to model complex metabolic networks. The conditions for the reactions to happen depend on many factors, such as initial quantity of input metabolites, changes in the physical-chemical environment surrounding the cell and many more. For this reason it is a hard task to observe all the states of the biological machinery under all the possible conditions and try to assign probabilities to reactions. Therefore there is a need for machine learning statistical methods that given certain conditions can learn distribution of probabilities from observations (the conditions here are meant as physical-chemical entities such as temperature, concentration of metabolites, entropy etc).

In order to model metabolic networks, two tasks must be performed. First, a relational model that describes the structure of the pathway must be build. There is already a large amount of accumulated knowledge about the structure of metabolic pathways such as that in KEGG and we can use all this background knowledge to skip the structure building process and concentrate on mining raw wet experimental-observational data. Indeed, graph structures are abundant but their main disadvantage in modeling cell's life is that they are static. This means that the pathway in Figure 1. does not express the stochastic dynamics in metabolic reactions. These graphs can be seen as useful static templates to interpret what can happen in the cell, but to faithfully reconstruct the cell's activity we must build a dynamic model that represents at a certain moment and under certain conditions what happens inside the cell. Thus, in order to mine biologically active patterns in the pathway under some conditions, we must first learn a dynamic-stochastic model from sequences of reactions that have been observed under those conditions. In order to confirm the feasibility of our approach of mining biological active patterns, we will proceed as follows. We will stochastically change the conditions for the reactions to happen (Section 5 describes how this is performed). Then, under each set of conditions, we stochastically generate sequences of reactions and finally after learning probability distributions for the reactions of the pathway, we perform mining for biological active patterns by querying the dynamic model we have built.

3. The Symbolic-Statistical Framework PRISM

PRISM (PRogramming In Statistical Modelling) [15] is a symbolic-statistical modeling language that integrates logic programming with learning algorithms for probabilistic programs. PRISM programs are not only just a probabilistic extension of logic programs but are also able to learn from examples through the EM (Expectation-Maximization) algorithm which is built-in in the language. PRISM represents a formal knowledge representation language for modeling scientific hypotheses about phenomena which are governed by rules and probabilities. The parameter learning algorithm [16], provided by the language, is a new EM algorithm called graphical EM algorithm that when combined with the tabulated search has the same time complexity as existing EM algorithms, i.e. the Baum-Welch algorithm for HMMs (Hidden Markov Models), the Inside-Outside algorithm for PCFGs (Probabilistic Context-Free Grammars), and the one for singly connected BNs (Bayesian Networks) that have been developed independently in each research field. Since PRISM programs can be arbitrarily complex (no restriction on the form or size), the most popular probabilistic modeling formalisms such as HMMs, PCFGs and BNs can be described by these programs. PRISM programs are defined as logic programs with a probability distribution given to facts that is called basic distribution. Formally a PRISM program is $P = F \cup R$ where R is a set of logical rules working behind the observations and F is a set of facts that models observations' uncertainty with a probability distribution. Through the built-in graphical EM algorithm the parameters (probabilities) of F are learned and through the rules this learned probability distribution over the facts induces a joint probability distribution over the set of least models of P, i.e. over the observations. This is called distributional semantics [17]. As an example, we present a hidden markov model with two states slightly modified from that in [16]:

$values(init, [s0, s1]).$ % State initialization
$values(out(\_), [a, b]).$ % Symbol emission
$values(tr(\_), [s0, s1]).$ % State transition
$hmm(L) :-$ % To observe a string L:
$str\_length(N),$ % Get the string length as N
$msw(init, S),$ % Choose an initial state randomly
$hmm(1, N, S, L).$ % Start stochastic transition (loop)
$hmm(T, N, \_, []) : -T > N, !.$ % Stop the loop
$hmm(T, N, S, [Ob|Y]) :-$ % Loop: current state is S, current time is T
$msw(out(S), Ob),$ % Output Ob at the state S
$msw(tr(S), Next),$ % Transit from S to Next.
$T1 is T + 1,$ % Count up time
$hmm(T1, N, Next, Y).$ % Go next (recursion)
$str\_length(10).$ % String length is 10
$set\_params : -set\_sw(init, [0.9, 0.1]), set\_sw(tr(s0), [0.2, 0.8]), set\_sw(tr(s1), [0.8, 0.2]), set\_sw(out(s0), [0.5, 0.5]), set\_sw(out(s1), [0.6, 0.4]).$

The most appealing feature of PRISM is that it allows the users to use random switches to make probabilistic choices. A random switch has a name, a space of possible outcomes, and a probability distribution. In the program above, $msw(init, S)$ probabilistically determines the initial state from which to start by tossing a coin. The predicate $set\_sw(init, [0.9, 0.1])$, states that the probability of starting from state $s0$ is 0.9 and from $s1$ is 0.1. The predicate learn in PRISM is used to learn from examples (a set of strings) the parameters (probabilities of init, out and tr) so that the ML (Maximum-Likelihood) is reached. For example, the learned parameters from a set of examples can be: $switchinit$: $s0(0.6570)$, $s1(0.3429)$; $switchout(s0)$: $a(0.3257), b(0.6742)$; $switchout(s1) : a(0.7048), b(0.2951)$; $switchtr(s0) : s0(0.2844)$, $s1(0.7155)$; $switchtr(s1)$: $s0(0.5703)$, $s1(0.4296)$. After learning these ML parameters, we can calculate the probability of a certain observation using the predicate prob: $prob(hmm([a, a, a, a, a, b, b, b, b, b]) = 0.000117528$. This way, we are able to define a probability distribution over the strings that we observe. Therefore from the basic distribution we have induced a joint probability distribution over the observations.

4. Modeling Bisphenol A Degradation Pathway in PRISM

Since PRISM is a logic-based language, we can easily represent the metabolic pathway presented in the previous section. Predicates that describe reactions remain unchanged from a language representation point of view. What we need to statistically model the metabolic pathway is the extension with random switches of the logic program that describes the pathway. We define for every reaction a random switch with its relative space outcome. For example, in the following we describe the random switches for the reactions in Figure 1.

$values(switch\_rea\_1\_97\_1, [rea\_1\_97\_1(yes, yes, yes, yes), rea\_1\_97\_1(yes, no, no, no)]).$
$values(switch\_rea\_1\_14\_13\_a, [rea\_1\_14\_13\_a(yes, yes), rea\_1\_14\_13\_a(yes, no)]).$
$values(switch\_rea\_1\_14\_13\_b, [rea\_1\_14\_13\_b(yes, yes), rea\_1\_14\_13\_b(yes, no)]).$
$values(switch\_rea\_1\_1\_3, [rea\_1\_1\_3(yes, yes), rea\_1\_1\_3(yes, no)]).$
$values(switch\_rea\_1\_14\_13\_b, [rea\_1\_14\_13\_c(yes, yes), rea\_1\_14\_13\_c(yes, no)]).$

For each of the three reactions there is a random switch that can take one of the stated values at a certain time. For example, the value $rea\_1\_97\_1(yes, yes)$ means that at a certain moment the metabolite $c13623$ is available and the reaction occurs producing the compounds $c13623, c13624$ and $c13625$. While the other value $rea\_1\_97\_1(yes, no, no, no)$ means that the input metabolite is present but the reaction stochastically did not occur, thus the products are not produced. Below we report the remaining part of the PRISM program for modeling the pathway in Figure 1. Together with the declarations in Section 2 for the possible reactions and those of the previous paragraph for the values of the random switches, the following logic program forms a model for stochastically modeling the pathway in Figure 1. (The complete PRISM code for the whole metabolic pathway can be requested to the authors).

$produces(Metabolites, Products) :-$
$produces(Metabolites, [], Products).$
$produces(Metabolites, Delayed, Products) :-$
$(reaction(Metabolites, Name, Inputs, Outputs, Rest)->$
$call\_reaction(Reaction, Inputs, Outputs, Call),$
$rand\_sw(Call, Value),$
$((Value == rea\_1\_97\_1(yes, yes, yes, yes);$
$Value == rea\_1\_14\_13\_a(yes, yes,);$
$Value == rea\_1\_14\_13\_b(yes, yes,);$

$Value == rea\_1\_14\_13\_c(yes, yes,);$
$Value == rea\_1\_1\_3(yes, yes)) - >$
$produces(Rest, Delayed, Products)$
;
$produces(Metabolites, [Reaction|Delayed], Products)$
;
$Products = Metabolites$
).
$rand\_sw(ReactAndArgs, Value) : -$
$ReactAndArgs = ..[Predicate|Arguments],$
$(Predicate == rea\_1\_97\_1 - > msw(switch\_rea\_1\_97\_1, Value);$
$(Predicate == rea\_1\_14\_13\_a - > msw(switch\_rea\_1\_14\_13\_a, Value);$
$(Predicate == rea\_1\_14\_13\_b - > msw(switch\_rea\_1\_14\_13\_b, Value);$
$(Predicate == rea\_1\_14\_13\_c - > msw(switch\_rea\_1\_14\_13\_c, Value);$
$(Predicate == rea\_1\_1\_3 - > msw(switch\_rea\_1\_1\_3, Value)$
;
$true)))))$. % do nothing

In the following, we trace the execution of the above logic program. The top goal to prove that represents the observations (sequences of reactions vastly produced by high-throughput technologies) for PRISM is $produces(Metabolites, Products)$. It will succeed if there is a pathway that leads from Metabolites to Products, in other words if there is a sequence of random choices (according to a probability distribution) that makes possible to prove the top goal. The predicate reaction controls among the first clauses of the program, if there is a possible reaction with Metabolites in input. Suppose that at a certain moment $Metabolites = [c13624]$ and thus two competing reactions can happen. Suppose one of the reaction is stochastically chosen and the variables Inputs and Outputs are bounded respectively to $[c13624]$ and $[c13629]$. The predicate $call\_reaction$ constructs the body of the reaction that is the predicate $Call$ which is in the form: $rea\_1\_14\_13\_a(\_, \_, \_, \_)$. This means that the next predicate $rand\_sw$ will perform a random choice for the switch $switch\_rea\_1\_14\_13\_a$. This random choice which is made by the built-in predicate $msw(switch\_rea\_1\_14\_13\_a, Value)$ of PRISM, determines the next step of the execution, since Value can be either $rea\_1\_14\_13\_a(yes, yes)$ or $rea\_1\_14\_13\_a(yes, no)$. In the first case it means the reaction has been probabilistically chosen to happen and the next step in the execution of the program which corresponds to the next reaction in the metabolic pathway is the call $produces(Rest, Delayed, Products)$. In the second case, the random choice $rea\_1\_14\_13\_a(yes, no)$ means that probabilistically the reaction did not occur and the sequence of the execution will be another, determined by the call $produces(Metabolites, [Reaction|Delayed], Products)$ which will try stochastically to choose the competing reaction catalyzed by the same enzyme $1.14.13.-$ that given the same input $c13624$ produces the compound $c13631$. If this reaction occurs, then the next reaction in the sequence will be one of the competing reactions with $c13631$ as input. In order to learn the probabilities of the reactions we need a set of observations of the form $produces(Metabolites, Products)$. These observations that represent metabolomic data, are being intensively collected through available high throughput instruments and stored in particular metabolomics databases. In the next section, we show that from these observations, PRISM is able to accurately learn reaction probabilities through the built-in graphical EM algorithm.

5. Reconstructing Pathways from Sequences of Reactions

A certain metabolic path becomes inactive or useless under certain conditions if a certain intermediate reaction in the path cannot occur under those conditions. In this paper we are not interested in the conditions themselves (these usually are stoichiometrics constraints). What is important for our purpose here, is that the conditions evolve stochastically. This means that by simulating various conditions we make possible a set of reactions instead of another, i.e. each set of conditions gives rise to a set of possible reactions that render some paths in the metabolic pathway biologically active and others biologically inactive under those conditions.

In order to simulate various conditions, for each experiment we randomly assign probabilities to reactions. These probabilities represent the switches probabilities in PRISM. Thus, we have for each single experiment a set of conditions under the form of assigned reactions' probabilities (as probabilities are randomly generated and some of them may be equal to zero or in the range $[0,9 - 0,999]$, among competing reactions one of them may not occur and this will cause some paths in the metabolic pathway to be inactive). The model constructed in this manner reflects the state of the biochemical environment under the given conditions at a certain moment. When the reactions happen, what is caught by a high-throughput instrument is a set of metabolites concentrations and their changes. For example, if a certain reaction happens then the concentration of the input metabolites decrease and that of the product compounds increase. This change is registered as a reaction, therefore catching all the time-series changes in concentration (this is actually performed intensively and accurately by current high-throughput technologies), means registering a time-series sequence of reactions. These constitute our mining data in order to re-construct biological active and inactive paths. By simulating the built model (this corresponds to simply running the PRISM program by calling the goal $produces(InputMetabolites, Products)$ where $InputMetabolites$ is a bounded list with the input compounds and $Products$ is a logic variable that will be bounded to the list of product compounds yielded by the series of reactions), we will have time-series sequences of reactions as if we were observing the model by high-throughput instruments.

In order to evaluate the validity of our approach we have proceeded as follows. For each experiment (each experiment has a different set of conditions, i.e. probabilities of random switches that are stochastically assigned) we have stochastically generated sequences of reactions by sampling from the previously defined model. This is made possible by the predicate sample of PRISM. Once the sequences have been generated, we launch the predicate learn of PRISM to learn the probability of each random switch from the sequences. Once the model has been reconstructed we query it over the sequences and mine biologically active paths with the predicate hindsight(Goal) where Goal is bounded to the top-goal $[InputMetabolites, Products]$. With this predicate we get the probabilities of all the sub-goals for the top-goal Goal. If any of these probabilities is equal to zero then the relative path of the sub-goal is biologically inactive under the given conditions. The relative path

Table 1. RMSE and learning time on average for 100 experiments, S: Number of sequences M-RMSE: Mean of RMSE on 100 experiments,MLT: Mean learning time on 100 experiments (seconds)

S	M-RMSE	MLT
100	0.13932	0.047
200	0.13593	0.068
500	0.12999	0.090
1000	0.10405	0.125
2000	0.09685	0.297
4000	0.08676	0.484
8000	0.06808	0.547
15000	0.05426	0.612
30000	0.03297	0.695
50000	0.02924	0.735
100000	0.02250	1.172

can be extracted by the predicate $probf(SubGoal, ExplGraph)$ where $ExplGraph$ (explanation graph in PRISM) represents the explanation paths for $SubGoal$. The accuracy of mining the sequences of reactions for biologically active patterns, depends on the ability to faithfully recontruct the model from the sequences. In order to assess the accuracy of learning the probabilities of the reactions and mining really biologically active paths we adopt the following method to evaluate the learning phase for the approach of the previous paragraph. We call the initial probability distribution (that represents the conditions) assigned to the clauses of the logic program the true probability distribution and call the M parameters the true parameters. Once the sequences have been stochastically generated by this model, we forget the true parameters and replace their probabilities by uniformly distributed ones. When learning starts, PRISM learns M new parameters, that represent the learned reaction probabilities from the sequences. In order to assess the accuracy of the learned P_i' towards P_i we use the RMSE (Root Mean Square Error) for each single experiment with S sequences.

$$RMSE = \sqrt{\sum_{i=1}^{M} \frac{(P_i - P_i')^2}{M}} \tag{1}$$

In this way we can measure the difference between the actual observations and the response predicted by the model. We have performed different experiments with a growing number S of sequences in order to evaluate how the number of sequences affects the accuracy and the learning time. Moreover, we wanted to test also large datasets of sequences in order to provide a robust methodology since real metabolomics datasets are in general highly voluminous. For each S we have performed 100 experiments where for each experiment the set of conditions is stochastically generated as presented above. Table 1. reports for each S the RMSE and the learning time on average for 100 experiments. We have used the version 1.10 of the system PRISM on a Pentium 4, 2.4GHz machine.

As Table 1 shows, the learning accuracy increases as more data are available and due to the tabulation techniques in PRISM, learning times increases reasonably as data dimension grows significantly. The accuracy of learning can be evaluated as good for a number of sequences between 1000 and 15.000 and excellent for a number of sequences greater than 15.000 considering that the range where probabilities fall is $[0, ..., 1]$ and the RMSE is under $0, 05$. This means that the paths have been faithfully reconstructed from the sequences and thus the predicates $hindsight$ and $probf$ in PRISM faithfully produce the biologically active paths in the pathway. Indeed, from empirical observations, we noted that all the queries performed by these two predicates reflected the real biological paths that are supposed to have produced the sequences. For instance, we noted that anytime the probability of the reaction catalyzed by the enzyme $1.14.13.-$ (with input the compound $c13624$ and output $c13631$) was stochastically assigned to be too low (from 0 to 0.05) by the conditions generation phase, then the path that involves one of the two next reactions, the one catalyzed by the enzyme $1.1.3.-$ and producing in output $c13633$, was mined as a biologically inactive path for the given conditions. Moreover, we noted for all the experiments that by slightly changing the conditions, many inactive paths became suddenly active and vice versa. This is quite interesting since it means that we can learn from sequences how conditions evolve in order to understand what changes them and what governs their randomness.

6. Related Work

The most important related work is that in [18] where a probabilistic relational formalism is used for modeling metabolic networks. The PRISM program we have presented here is syntactically quite similar to the logic program in [18], but is semantically different in the way probability distributions are defined. Stochastic Logic Programs (SLPs) [19], used in [18], assign probabilities to clauses and define probability distributions on Prolog proof trees, while PRISM programs are based on the distributional semantics [17] and assign probabilities to atoms as we explained in Section 3. Most of other related work is not based on symbolic-statistical approaches. In [20, 21], graph-theory based approaches are used to find common or unique sub-graphs in different pathway graphs to understand better why and how pathways differ or are similar. Other approaches are those that focus on text mining for metabolic pathways [22]. These methods have been applied to the voluminous literature on metabolic pathways to discover knowledge about the structure of the pathways. Text mining techniques focus on the structure building process trying to identify, in the accumulated experience about metabolic pathways, significant structural properties. Other approaches attempt to only stochastically simulate biochemical processes such [23] or [24]. These are powerful tools to model the dynamic nature of cells for simulation purposes but lack machine learning abilities to infer knowledge from observations.

7. Conclusion and Future Work

We have applied the hybrid symbolic-statistical framework PRISM to a problem of modeling metabolic pathways and have shown through experiments the feasibility of learning reaction probabilities from metabolomics data and mining biologically active paths from time-series sequences of reactions. The power of the proposed approach stands in the description language that allows to model relations and in the ability to model uncertainty in a robust manner. Moreover, we have also shown that the symbolic-statistical framework PRISM can be used as a stochastic simulator for biochemical reactions.

Although we have been able to reconstruct the model from the sequences of reactions, our approach is far from completing the real picture of a biochemical network. Much work remains to be done. First of all, we have not considered stoiochiometrics constraints which express quantitative relationships of the reactants and products in chemical reactions. We believe that adding these constraints to our approach will help reproduce better models. Another direction for future work regards plugging in the model other sources of data. Considering multiple sources of data can lead to better models in modeling metabolic pathways [25]. In PRISM this is straightforward because relational problems can be easily modeled due to the logic-based language. Another challenge is learning from incomplete raw metabolomic data. EM algorithms [26] are the state-of-the art for learning in the presence of missing data and since the graphical EM algorithm [16] that PRISM uses, is a version of this class of learning algorithms, we believe this will help in dealing with incomplete real datasets. In addition, in this paper we have considered a medium-sized metabolic pathway. For future work we intend to model very large metabolic pathways and hierarchical metabolic networks to see how the learning algorithms in PRISM scales for large datasets. We think the tabulation techniques used in PRISM will greatly help in dealing with a high number of paths to be explored. We also plan to investigate other important problems using the symbolic-statistical framework PRISM and other learning capabilities such as inductive relational learning for inferring missing pathways in existing metabolic networks or reconstructing whole novel pathways from sequences of observations.

References

[1] Harrigan, G.G., Goodacre, R.e.: *Metabolic Profiling: Its Role in Biomarker Discovery and Gene Function Analysis.* Kluwer Academic Publishers, Boston (2003)

[2] Oliver, S.G., Winson, M.K., Kell, D.B., Baganz, F.: Systematic functional analysis of the yeast genome. *Trends Biotechnol.* **16**(10) (1998) 373–378

[3] Kitano, H.e.: Foundations of Systems Biology. *MIT Press* (2001)

[4] Weckwerth, W.: Metabolomics in systems biology. *Annu. Rev. Plant Biol.* **54** (2003) 669–689

[5] Kriete, A., Eils, R.: *Computational Systems Biology.* Elsevier - Academic Press (2005)

[6] Getoor, L., Taskar, B.: *Introduction to Statistical Relational Learning* . MIT (2007)

[7] Bacchus, F.: *Representing and Reasoning with Probabilistic Knowledge.* Cambridge, MA: MIT Press (1990)

[8] Halpern, J.: An analysis of first-order logics of probability. *Artificial Intelligence* **46** (1990) 311–350

[9] Nilsson, N.: Probabilistic logic. *Artificial Intelligence* **28** (1986) 71–87

[10] Wellman, M. Breese, J.S., Goldman, R.P.: From knowledge bases to decision models. *Knowledge Engineering Review* **7** (1992)

[11] Getoor, L., Friedman, N., Koller, D., Taskar, B.: Learning probabilistic models of link structure. *Journal of Machine Learning Research* **3** (2002) 679–707

[12] Natarajan, S., Tadepalli, P., Altendorf, E., Dietterich, T.G., Fern, A., Restificar, A.C.: Learning first-order probabilistic models with combining rules. In: *ICML.* (2005) 609–616

[13] Neville, J., Jensen, D.: Dependency networks for relational data. In: *Proc. 4th IEEE Int'l Conf. on Data Mining*, IEEE Computer Society Press. (2004) 170–177

[14] Jaeger, M.: Parameter learning for relational bayesian networks. In: *ICML.* (2007) 369–376

[15] Sato, T., Kameya, Y.: Prism: A symbolic-statistical modeling language. In: *Proceedings of the Fifteenth International Joint Conference on Artificial Intelligence* , Nagoya, Japan: Morgan Kaufmann (1997) 1330–1335

[16] Sato, T., Kameya, Y.: Parameter learning of logic programs for symbolic-statistical modeling. *Journal of Artificial Intelligence Research* **15** (2001) 391–454

[17] Sato, T.: A statistical learning method for logic programs with distribution semantics. In: In Leon Sterling, editor, *Proc. Twelfth International Conference on Logic Programming*, MIT Press. (1995) 715–729

[18] N., A., S.H., M.: Machine learning metabolic pathway descriptions using a probabilistic relational representation. *Electronic Transactions in Artificial Intelligence* **6** (2002)

[19] Muggleton, S.: Stochastic logic programs. In: In L. De Raedt (Ed.), *Advances in inductive logic programming.* IOS Press, Amsterdam (1996)

[20] Koyuturk, M., Grama, A., Szpankowski, W.: An efficient algorithm for detecting frequent subgraphs in biological networks. In: Bioinformatics, Suppl. 1: *Proc. 12th Intl. Conf. Intelligent Systems for Molecular Biology (ISMB'04).* (2004) 200–207

[21] You, C., Holder, L., Cook, J.: Application of graph-based data mining to metabolic pathways. In: *Workshop on Data Mining in Bioinformatics,* ICDM,. (2006)

[22] Hoffmann, R., Krallinger, M., Andres, E., Tamames, J., Blaschke, C., Valencia, A.: Text mining for metabolic pathways, signaling cascades, and protein networks. *Sci STKE 283* **21** (2005)

[23] Le Novre, N., Shimizu, T.S.: Stochsim: modelling of stochastic biomolecular processes. *Bioinformatics* **17** (2001) 575–576

[24] Klamt S, Stelling J, G.M., ED., G.: Fluxanalyzer: exploring structure, pathways, and flux distributions in metabolic networks on interactive flux maps. *Bioinformatics* **19(2)** (2003) 261–269

[25] Fiehn, O.: Combining genomics, metabolome analysis, and biochemical modelling to understand metabolic networks. *Comp. Funct. Genomics* **2(3)** (2001) 155–168

[26] Dempster, A.P., Laird, N.M., Rubin, D.B.: Maximum likelihood from incomplete data via the em algorithm. *Royal Statistical Society* **B39(1)** (1977) 1–38

In: Metabolomics: Metabolites, Metabonomics... ISBN: 978-1-61668-006-0
Editors: J.S. Knapp and W.L. Cabrera, pp. 229-241

Chapter 8

METABOLOMICS

Viroj Wiwanikit
Chulalongkorn University, Bangkok, Thailand

Introduction to Metabolomics

A. General Information on Metabolomics

Generally, a large proportion of the genes in any genome encode enzymes of primary and specialized (secondary) metabolism [1]. Not all primary metabolites, those that are found in all or most species, have been identified and only a small portion of the estimated hundreds of thousand specialized metabolites, those found only in restricted lineages, have been studied in any species [1]. Fridman and Pichersky [1] noted that the correlative analysis of extensive metabolic profiling and gene expression profiling had proven a powerful approach for the identification of candidate genes and enzymes, particularly those in secondary metabolism [2]. It is rapidly becoming possible to measure hundreds or thousands of metabolites in small samples of biological fluids or tissues. Arita [3] said that metabolomics, a comprehensive extension of traditional targeted metabolite analysis, had recently attracted much attention as the biological missing pieces that can complement transcriptome and proteome analysis. Metabolic profiling applied to functional genomics (metabolomics) is in an early stage of development [4]. Fridman and Pichersky [1] said that the final characterization of substrates, enzymatic activities, and products requires biochemical analysis, which had been most successful when candidate proteins have homology to other enzymes of known function. To facilitate the analysis of experiments using post-genomic technologies, new concepts for linking the vast amount of raw data to a biological context have to be developed [5]. Visual representations of pathways help biologists to understand the complex relationships between components of metabolic network [5].

Organ function can only be completely understood through knowledge of molecular and cellular processes within the constraints of structure-function relations at the tissue level [6]. Knowledge on integrative computational physiology is required. Cellular components interact with each other to form networks that process information and evoke biological responses [7]. Today different database systems for molecular structures (genes and proteins) and metabolic pathways are available. All these systems are characterized by the static data representation [8]. For progress in biotechnology the dynamic representation of this data is important. The metabolism can be characterized as a complex biochemical network [8]. A deep understanding of the behavior of these networks requires the development and analysis of mathematical models [7]. Computer modeling of metabolic networks can help better understand complex metabolism [9 - 10]. As previously mentioned, mathematical modeling is one of the key methodologies of metabolic engineering [11]. Based on a given metabolic model different computational tools for the simulation, data evaluation, systems analysis, prediction, design and optimization of metabolic systems have been developed [11]. More details on mathematical modeling can be seen in another specific chapter in this book. In additional to mathematical model, graph-based analysis of metabolic networks is another widely used technique in metabolomics [12].

B. Database and Tool in Metabolomics

Since metabolomics is new, the database and tool as well as application of metabolomic database in medicine is still limited. German et al [2] noted that metabolomics made it possible to assess the metabolic component of nutritional phenotypes and allow individualized dietary recommendations. German et al [2] proposed that the American Society for Nutritional Science (ASNS) had to take action to ensure that appropriate technologies were developed and that metabolic databases were constructed with the right inputs and organization. German et al [2] also mentioned that the relations between diet and metabolomic profiles and between those profiles and health and disease should be established. The details of important databases and tools in metabolomics and their application are hereby presented.

- **MSFACTs [13]**

This tool is for metabolomics spectral formatting, alignment and conversion [13]. It covers metabolomics spectral formatting, alignment and conversion [13].

- **HybGFS [14]**

HybGFS is a hybrid method for genome-fingerprint scanning [14]. This technique combines genome sequence-based peptide MS/MS ion searching with liquid-chromatography elution-time (LC-ET) prediction, to improve the reliability of identification [14]. This hybrid method allows the simultaneous identification and mapping of proteins without a priori information about their coding sequences [14].

- **HMDB [15]**

The Human Metabolome Database (HMDB) is currently the most complete and comprehensive curated collection of human metabolite and human metabolism data in the world [5]. It contains records for more than 2180 endogenous metabolites with information gathered from thousands of books, journal articles and electronic databases [15]. In addition to its comprehensive literature-derived data, the HMDB also contains an extensive collection of experimental metabolite concentration data compiled from hundreds of mass spectra (MS) and Nuclear Magnetic resonance (NMR) metabolomic analyses performed on urine, blood and cerebrospinal fluid samples [15]. This is further supplemented with thousands of NMR and MS spectra collected on purified, reference metabolites. Each metabolite entry in the HMDB contains an average of 90 separate data fields including a comprehensive compound description, names and synonyms, structural information, physico-chemical data, reference NMR and MS spectra, biofluid concentrations, disease associations, pathway information, enzyme data, gene sequence data, SNP and mutation data as well as extensive links to images, references and other public databases [15].

- **aMAZE LightBench [16]**

The aMAZE LightBench (http://www.amaze.ulb. ac.be/) is a web interface to the aMAZE relational database, which contains information on gene expression, catalysed chemical reactions, regulatory interactions, protein assembly, as well as metabolic and signal transduction pathways [16]. It allows the user to browse the information in an intuitive way, which also reflects the underlying data model [16].

- **BioSilico [17]**

BioSilico is a web-based database system that facilitates the search and analysis of metabolic pathways [17]. Heterogeneous metabolic databases including LIGAND, ENZYME, EcoCyc and MetaCyc are integrated in a systematic way, thereby allowing users to efficiently retrieve the relevant information on enzymes, biochemical compounds and reactions [17]. In addition, it provides well-designed view pages for more detailed summary information [17].

- **Eco Cyc [18 - 21]**

The EcoCyc database describes the genome and gene products of Escherichia coli, its metabolic and signal-transduction pathways, and its tRNAs [18]. The database describes 4391 genes of *E.coli*, 695 enzymes encoded by a subset of these genes, 904 metabolic reactions that occur in *E.coli*, and the organization of these reactions into 129 metabolic pathways [18]. EcoCyc is available at URL http://ecocyc.PangeaSystems.com/ecocyc/ [18].

- **Patikaweb [22]**

Patikaweb provides a Web interface for retrieving and analyzing biological pathways in the Patika database, which contains data integrated from various prominent public pathway databases [22].

- **PathAligner [23]**

PathAligner extracts metabolic information from biological databases via the Internet and builds metabolic pathways with data sources of genes, sequences, enzymes, metabolites etc [23]. It provides an easy-to-use interface to retrieve, display and manipulate metabolic information [23]. PathAligner also provides an alignment method to compare the similarity between metabolic pathways [23]. PathAligner is available at http://bibiserv.techfak.uni-bielefeld.de/pathaligner [23].

- **MetaCys [24 – 26]**

MetaCyc is a database of metabolic pathways and enzymes located at http://MetaCyc.org/. Its goal is to serve as a metabolic encyclopedia, containing a collection of non-redundant pathways central to small molecule metabolism, which have been reported in the experimental literature [24 – 25]. Most of the pathways in MetaCyc occur in microorganisms and plants, although animal pathways are also represented [24 – 25]. MetaCyc contains metabolic pathways, enzymatic reactions, enzymes, chemical compounds, genes and review-level comments [24 – 25]. Enzyme information includes substrate specificity, kinetic properties, activators, inhibitors, cofactor requirements and links to sequence and structure databases. Data are curated from the primary literature by curators with expertise in biochemistry and molecular biology [24 – 25].

- **Golm Metabolome Database [27]**

The Golm Metabolome Database (GMD) is an open access metabolome database, which should enable these processes [26]. GMD provides public access to custom mass spectral libraries, metabolite profiling experiments as well as additional information and tools with regard to methods, spectral information or compounds [27]. The main goal will be the representation of an exchange platform for experimental research activities and bioinformatics to develop and improve metabolomics by multidisciplinary cooperation [27]. GMD is available at http:// csbdb.mpimp-golm.mpg.de/gmd.html [27].

C. Application of Metabolomics

Metabolomics is the newest "omics" science. It focuses on a dynamic portrait of the metabolic status of living systems. Metabolomics can bring enomous new insights on metabolic fluxes and a more comprehensive and holistic understanding of a cell's environment. Metabolomics, in particular gas chromatography-mass spectrometry (GC-MS) based metabolite profiling of biological extracts, is rapidly becoming one of the cornerstones of functional genomics and systems biology [27]. Metabolite profiling has profound applications in discovering the mode of action of drugs or herbicides, and in unravelling the effect of altered gene expression on metabolism and organism performance in biotechnological applications [27].

1. Application in Oncology

The tumor metabolome is characterized by high glycolytic and glutaminolytic capacities, high phosphometabolite levels and a high channelling of glucose carbons to synthetic processes [28]. This allows tumor cells to proliferate under strong variations in oxygen and glucose supply (http://www.metabolic-database.com) [28]. The main current applications and challenges of metabolomics in cancer research, including a) protein expression profiling of tumours, tumour fluids and tumour cells; b) protein microarrays; c) mapping of cancer signalling pathways; d) pharmacoproteomics; e) biomarkers for diagnosis, staging and monitoring of the disease and therapeutic response; and f) the immune response to cancer [29]. All these applications continue to benefit from further technological advances, such as the development of quantitative proteomics methods, high-resolution, high-speed and high-sensitivity MS, functional protein assays, and advanced bioinformatics for data handling and interpretation [29].

The best example of metabolomics application in oncology is the case of breast cancer. The metabolomics technology permits simultaneous monitoring of many hundreds, or thousands, of macro- and small molecules, as well as functional monitoring of multiple pivotal cellular pathways [30]. In addition, elucidation of cellular responses to molecular damage, including evolutionarily conserved inducible molecular defense systems, could be achieved with metabolomics and could lead to the discovery of new biomarkers of molecular responses to functional perturbations [30].

2. Application in Pharmacology

Metabolomics is the study of global metabolite profiles in a system (cell, tissue, or organism) under a given set of conditions [31]. The analysis of the metabolome is particularly challenging due to the diverse chemical nature of metabolites [31]. The potential of metabolomics for natural product drug discovery and functional food analysis, primarily as incorporated into broader "omic" data sets, is widely discussed [31 - 32]. In the past, new drug design especially for new mixture is very hard. Rational design of drug mixtures has been nearly impossible due to the lack of information about in vivo cell regulation, mechanisms of pathway activation, and interactions between different pathways in vivo [33]. However, with the advent in metabolomics, this gap can be solved [33]. Metabolomics experiments aim to quantify all metabolites in a cellular system (cell or tissue) under defined states and at different time points so that the dynamics of any biotic, abiotic, or genetic perturbation can be accurately assessed [34]. This can help develop new drug for hopeless diseases such as cancer and new emerging untreatable infectious diseases. Metabolomics incorporates the most advanced approaches to molecular phenotype system readout and provides the ideal theranostic technology platform for the discovery of biomarker patterns associated with healthy and diseased states, for use in personalized health monitoring programs and for the design of individualized interventions [35].

The inducibility of drug-metabolizing enzymes and transporters by numerous xenobiotics has become a vital issue to be considered in the drug development process [36]. Activation of so-called orphan nuclear receptors has been identified to result in increased expression of these detoxifying systems and consequently altered drug levels in the human body [36]. The computational assessment of drug metabolism has gained considerable interest in pharmaceutical research [37]. Amongst others, machine learning techniques have been employed to model relationships between the chemical structure of a compound and its metabolic fate [37]. Examples for these techniques, which were originally developed in fields far from drug discovery, are artificial neural networks or support vector machines [37]. Newer computational technologies are also being applied in order to attempt to predict induction from the molecular structure alone before a molecule is even synthesized or tested [38]. Prediction of human drug metabolizing enzyme induction can also be performed as in silico study [38].

3. Application in Genetics

Metabolomics is an important “omic” science to fill the gap between genomics and proteomics. Pharmacogenetics can be accepted as a variant of metabolomics in term. Metabonomics involves the determination of multiple metabolites simultaneously in biofluids, tissues and tissue extracts and these are all have some levels of genetic involvement [39]. Recently, Fu et al described the MetaNetwork protocol to reconstruct metabolic networks using metabolite abundance data from segregating populations [40]. MetaNetwork maps metabolite quantitative trait loci (mQTLs) underlying variation in metabolite abundance in individuals of a segregating population using a two-part model to account for the often observed spike in the distribution of metabolite abundance data [40]. MetaNetwork predicts and visualizes potential associations between metabolites using correlations of mQTL profiles, rather than of abundance profiles [40]. In addition, MetaNetwork is able to integrate high-throughput data from subsequent metabolomics, transcriptomics and proteomics experiments in conjunction with traditional phenotypic data [40].

To help the reader get a better view on this topic, the author will discuss the application of metabolomics in the case of preterm parturition syndrome. For this syndrome, the application of metabolomics is to identify the metabolic footprints of women with preterm labor likely to deliver preterm and those who will deliver at term [41].

Pathway and Metabolism

A. What is Metabolism?

Metabolism is a set of chemical reactions that occur in living organisms in order to maintain life. There are two main types of metabolism, anabolism and catabolism. Anabolism means the constructive aspects of metabolism while catabolism means the destructive aspects of metabolism. As a consequence, there are three main parts of metabolism, input, process and output. Input, are any biomolecules called substrates. The process is the reaction. The output or result of metabolism is product of metabolite. The resulted metabolite is the target of any metabolism in living things. Metabolomics is an “omic” science directly involving metabolism.

B. Pathway Drawing

Metabolism consists of many reactions or pathways. To simplify and make metabolism understandable, scientists make use of pathway drawing to demonstrated numerous words describing metabolism. This is a basic knowledge in biochemistry. The symbols “+” and “→ ” are the two most commonly used in the pathway. A symbols “+” means react between molecules. The symbols “ → ” means result into. It should be noted that the symbol “→” has directional meaning. A “ → ” means a forward direction while a ←“” means backward direction. The combination “ ⇄ ” means reversible process.

In bioinformatics, there are many new pathway drawing tools that can help to make better drawings. Details of some important pathway drawing tools are hereby presented.

1. PathFinder [42]

PathFinder is a tool for the dynamic visualization of metabolic pathways based on annotation data [42]. Pathways are represented as directed acyclic graphs [42], graph layout algorithms accomplish the dynamic drawing and visualization of the metabolic maps [42]. A more detailed analysis of the input data on the level of biochemical pathways helps to identify genes and detect improper parts of annotations [42]. As an Relational Database Management System (RDBMS) based internet application PathFinder reads a list of EC-numbers or a given annotation in EMBL- or Genbank-format and dynamically generates pathway graphs [42].

2. MetaViz [43]

MetaViz enables to draw a genome-scale metabolic network and that also takes into account its structuration into pathways [43]. This method consists in two steps: a clustering step which addresses the pathway overlapping problem and a drawing step which consists in drawing the clustered graph and each cluster [43]. The method we propose is original and addresses new drawing issues arising from the no-duplication constraint [43].

3. FluxAnalyzer [44]

The FluxAnalyzer is a package for MATLAB and facilitates integrated pathway and flux analysis for metabolic networks within a graphical user interface [44]. Arbitrary metabolic network models can be composed by instances of four types of network elements [44]. The abstract network model is linked with network graphics leading to interactive flux maps which allow for user input and display of calculation results within a network visualization [44]. Therein, a large and powerful collection of tools and algorithms can be applied interactively including metabolic flux analysis, flux optimization, detection of topological features and pathway analysis by elementary flux modes or extreme pathways [44].

4. ePath3D

ePath3D is an easy-to-use, powerful software tool for creating and managing illustrated 3D pathways for publications and presentations. This new desktop software includes a powerful drawing feature that allows for the easy creation and management of dramatic 3D signaling and metabolic pathways ideal for teaching, presentations, publications and posters.

C. Usefulness of Pathway Analysis

Metabolic pathways are a central paradigm in biology [45]. Classically, they have been defined on the basis of their step-by-step discovery [45]. However, the genome-scale metabolic networks now being reconstructed from annotation of genome sequences demand new network-based definitions of pathways to facilitate analysis of their capabilities and functions, such as metabolic versatility and robustness, and optimal growth rates [45]. This demand has led to the development of a new mathematically based analysis of complex, metabolic networks that enumerates all their unique pathways that take into account all requirements for cofactors and byproducts [45]. The ability to visualise the complex data dynamically would be useful for building more powerful research tools to access the databases [46]. Metabolic pathways are typically modelled as graphs in which nodes represent chemical compounds, and edges represent chemical reactions between compounds [46]. Thus, the problem of visualising pathways can be formulated as a graph layout problem [46]. The automatic generation of drawings of metabolic pathways is a challenging problem that depends intimately on exactly what information has been recorded for each pathway and on how that information is encoded [47].

Table 1. Some interesting reports on usefulness of pathway analysis

Authors	Details
Papin et al [49]	Genome-scale extreme pathways associated with the production of non-essential amino acids in *Haemophilus influenzae* were computed [49]. Three key results were obtained [49]. First, there were multiple internal flux maps corresponding to externally indistinguishable states. It was shown that there was an average of 37 internal states per unique exchange flux vector in *H. influenzae* when the network was used to produce a single amino acid while allowing carbon dioxide and acetate as carbon sinks [49]. Second, an analysis of the carbon fates illustrated that the extreme pathways were non-uniformly distributed across the carbon fate spectrum [49]. Third, this distribution fell between distinct systemic constraints [49].
Price et al [50]	The first study of genome-scale extreme pathways for the simultaneous formation of all nonessential amino acids or ribonucleotides in *Helicobacter pylori* was presented [50]. First, the extreme pathways for the production of individual amino acids in H. pylori showed far fewer internal states per external state than previously found in *H. influenza* [50]. Second, the degree of pathway redundancy in *H. pylori* was essentially the same for the production of individual amino acids and linked amino acid sets, but was approximately twice that of the production of the ribonucleotides [50]. Third, the metabolic network of *H. pylori* was unable to achieve extensive conversion of amino acids consumed to the set of either nonessential amino acids or ribonucleotides [50]
Wiback and Palsson [51]	In this work, extreme pathways of the well-characterized human red blood cell metabolic network were calculated and interpreted in a biochemical and physiological context [51]. These extreme pathways were divided into groups based on such criteria as their cofactor and by-product production, and carbon inputs including those that 1) convert glucose to pyruvate; 2) interchange pyruvate and lactate; 3) produce 2,3-diphosphoglycerate that binds to hemoglobin; 4) convert inosine to pyruvate; 5) induce a change in the total adenosine pool; and 6) dissipate ATP [51].
Papin and Palsson [52]	A reconstruction of the JAK-STAT signaling system in the human B-cell was described and a scalable framework for its network analysis was presented [52]. From the extreme signaling pathways, emergent systems properties of the JAK-STAT signaling network had been characterized, including 1), a mathematical definition of network crosstalk; 2), an analysis of redundancy in signaling inputs and outputs; 3), a study of reaction participation in the network; and 4), a delineation of 85 correlated reaction sets, or systemic signaling modules [52].

A useful approach to unraveling and understanding complex biological networks is to decompose networks into basic functional and structural units [48]. Recent application of convex analysis to metabolic networks leads to the development of network-based metabolic pathway analysis and the decomposition of metabolic networks into metabolic extreme pathways that are true functional units of metabolic systems [48]. Metabolic extreme pathways are derived from limited knowledge of the metabolic networks, but provide an integrated predictive description of metabolic networks [48]. Some interesting reports on usefulness of pathway analysis are presented in Table 1.

D. Pathway Analysis Tool

There are also many pathway analysis tools at present. These tools are very useful. Details of some important tools will be hereby presented.

1. Pathway Miner [53]

Pathway Miner catalogs genes based on their role in metabolic, cellular and regulatory pathways [53]. A Fisher exact test is provided as an option to rank pathways [53]. The genes are mapped onto pathways and gene product association networks are extracted for genes that co-occur in pathways [53]. Pathway Miner is a freely available web accessible tool at http://www.biorag.org/pathway.html [53].

2. WholePathwayScope [54]

WholePathwayScope (WPS) is for deriving biological insights from analysis of High Throughput data [54]. WPS extracts gene lists with shared biological themes through color cue templates [54]. WPS statistically evaluates global functional category enrichment of gene lists and pathway-level pattern enrichment of data [54]. WPS incorporates well-known biological pathways from KEGG (Kyoto Encyclopedia of Genes and Genomes) and Biocarta, GO (Gene Ontology) terms as well as user-defined pathways or relevant gene clusters or groups, and explores gene-term relationships within the derived gene-term association networks (GTANs) [54]. WPS simultaneously compares multiple datasets within biological contexts either as pathways or as association networks. WPS also integrates Genetic Association Database and Partial MedGene Database for disease-association information. We have used this program to analyze and compare microarray and proteomics datasets derived from a variety of biological systems [54]. The tool is freely available at http://www.abcc.ncifcrf.gov/wps/wps_index.php [54].+

3. ArrayxPath [55]

ArrayXPath (http://www.snubi.org/software/ArrayXPath/) is a web-based service for mapping and visualizing microarray gene-expression data for integrated biological pathway resources using Scalable Vector Graphics [55]. By integrating major bio-databases and searching pathway resources, ArrayXPath automatically maps different types of identifiers from microarray probes and pathway elements [55]. When one inputs gene-expression clusters, ArrayXPath produces a list of the best matching pathways for each cluster [55].

4. Genome Expression Pathway Analysis Tool [56]

Genome Expression Pathway Analysis Tool (GEPAT) offers an analysis of gene expression data under genomic, proteomic and metabolic context [56]. GEPAT offers various statistical data analysis methods, as hierarchical, k-means and PCA clustering, a linear model based t-test or chromosomal profile comparison [56]. GEPAT offers no linear work flow, but allows the usage of any subset of probes and samples as a start for a new data analysis [56]. GEPAT relies on established data analysis packages, offers a modular approach for an easy extension, and can be run on a computer grid to allow a large number of users [56]. It is freely available under the LGPL open source license for academic and commercial users at http://gepat.sourceforge.net [56].

Introduction to Metabolonics

Metabolonics is a science to study cellular metabolic activities. The aim of metabolonics is to study the complete metabolic response of living things to genetic modifications or environmental stimuli. The knowledge in this new "omic" science is still limited but carries great hope in science and medicine.

References

[1] Fridman, E; Pichersky, E. Metabolomics, genomics, proteomics, and the identification of enzymes and their substrates and products. *Curr Opin Plant Biol.,* 2005, 8(3), 242-8.

[2] German, JB; Bauman, DE; Burrin, DG; Failla, ML; Freake, HC; King, JC; Klein, S; Milner, JA; Pelto, GH; Rasmussen, KM; Zeisel, SH. Metabolomics in the opening decade of the 21st century: building the roads to individualized health. *J Nutr.,* 2004, 134(10), 2729-32.

[3] Arita, M. Additional paper: computational resources for metabolomics. *Brief Funct Genomic Proteomic.,* 2004, 3(1), 84-93.

[4] Mendes, P. Emerging bioinformatics for the metabolome. *Brief Bioinform.,* 2002, 3(2), 134-45.

[5] Lange, BM; Ghassemian, M. Comprehensive post-genomic data analysis approaches integrating biochemical pathway maps. *Phytochemistry.,* 2005, 66(4), 413-51.

[6] Hunter, P; Nielsen, P. A strategy for integrative computational physiology. *Physiology (Bethesda).,* 2005, Oct; 20, 316-25.

[7] Eungdamrong, NJ; Iyengar, R. Computational approaches for modeling regulatory cellular networks. *Trends Cell Biol.,* 2004, Dec;14(12), 661-9.

[8] Hofestädt, R; Thelen, S. Quantitative modeling of biochemical networks. *In Silico Biol.,* 1998, 1(1), 39-53.

[9] Cabrera, ME; Saidel, GM; Kalhan, SC. Modeling metabolic dynamics. From cellular processes to organ and whole body responses. *Prog Biophys Mol Biol.,* 1998, 69(2-3), 539-57.

[10] Arita, M. Computer modeling of metabolic networks. *Tanpakushitsu Kakusan Koso.,* 2003, Jun; 48(7), 823-8

[11] Wiechert, W. Modeling and simulation: tools for metabolic engineering. *J Biotechnol.,* 2002, Mar, 14, 94(1), 37-63.

[12] van Helden, J; Wernisch, L; Gilbert, D; Wodak, SJ. Graph-based analysis of metabolic networks. *Ernst Schering Res Found Workshop.,* 2002, (38), 245-74.

[13] Duran, AL; Yang, J; Wang, L; Sumner, LW. Metabolomics spectral formatting, alignment and conversion tools (MSFACTs). *Bioinformatics.,* 2003, 19(17), 2283-93.

[14] Shinoda, K; Yachie, N; Masuda, T; Sugiyama, N; Sugimoto, M; Soga, T; Tomita, M. HybGFS: a hybrid method for genome-fingerprint scanning. *BMC Bioinformatics.,* 2006, Oct 29, 7, 479.

[15] Wishart, DS; Tzur, D; Knox, C; Eisner, R; Guo, AC; Young, N; Cheng, D; Jewell, K; Arndt, D; Sawhney, S; Fung, C; Nikolai, L; Lewis, M; Coutouly, MA; Forsythe, I; Tang, P; Shrivastava, S; Jeroncic, K; Stothard, P; Amegbey, G; Block, D; Hau, DD; Wagner, J; Miniaci, J; Clements, M; Gebremedhin, M; Guo, N; Zhang, Y; Duggan, GE; Macinnis, GD; Weljie, AM; Dowlatabadi, R; Bamforth, F; Clive, D; Greiner, R; Li, L; Marrie, T; Sykes, BD; Vogel, HJ; Querengesser, L. HMDB: the Human Metabolome Database. *Nucleic Acids Res.* 2007 Jan, 35, (Database issue), D521-6.

[16] Lemer, C; Antezana, E; Couche, F; Fays, F; Santolaria, X; Janky, R; Deville, Y; Richelle, J; Wodak, SJ. The aMAZE LightBench: a web interface to a relational database of cellular processes. *Nucleic Acids Res.,* 2004, Jan 1, 32(Database issue):D443-8.

[17] Hou, BK; Kim, JS; Jun, JH; Lee, DY; Kim, YW; Chae, S; Roh, M; In, YH; Lee, SY. BioSilico: an integrated metabolic database system. *Bioinformatics.,* 2004, Nov 22, 20(17), 3270-2.

[18] Karp, PD; Riley, M; Paley, SM; Pellegrini-Toole, A; Krummenacker, M. EcoCyc: Enyclopedia of Escherichia coli Genes and Metabolism. *Nucleic Acids Res.*, 1997, Jan 1, 25(1), 43-51.

[19] Karp, PD; Riley, M; Paley, SM; Pellegrini-Toole, A; Krummenacker, M. Eco Cyc: encyclopedia of Escherichia coli genes and metabolism. *Nucleic Acids Res.*, 1999, Jan 1, 27(1), 55-8.

[20] Karp, PD; Riley, M; Paley, SM; Pellegrini-Toole, A; Krummenacker, M. yc: Encyclopedia of Escherichia coli genes and metabolism. *Nucleic Acids Res.,* 1998, Jan 1, 26(1), 50-3.

[21] Karp, PD; Riley, M; Paley, SM; Pelligrini-Toole, A. EcoCyc: an encyclopedia of Escherichia coli genes and metabolism. *Nucleic Acids Res.,* 1996, Jan 1, 24(1), 32-9.

[22] Dogrusoz, U; Erson, EZ; Giral, E; Demir, E; Babur, O; Cetintas, A; Colak, R. PATIKAweb: a Web interface for analyzing biological pathways through advanced querying and visualization. *Bioinformatics.,* 2006, Feb 1, 22(3), 374-5.

[23] Chen, M; Hofestädt, R. PathAligner: metabolic pathway retrieval and alignment. *Appl Bioinformatics.,* 2004, 3(4), 241-52.

[24] Caspi, R; Foerster, H; Fulcher, CA; Hopkinson, R; Ingraham, J; Kaipa, P; Krummenacker, M; Paley, S; Pick, J; Rhee, SY; Tissier, C; Zhang, P; Karp, PD. MetaCyc: a multiorganism database of metabolic pathways and enzymes. *Nucleic Acids Res.* 2006, Jan 1, 34(Database issue):D511-6.

[25] The MetaCyc Database. Karp, PD; Riley, M; Paley, SM; Pellegrini-Toole, A. *Nucleic Acids Res.,* 2002, Jan 1, 30(1), 59-61.

[26] Zhang, P; Foerster, H; Tissier, CP; Mueller, L; Paley, S; Karp, PD; Rhee, SY. MetaCyc and AraCyc. Metabolic pathway databases for plant research. *Plant Physiol.,* 2005, May;138(1), 27-37.

[27] Kopka, J; Schauer, N; Krueger, S; Birkemeyer, C; Usadel, B; Bergmüller, E; Dörmann, P; Weckwerth, W; Gibon, Y; Stitt, M; Willmitzer, L; Fernie, AR; Steinhauser, D. GMD@CSB.DB: the Golm Metabolome Database. *Bioinformatics.,* 2005, Apr 15, 21(8), 1635-8.

[28] Mazurek, S; Eigenbrodt, E. The tumor metabolome. *Anticancer Res.,* 2003, Mar-Apr, 23(2A), 1149-54.

[29] Kolch, W; Mischak, H; Pitt, AR. The molecular make-up of a tumour: proteomics in cancer research. *Clin Sci (Lond).,* 2005, May, 108(5), 369-83.

[30] Claudino, WM; Quattrone, A; Biganzoli, L; Pestrin, M; Bertini, I; Di Leo, A. Metabolomics: available results, current research projects in breast cancer, and future applications. *J Clin Oncol.,* 2007, Jul 1, 25(19), 2840-6

[31] Rochfort, S. Metabolomics reviewed: a new "omics" platform technology for systems biology and implications for natural products research. *J Nat Prod.,* 2005, Dec, 68(12), 1813-20.

[32] Xu, M; Lin, DH; Liu, CX. Current status and prospect of metabonomics. *Yao Xue Xue Bao.,* 2005, Sep, 40(9), 769-74.

[33] Sivachenko, A; Kalinin, A; Yuryev, A. Pathway analysis for design of promiscuous drugs and selective drug mixtures. *Curr Drug Discov Technol.*, 2006, Dec, 3(4), 269-77.

[34] Goodacre, R. Metabolomics of a superorganism. *J Nutr.,* 2007, Jan; 137(1 Suppl), 259S-266S.

[35] van der Greef, J; Hankemeier, T; McBurney, RN. Metabolomics-based systems biology and personalized medicine: moving towards n = 1 clinical trials? *Pharmacogenomics.,* 2006, Oct;7(7), 1087-94.

[36] Schuster, D; Steindl, TM; Langer, T. Predicting drug metabolism induction in silico. *Curr Top Med Chem.,* 2006, 6(15), 1627-40.

[37] Fox, T; Kriegl, JM. Machine learning techniques for in silico modeling of drug metabolism. *Curr Top Med Chem.,* 2006, 6(15), 1579-91.

[38] Mankowski, DC; Ekins, S. Prediction of human drug metabolizing enzyme induction. *Curr Drug Metab.,* 2003, Oct; 4(5), 381-91.

[39] Lindon, JC; Holmes, E; Nicholson, JK. Metabonomics in pharmaceutical R&D. *FEBS J.,* 2007, Mar; 274(5), 1140-51.

[40] Fu, J; Swertz, MA; Keurentjes, JJ; Jansen, RC. MetaNetwork: a computational protocol for the genetic study of metabolic networks. *Nat Protoc.,* 2007, 2(3), 685-94.

[41] Romero, R; Espinoza, J; Gotsch, F; Kusanovic, JP; Friel, LA; Erez, O; Mazaki-Tovi, S; Than, NG; Hassan, S; Tromp, G. The use of high-dimensional biology (genomics, transcriptomics, proteomics, and metabolomics) to understand the preterm parturition syndrome. *BJOG.,* 2006, Dec;113, Suppl, 3, 118-35.

[42] Goesmann, A; Haubrock, M; Meyer, F; Kalinowski, J; Giegerich, R. PathFinder: reconstruction and dynamic visualization of metabolic pathways. *Bioinformatics.,* 2002, Jan;18(1), 124-9.

[43] Bourqui, R; Cottret, L; Lacroix, V; Auber, D; Mary, P; Sagot, MF; Jourdan, F. Metabolic network visualization eliminating node redundance and preserving metabolic pathways. *BMC Syst Biol.,* 2007, Jul 3, 1, 29.

[44] Klamt, S; Stelling, J; Ginkel, M; Gilles, ED. FluxAnalyzer: exploring structure, pathways, and flux distributions in metabolic networks on interactive flux maps. *Bioinformatics.,* 2003, Jan 22, 19(2):261-9.

[45] Papin, JA; Price, ND; Wiback, SJ; Fell, DA; Palsson, BO. Metabolic pathways in the post-genome era. *Trends Biochem Sci.,* 2003, May;28(5), 250-8.

[46] Becker, MY; Rojas, I. A graph layout algorithm for drawing metabolic pathways. *Bioinformatics.,* 2001, May; 17(5), 461-7.

[47] Karp, PD; Paley, SM. Representations of metabolic knowledge: pathways. *Proc Int Conf Intell Syst Mol Biol.,* 1994, 2, 203-11.

[48] Xiong, M; Zhao, J; Xiong, H. Network-based regulatory pathways analysis. *Bioinformatics.,* 2004, Sep 1, 20(13), 2056-66.

[49] Papin, JA; Price, ND; Edwards, JS; Palsson, B. BØ. The genome-scale metabolic extreme pathway structure in Haemophilus influenzae shows significant network redundancy. *J Theor Biol.,* 2002, Mar 7, 215(1), 67-82.

[50] Price, ND; Papin, JA; Palsson, BØ. Determination of redundancy and systems properties of the metabolic network of Helicobacter pylori using genome-scale extreme pathway analysis. *Genome Res.,* 2002, May, 12(5), 760-9.

[51] Wiback, SJ; Palsson, BO. Extreme pathway analysis of human red blood cell metabolism. *Biophys J.,* 2002, Aug, 83(2), 808-18.

[52] Papin, JA; Palsson, BO. The JAK-STAT signaling network in the human B-cell: an extreme signaling pathway analysis. *Biophys J.,* 2004, Jul, 87(1), 37-46.

[53] Pandey, R; Guru, RK; Mount, DW. Pathway Miner: extracting gene association networks from molecular pathways for predicting the biological significance of gene expression microarray data. *Bioinformatics.,* 2004, Sep, 1, 20(13), 2156-8.

[54] Yi, M; Horton, JD; Cohen, JC; Hobbs, HH; Stephens, RM. WholePathwayScope: a comprehensive pathway-based analysis tool for high-throughput data. *BMC Bioinformatics.,* 2006, Jan 19, 7, 30.

[55] Chung, HJ; Kim, M; Park, CH; Kim, J; Kim, JH. ArrayXPath: mapping and visualizing microarray gene-expression data with integrated biological pathway resources using Scalable Vector Graphics. *Nucleic Acids Res.,* 2004, Jul 1, 32(Web Server issue):W460-4.

[56] Weniger, M; Engelmann, JC; Schultz, J. Genome Expression Pathway Analysis Tool--analysis and visualization of microarray gene expression data under genomic, proteomic and metabolic context. *BMC Bioinformatics.,* 2007, Jun 2, 8, 179.

In: Metabolomics: Metabolites, Metabonomics…
Editors: J.S. Knapp and W.L. Cabrera, pp. 243-251
ISBN: 978-1-61668-006-0

Chapter 9

THE ROLE OF SPECIFIC ESTROGEN METABOLITES IN THE INITIATION OF BREAST AND OTHER HUMAN CANCERS

Eleanor G. Rogan [*] ***and Ercole L. Cavalieri***
Eppley Institute for Research in Cancer and Allied Diseases,
University of Nebraska Medical Center, 986805 Nebraska Medical Center,
Omaha, NE, USA

Keywords: Breast cancer, cancer initiation, catechol estrogen quinones, depurinating DNA adducts, estrogens.

Introduction

Various types of evidence have implicated estrogens in the etiology of human breast cancer [1-8]. They are generally thought to cause proliferation of breast epithelial cells through estrogen receptor-mediated processes [4]. Rapidly proliferating cells are susceptible to genetic errors during DNA replication, which, if uncorrected, can ultimately lead to malignancy. While receptor-mediated processes may play an important role in the development and growth of tumors, accumulating evidence suggests that specific oxidative metabolites of estrogens, if formed, can be endogenous ultimate carcinogens that react with DNA to cause the mutations leading to initiation of cancer [6-9]. Thus, estrogen metabolites, specifically catechol estrogen-3,4-quinones, are hypothesized to be endogenous initiators of breast, prostate and other human cancers.

Several lines of evidence, including metabolism and carcinogenicity studies by Liehr and coworkers, led to the recognition that the 4-hydroxylated estrogens play a major role in the genotoxic properties of estrogens [1-3]. We have hypothesized that the estrogens estrone (E_1) and estradiol (E_2) initiate breast and other human cancers by reaction of their electrophilic

[*] E-mail address: egrogan@unmc.edu. Tel: 402-559-4095, Fax: 402-559-8068. (Corresponding author)

metabolites, catechol estrogen-3,4-quinones [$E_1(E_2)$-3,4-Q], with DNA to form depurinating adducts [5-8]. These adducts generate apurinic sites leading to mutations that may initiate breast, prostate and other human cancers [6-9].

The estrogens, E_1 and E_2, are obtained via aromatization of 4-androstene-3,17-dione and testosterone, respectively, catalyzed by cytochrome P450(CYP)19, aromatase (Figure 1). E_1 and E_2, which are biochemically interconvertible by the enzyme 17β-estradiol dehydrogenase, are metabolized to the 2-catechol estrogens, 2-OH$E_1(E_2)$, and 4-OH$E_1(E_2)$, predominantly catalyzed by the activating enzymes CYP1A1 [10] and 1B1 [10-13], respectively, in extrahepatic tissues. The estrogens are also metabolized, to a lesser extent, to 16α-hydroxy derivatives (not shown). The catechol estrogens are further easily oxidized to the catechol estrogen quinones, $E_1(E_2)$-2,3-Q and $E_1(E_2)$-3,4-Q (Figure 1) by metal ions, peroxidases and cytochrome P450. In general, the catechol estrogens are inactivated by conjugating reactions, such as glucuronidation and sulfation. A common pathway of inactivation in extrahepatic tissues, however, occurs by *O*-methylation catalyzed by the ubiquitous catechol-*O*-methyltransferase (COMT) [14]. If formation of E_1 or E_2 is excessive, due to overexpression of aromatase and/or the presence of excess sulfatase that converts the stored E_1 sulfate to E_1, increased formation of catechol estrogens is expected. In particular, the presence and/or induction of CYP1B1 and other 4-hydroxylases could render the 4-OH$E_1(E_2)$, which are usually minor metabolites, as the major metabolites [15-17]. Thus, conjugation of 4-OH$E_1(E_2)$ via methylation in extrahepatic tissues might become insufficient, and competitive catalytic oxidation of 4-OH$E_1(E_2)$ to $E_1(E_2)$-3,4-Q could occur. (Figure 1)

Protection at the quinone level can occur by conjugation of $E_1(E_2)$-Q with glutathione (GSH, Figure 1). A second inactivating process for $E_1(E_2)$-Q is their reduction to catechol estrogens by quinone reductase. If these two inactivating processes are not effective, $E_1(E_2)$-Q may react with DNA to form stable and depurinating adducts [5-8,18-20]. Imbalances in estrogen homeostasis [17,20], that is the equilibrium between activating and protective enzymes with the scope of avoiding formation of catechol estrogen semiquinones and quinones, could lead to initiation of cancer by estrogens.

Catechol Estrogen Quinones as Mutagens Initiating Breast, Prostate and Other Human Cancers

Experiments on estrogen metabolism [17,19-22], formation of DNA adducts [5-8], carcinogenicity [23-25], and mutagenicity [9,26,27] provide a basis for the hypothesis that reaction of certain estrogen metabolites, predominantly catechol estrogen-3,4-quinones, with DNA can generate the critical mutations initiating breast, prostate and other cancers [7-9,28].

Imbalance in Estrogen Homeostasis

Estrogen metabolism involves a balance between activating and deactivating (protective) pathways. There are several factors that can unbalance estrogen homeostasis, that is, the equilibrium between activating and deactivating pathways, to limit formation and/or reaction of the endogenous carcinogenic $E_1(E_2)$-Q with DNA. The first imbalancing factor could be excessive synthesis of E_2 by high expression of aromatase (CYP19) in target tissues [29-31] and/or the presence of sulfatase that excessively converts stored E_1 sulfate to E_1 [32,33]. A striking result of *in situ* production of E_2 in human breast tissue is the similar levels of E_2 in breast tissue in pre-menopausal and post-menopausal women, even though plasma levels are 10-50 fold higher in pre-menopausal than post-menopausal women [34]. Both aromatase and sulfatase contribute to *in situ* estrogen production [32,33].

A second critical factor leading to imbalances in estrogen homeostasis might be high levels of 4-OHE$_1$(E$_2$) due to high expression of CYP1B1, which metabolizes E_2 predominantly to form 4-OHE$_2$ [10,12,35]. This could result in relatively large amounts of 4-OHE$_1$(E$_2$) that, in turn, can lead to more extensive oxidation to the carcinogenic E$_1$(E$_2$)-3,4-Q. A third factor could be a lack or a low level of activity of the protective COMT enzyme. If this enzyme is insufficient, 4-OHE$_1$(E$_2$) will not be effectively methylated in extrahepatic tissues, but will be oxidized to the ultimate carcinogenic metabolites E$_1$(E$_2$)-3,4-Q. A fourth factor could be a low level of GSH and/or low levels of quinone oxidoreductase and/or CYP reductase, which could leave available higher levels of E$_1$(E$_2$)-3,4-Q that may react with DNA.

Imbalances in estrogen homeostasis have been observed in laboratory animals and in breast tissue from women with breast cancer:

The Kidney of Syrian Golden Hamsters

The hamster provides an excellent model for studying estrogen homeostasis because implantation of E_1 or E_2 in male Syrian golden hamsters induces 100% of renal carcinomas, but does not induce liver tumors [36]. Therefore, comparison of the profile of estrogen metabolites, conjugates and DNA adducts in the two organs, after treatment of hamsters with E_2, should provide information on the relative imbalance in estrogen homeostasis in the two tissues [20]. In the liver, more *O*-methylation of 2-OHE$_1$(E$_2$) was observed, whereas more formation of E$_1$(E$_2$)-Q was detected in the kidney. These results suggest greater oxidation of catechol estrogens to E$_1$(E$_2$)-Q and less protective methylation of 2-OHE$_1$(E$_2$) in the kidney. When normal levels of GSH were depleted before hamsters were treated with E_2, very low levels of catechol estrogens and methoxy catechol estrogens were observed in the kidney compared to the liver, suggesting little protective reduction of E$_1$(E$_2$)-Q to catechol estrogens in the kidney. More importantly, the 4-OHE$_1$(E$_2$)-1-N7Gua depurinating adduct arising from reaction of E$_1$(E$_2$)-3,4-Q with DNA was detected in the kidney, but not in the liver [20].

These results suggest that tumor initiation in the kidney occurs because of poor methylation of catechol estrogens, rendering more likely competitive oxidation of catechol estrogens to E$_1$(E$_2$)-Q, as well as poor quinone reductase activity to remove the E$_1$(E$_2$)-Q. These two effects produce a large amount of E$_1$(E$_2$)-Q, which can react with the nucleophilic groups of DNA.

The Mammary Gland of ERKO/Wnt-1 Mice

Mammary tumors develop in female estrogen receptor-α knock-out (ERKO)/Wnt-1 mice despite their lack of functional estrogen receptor-α [37]. Extracts of hyperplastic mammary tissue and mammary tumors from these mice were analyzed by HPLC interfaced with an electrochemical detector [21]. Picomole amounts of the 4-catechol estrogens were detected, but their methoxy conjugates were not. Neither the 2-catechol estrogens nor 2-methoxy catechol estrogens were detected. 4-$OHE_1(E_2)$-GSH conjugates or their hydrolytic products (conjugates of cysteine and *N*-acetylcysteine) were detected in picomole amounts in both tumors and hyperplastic mammary tissue, demonstrating the formation of $E_1(E_2)$-3,4-Q. These preliminary findings indicate that estrogen homeostasis is unbalanced in the mammary tissue, in that the normally minor 4-catechol estrogen metabolites were detected in the mammary tissue, but not the normally predominant 2-catechol estrogens. Furthermore, methylation of catechol estrogens was not detected, whereas formation of 4-$OHE_1(E_2)$-GSH conjugates was. These results are consistent with the hypothesis that mammary tumor development is primarily initiated by metabolism of estrogens to $E_1(E_2)$-3,4-Q, which may react with DNA to induce oncogenic mutations.

The Prostate of Noble Rats

Estrogen metabolites and conjugates were analyzed in the ventral and anterior lobes of the rat prostate, which are not susceptible to estrogen-induced carcinogenesis, and in the susceptible dorsolateral and periurethral prostate of rats treated with 4-OHE_2 or E_2-3,4-Q [22]. The analyses revealed that the areas of the prostate susceptible to induction of carcinomas have less protection by COMT, quinone reductase and GSH, thereby favoring reaction of $E_1(E_2)$-3,4-Q with DNA.

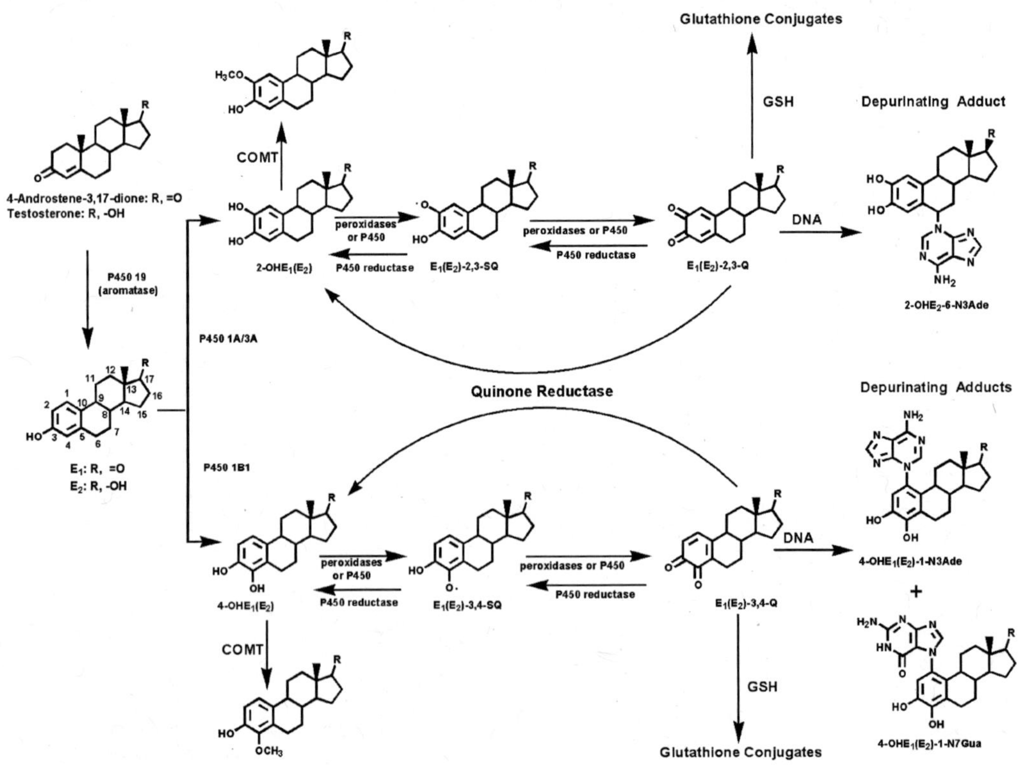

Figure 1. Formation, metabolism, conjugation and DNA adducts of estrogens.

The Breast of Women with Breast Carcinoma

A study of breast tissue from women with and without breast cancer provides key evidence in support of the concept of estrogen homeostasis [17]. In fact, relative imbalances in estrogen homeostasis were observed in analysis of women with breast cancer (Figure 2). Levels of E_1 and E_2 in women with carcinoma were higher than in controls. In women without cancer, a larger amount of 2-OHE$_1$(E$_2$) than 4-OHE$_1$(E$_2$) was observed. In women with carcinoma, the 4-OHE$_1$(E$_2$) were three times more abundant than the 2-OHE$_1$(E$_2$). The 4-OHE$_1$(E$_2$) were also four times higher than in women without cancer. Furthermore, a lower level of methylation was observed for the catechol estrogens in cancer cases *vs* the controls. Levels of E$_1$(E$_2$)-Q conjugates in women with cancer were three times those in controls, suggesting a larger probability for the E$_1$(E$_2$)-Q to react with DNA in the breast tissue of women with carcinoma. Levels of 4-OHE$_1$(E$_2$) ($p<0.01$) and quinone conjugates ($p<0.003$) appear to be highly significant predictors of breast cancer [17]. Further support for this concept is provided by detection of the 4-OHE$_2$-1-N3Ade adduct in non-tumor breast tissue from a woman with breast carcinoma at a level 30 times higher than in breast tissue from a woman without breast cancer [38].

In summary, it appears from these animal and human studies that the formation of E$_1$(E$_2$)-3,4-Q from catechol estrogens is the result of an imbalance of one or more enzymes involved in the maintenance of estrogen homeostasis.

Estrogen-Induced Mutations and Cell Transformation

Mutations are induced in the Harvey (H)-*ras* oncogene in the skin of female SENCAR mice following topical treatment with E$_2$-3,4-Q [9]. Mutations are also induced in the H-*ras* oncogene in the mammary gland of female ACI rats [28], which develop mammary tumors when implanted with E$_2$ [39]. These studies demonstrate that E$_2$-3,4-Q is mutagenic. This mutagenicity has been correlated with formation of depurinating DNA adducts, in particular the rapidly depurinating 4-OHE$_2$-1-N3Ade [9,19,28].

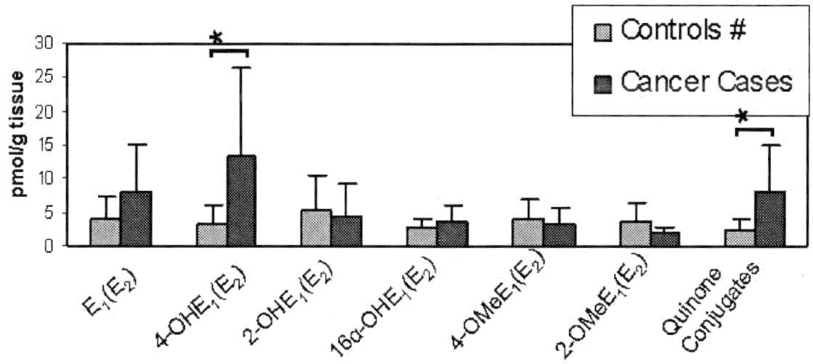

Figure 2. Analysis of estrogen metabolites and conjugates in human breast tissue from women with and without breast cancer. Controls are benign fatty breast tissue and benign fibrocystic changes. Quinone conjugates are 4-OHE$_1$(E$_2$)-2-NAcCys, 4-OHE$_1$(E$_2$)-2-Cys, 2-OHE$_1$(E$_2$)-(1+4)-NAcCys and 2-OHE$_1$(E$_2$)-(1+4)-Cys. *Statistically significant differences were determined using the Wilcoxon rank sum test, p<0.01 [4-OHE$_1$(E$_2$)] and p<0.003 (quinone conjugates).

Both E_2 and the catechol estrogen 4-OHE_2 also induce cell transformation in the human breast epithelial MCF-10F cell line [40,41]. It is significant to note that the neoplastic transformation of MCF-10F cells by E_2 or 4-OHE_2 is not blocked by the antiestrogen ICI 182-780, indicating that this event is occurring by a non-receptor mediated process [42]. These data suggest that the initiating step leading to cell transformation derives from the DNA damage produced by E_2-3,4-Q, the oxidative metabolite of 4-OHE_2.

Conclusions

A growing body of evidence from studies with laboratory animals, cultured cells and human tissues supports the hypothesis that estrogens can initiate cancer by formation of specific DNA adducts leading to mutations in critical genes. The $E_1(E_2)$-3,4-Q are the predominant estrogen metabolites that react with DNA to form depurinating N7Gua and N3Ade adducts, generating apurinic sites and subsequent mutations. This approach to studying estrogen-induced cancer not only guides the study of the role of estrogens in initiating cancer, but also provides candidate biomarkers that may be used to determine risk of developing breast, prostate or other types of cancer. These studies also suggest possible strategies to prevent the development of cancer.

Acknowledgments

Preparation of this article was supported by U.S. Public Health Service grants P01 CA49210 and R01 CA49917 from the National Cancer Institute. Core support in the Eppley Institute is provided by grant P30 CA36727 from the National Cancer Institute.

References

[1] Liehr, J. G. (1990). Genotoxic effects of estrogens. *Mutat Res*, **238**, 269-276.

[2] Liehr, J. G. (2000). Is estradiol a genotoxic mutagenic carcinogen? *Endocr Rev*, **21**, 40-54.

[3] Liehr, J. G. (2001). Genotoxicity of the steroidal oestrogens oestrone and oestradiol: Possible mechanism of uterine and mammary cancer development. *Human Repro Update*, **7**, 273-281.

[4] Feigelson, H. S. & Henderson, B. E. (1996). Estrogens and breast cancer. *Carcinogenesis*, **17**, 2279-2284.

[5] Cavalieri, E. L., Stack, D. E., Devanesan, P. D., et al. (1997). Molecular origin of cancer: Catechol estrogen-3,4-quinones as endogenous tumor initiators. *Proc Natl Acad Sci.*, USA, **94**, 10937-10942.

[6] Cavalieri, E., Frenkel, K., Liehr, J. G., Rogan, E. & Roy, D. (2000). Estrogens as endogenous genotoxic agents: DNA adducts and mutations. In: *JNCI Monograph* **27**: *Estrogens as Endogenous Carcinogens in the Breast and Prostate*. E. Cavalieri, & E. Rogan (Eds.), Oxford Press, 75-93.

[7] Cavalieri, E. L., Rogan, E. G. & Chakravarti, D. (2002). Initiation of cancer and other diseases by catechol ortho-quinones: A unifying mechanism. *Cell & Mol Life Sci*, **59**, 665-681.

[8] Cavalieri, E., Rogan, E. & Chakravarti, D. (2004). The role of endogenous catechol quinones in the initiation of cancer and neurodegenerative diseases. In: *Methods in enzymology, quinones and quinone enzymes, part, B*. In H. Sies, & L. Packer (Eds.), Elsevier, Duesseldorf, Germany, 293-319.

[9] Chakravarti, D., Mailander, P., Li, K. M, et al. (2001). Evidence that a burst of DNA depurination in SENCAR mouse skin induces error-prone repair and form mutations in the H-*ras* gene. *Oncogene*, **20**, 7945-7953.

[10] Spink, D. C., Spink, B. C., Cao, J. Q., et al. (1998). Differential expression of CYP1A1 and CYP1B1 in human breast epithelial cells and breast tumor cells. *Carcinogenesis*,, **19**, 291-298.

[11] Spink, D. C., Hayes, C. L., Young, N. R., et al. (1994). The effects of 2,3,7,8-tetrachlorodibenzo-*p*-dioxin on estrogen metabolism in MCF-7 breast cancer cells: Evidence for induction of a novel 17β-estradiol 4-hydroxylase. *J Steroid Biochem Mol Biol*, **51**, 251-258.

[12] Hayes, C. L., Spink, D. C., Spink, B. C., Cao, J. Q., Walker, N. J. & Sutter, T. R. (1996). 17β-estradiol hydroxylation catalyzed by human P450 1B1. *Proc Natl Acad Sci* USA, **93**, 9776-9781.

[13] Spink, D. C., Spink, B. C., Cao, J. Q., et al. (1997). Induction of cytochrome P450 1B1 and catechol estrogen metabolism in ACHN human renal adenocarcinoma cells. *J Steroid Biochem Mol Biol*, **62**, 223-232.

[14] Ball, P. & Knuppen, R. (1980). Catechol oestrogens (2- and 4-hydroxyestrogens): Chemistry, biogenesis, meta-bolism, occurrence and physiological significance. *Acta Endocrinol* (Copenhagen), **93**, (Suppl 232), 1-127.

[15] Castagnetta, L. A., Granata, O. M., Arcuri, F. P., Polito, L. M., Rosati, F., Cartoni, G. P. (1992). Gas chromatography/mass spectrometry of catechol estrogens. *Steroids*, **57**, 437-443.

[16] Liehr, J. G., Ricci, M. J. (1996). 4-Hydroxylation of estrogens as marker of human mammary tumors. *Proc Natl Acad Sci.*, USA, **93**, 3294-3296.

[17] Rogan, E. G., Badawi, A. F., Devanesan, P. D., et al. (2003). Relative imbalances in estrogen metabolism and conjugation in breast tissue of women with carcinoma: Potential biomarkers of susceptibility to cancer. *Carcinogenesis*, **24**, 697-702.

[18] Dwivedy, I., Devanesan, P., Cremonesi, P., Rogan, E. & Cavalieri, E. (1992). Synthesis and characterization of estrogen 2,3- and 3,4-quinones. Comparison of DNA adducts formed by the quinones *versus* horseradish peroxidase-activated catechol estrogens. *Chem Res Toxicol*, **5**, 828-833.

[19] Li, K. M., Todorovic, R., Devanesan, P., et al. (2004). Metabolism and DNA binding studies of 4-hydroxyestradiol and estradiol-3,4-quinone *in vitro* and in Female ACI rat mammary gland *in vivo*. *Carcinogenesis*, **25**, 289-297.

[20] Cavalieri, E. L., Kumar, S., Todorovic, R., Higginbotham, S., Badawi, A. F. & Rogan, E. G. (2001). Imbalance of estrogen homeostasis in kidney and liver of hamsters treated with estradiol: Implications for estrogen-induced initiation of renal tumors. *Chem Res Toxicol*, **14**, 1041-1050.

[21] Devanesan, P., Santen, R. J., Bocchinfuso, W. P., Korach, K. S., Rogan, E. G. & Cavalieri, E. L. (2001). Catechol estrogen metabolites and conjugates in mammary tumors and hyperplastic tissue from estrogen receptor-α knock-out (ERKO)/Wnt-1 mice: Implications for initiation of mammary tumors. *Carcinogenesis*, **22**, 1573-1576.

[22] Cavalieri, E. L., Devanesan, P., Bosland, M. C., Badawi, A. F. & Rogan, E. G. (2002). Catechol estrogen metabolites and conjugates in different regions of the prostate of Noble rats treated with 4-hydroxyestradiol: Implications for estrogen-induced initiation of prostate cancer. *Carcinogenesis*, **23**, 329-333.

[23] Liehr, J. G., Fang, W. F., Sirbasku, D. A. & Ari-Ulubelen, A. (1986). Carcinogenicity of catecholestrogens in Syrian hamsters. *J Steroid Biochem*, **24**, 353-356.

[24] Li, J. J. & Li, S. A. (1987). Estrogen carcinogenesis in Syrian hamster tissue: Role of metabolism. *Fed Proc*, **46**, 1858-1863.

[25] Newbold, R. R. & Liehr, J. G. (2000). Induction of uterine adenocarcinoma in CD-1 mice by cagechol estrogens. *Cancer Res*, **60**, 235-237.

[26] Rajah, T. T. & Pento, J. T. (1995). The mutagenic potential of antiestrogens in the *HPRT* locus of V79 cells. *Res Comm Molecul Pathol & Pharmacol*, **89**, 85-92.

[27] Kong, L. Y., Szaniszlo, P., Albrecht, T. & Liehr, J. G. (2000). Frequency and molecular analysis of *HPRT* mutations induced by estradiol in Chinese hamster V79 cells. *Intl J Oncol*, **17**, 1141-1149.

[28] Chakravarti, D., Mailander, P. C., Higginbotham, S., Cavalieri, E. L. & Rogan, E. G. (2003). The catechol estrogen-3,4-quinone metabolites induces mutations in the mammary gland of ACI rats. *Proc Amer Assoc Cancer Res*, **44**, (2nd ed.): 180.

[29] Miller, W. R. & O'Neill, J. (1987). The importance of local synthesis of estrogen within the breast. *Steroids*, **50**, 537-548.

[30] Simpson, E. R., Mahendroo, M. S., Means, G. D., Kilgore, M. W., Hinshelwood, M. M., Graham-Lorence, S., Amarneh, B., Ito, Y., Fisher, C. R., Michael, M. D., Mendelson, C. R. & Bulun, S. E. (1994). Aromatase cytochrome P450, the enzyme responsible for estrogen biosynthesis. *Endocrine Rev*, **15**, 342-355.

[31] Jefcoate, C. R., Liehr, J. G., Santen, R. J., Sutter, T. R., Yager, J. D., Yue, W., Santner, S. J., Tekmal, R., Demers, L., Pauley, R., Naftolin, F., Mor, G. & Berstein, L. (2000). In: *Estrogens as Endogenous Carcinogens in the Breast and Prostate*. E. Cavalieri, & E. Rogan (Eds.), Oxford Press, 95-112.

[32] Santner, S. J., Feil, P. D. & Santen, R. J. (1984). *In situ* estrogen production via the estrone sulfatase pathway in breast tumors: Relative importance versus the aromatase pathway. *J Clin Endocrinol Metab,* **59**, 29-33.

[33] Pasqualini, J. R., Chetrite, G., Blacker, C., Feinstein, M. C., Delalonde, L., Talbi, M. & Maloche, C. (1996). Concentrations of estrone, estradiol and estrone sulfate and evaluation of sulfatase and aromatase activities in pre and postmenopausal breast cancer patients. *J Clin Endo Metab*, **81**, 1460-1464.

[34] Van Landeghem, A. A., Poortman, J., Nabuurs, M. & Thijssen, J. H. (1985). Endogenous concentration and subcellular distribution of estrogens in normal and malignant human breast tissue. *Cancer Res*, **45**, 2900-2906.

[35] Savas, U., Bhattacharya, K. K., Christou, M., Alexander, D. L. & Jefcoate, C. R. (1994). Mouse cytochrome P-450EF, representative of a new 1B subfamily of cytochrome P-450s. Cloning, sequence determination, and tissue expression. *J Biol Chem*, **269**, 14905-14911.

[36] Li, J. J., Li, S. A., Klicka, J. K., Parsons, J. A. & Lam, L. K. (1983). Relative carcinogenic activity of various synthetic and natural estrogens in the Syrian hamster kidney *Cancer Res*, **43**, 5200-5204.

[37] Bocchinfuso, W. P., Hively, W. P., Couse, J. F., Varmus, H. E. & Korach, K. S. (1999). A mouse mammary tumor virus-*Wnt*-**1** transgene induces mammary gland hyperplasia and tumorigenesis in mice lacking estrogen receptor-α. *Cancer Res.*, **59**, 1869-1876.

[38] Markushin, Y., Zhong, W., Cavalieri, E. L., Rogan, E. G., Small, G. J., Yeung, E. S. & Jankowiak, R. (2003). Spectral characterization of catechol estrogen quinone (CEQ)-derived DNA adducts and their identification in human breast tissue extract. *Chem Res Toxicol*, **16**, 1107-1117.

[39] Shull, J. D., Spady, T. J., Snyder, M. D., Johansson, S. L. & Pennington, K. L. (1997). Ovary intact, but not ovariectomized, female ACI rats treated with 17β-estradiol rapidly develop mammary carcinoma. *Carcinogenesis*, **18**, 1595-1601.

[40] Russo, J., Lareef, M. H., Tahin, Q., Hu, Y. F., Slater, C., Ao, X. & Russo, I. H. (2002). 17β-Estradiol is carcinogenic in human breast epithelial cells. *J Steroid Biochem Mol Biol*, **1656**, 1-14.

[41] Russo, J., Hasan Lareef, M., Balogh, G., Guo, S. & Russo, I. H. (2003). Estrogen and its metabolites are carcinogenic agents in human breast epithelial cells. *J Steroid Biochem Mol Biol*, **87**, 1-25.

[42] Lareef, M. H., Heulings, R. C., Russo, P. A., Garber, J., Russo, I. H. & Russo, J. (2004). The estrogen antagonist ICI182-780 does not inhibit the proliferative activity and invasiveness induced in human breast epithelial cells by estradiol and its metabolite 4-OH estradiol. *Proc Amer Assoc Cancer Res*, 95th AACR, **45**, 11.

INDEX

A

B

C

D

G

H

I

N

O

P

Q

R

S

T

U

V

W

X

Y